Latin-American Symposium on Mathematical Logic 1976 :

Non-classical logics, model theory, and computability

NON-CLASSICAL LOGICS, MODEL THEORY, AND COMPUTABILITY

STUDIES IN LOGIC

AND

THE FOUNDATIONS OF MATHEMATICS

VOLUME 89

NORTH-HOLLAND PUBLISHING COMPANY–AMSTERDAM • NEW YORK • OXFORD

NON-CLASSICAL LOGICS, MODEL THEORY, AND COMPUTABILITY

Proceedings of the Third Latin-American Symposium on Mathematical Logic, Campinas, Brazil, July 11-17, 1976

Edited by

A. I. ARRUDA
Universidade Estadual de Campinas
Brazil

N. C. A. da COSTA
Universidade de São Paulo
Brazil

R. CHUAQUI
Universidad Católica de Chile
Chile

1977

NORTH-HOLLAND PUBLISHING COMPANY – AMSTERDAM • NEW YORK • OXFORD

North-Holland ISBN: 0 7204 0752 4

Published by:

North-Holland Publishing Company – Amsterdam • New York • Oxford

Sole distributors for the U.S.A. and Canada:

Elsevier North-Holland, Inc.
52 Vanderbilt Avenue
New York, N.Y. 10017

PRINTED IN THE NETHERLANDS

PREFACE

This volume contitutes the Proceedings of the Third Latin American Symposium on Mathematical Logic which was held at the State University of Campinas, Campinas, São Paulo, Brazil from July 11 to July 17, 1976. The meeting was sponsored by the State University of Campinas, the Association for Symbolic Logic, and the Division of Logic, Methodology, and Philosophy of Science of the International Union of History and Philosophy of Science. The meeting was organized by an executive committee consisting of A. I. Arruda (Chairman), R. B. Chuaqui, N. C. A. da Costa, and F. Mirō Quesada.

The Symposium was divided into three sections: Non-classical Logics, Model Theory, and Computability. The papers which appear in this volume are the texts, at time considerably expanded and revised, of most of the addresses presented by invitees to the meeting. Also included are expanded versions of short communications which the editors thought would give an idea of the present topics of research in Latin America.

Abstracts of all addresses and contributed papers were issued to members at the beginning of the Symposium and appeared in the Journal of Symbolic Logic. The titles of all papers are listed in the end of the introduction to this volume.

The editors would like to acknowledge the financial support given to the meeting by the following institutions: State University of Campinas,

Fundação de Amparo à Pesquisa do Estado de São Paulo (FAPESP), Organization of American States, International Union of History and Philosophy of Science, Banco do Estado de São Paulo, and Ministry of Foreign Relations of Brazil.

The editors would also to thank North-Holland Publishing Co. for the inclusion of this volume in the series Studies in Logic and the Foundations of Mathematics.

The Editors.

Departamento de Matemática
Universidade Estadual de Campinas
February 1977.

The editors N. C. A. da Costa and R. B. Chuaqui wish to express their appreciation for the work of A. I. Arruda. Her efforts were the main force behind these Proceedings.

CONTENTS

A Short History of the Latin American Logic Symposia

We would like to give an idea of part of the development of Mathematical Logic in Latin America through a short history of the three symposia sponsored by the Association for Symbolic Logic. Because it is mainly based on the recollections of the editors, who were participants or organizers of the three symposia, it will not be a completely balanced history.

Since the early sixties there had been the intention of having a meeting of the A.S.L. in South America. The good experience with the European meetings encouraged the idea of their being held in other continents. William Craig, when President of the A.S.L., for instance, talked about it with several people.

The first concrete step, however, was taken in 1967 by A.S.L. President Abraham Robinson. While in attendance at the logic year at the University of California, Los Angeles, Professor Robinson approached Rolando Chuaqui, who was visiting from the University of Chile, with the idea of having a first Latin Americam meeting in Chile. This was immediately accepted. Robinson then named the first Advisory Committee for Logic in Latin America formed by himself, as A.S.L. President, David Kaplan from UCLA, Antonio Monteiro, from Universidad Nacional del Sur, Bahía Blanca, Argentina, and Rolando Chuaqui (Chairman) from University of Chile, Santiago.

The meeting was first planned for January 1969. Due to internal difficulties in the University of Chile it could not take place at this time. When Chuaqui moved to the Catholic University of Chile (Santiago), in 1969, he found enthusiastic support for the idea from the Rector Fernando Castillo Velasco and other university authorities, and it was decided to hold the First Latin American Symposium on Mathematical Logic (I SLALM) in July 1970.

Abraham Robinson was an indefatigable worker for the success of this meeting. He sent letters to all possible organizations asking for financial support. We reproduce below Robinson's proposal for I SLALM:

A PROPOSAL FOR THE ORGANIZATION OF A SEMINAR AND COLLOQUIUM IN MATHEMATICAL LOGIC, TO BE HELD IN CHILE.

(Revised Version, June 3, 1969)

1. There is, at present, a small but growing group of mathematical logicians in South America. The following proposal is designed to encourage the further development of interest in Mathematical Logic in that region. It is patterned after a format which has been employed repeatedly in recent years in Western Europe and has proved successful in strengthening Mathematical Logic there. The Association for Symbolic Logic (ASL), which has co-sponsored some of these activities, stands ready to give its backing also to the plan outlined below. While the proposal is stated in rather precise terms, modifications would of course be in order.

2. The event is to take place at the Universidad Católica de Chile in July 1970. It is to include:

(i) A seminar of about three weeks'duration, to consist of three courses, at an intermediate level, in Model Theory,in Set Theory, and in Recursion Theory.

(ii) A colloquium of 3-5 days' duration, to be held at the end of the seminar, to include invited addresses and contributed papers. The scientific standards of the colloquium would be strengthned by including the seminar course instructors in the list of invited speakers. The colloquium will be sponsored or co-sponsored by the Association for Symbolic Logic. This will ensure that an account of the colloquium, including abstracts, will be published in the Journal of Symbolic Logic. It is hoped that the Mathematical Society of Chile and the Universidad Católica de Chile will join us in co-sponsoring the meeting.

3. The Association for Symbolic Logic will offer advice and moral support to the organizers of the seminar and colloquium. We are applying to the Pan American Union for financial su-

port. We also hope that it may be possible to obtain financial support from ICSU through International Union ofthe History and Philosophy of Science (IUHPS). In case such suppot is forthcoming IUHPS will be invited to co-sponsor the meeting.

4. The Association for Symbolic Logic has created an Advisory Committee on Logic in Latin America. Its members are Professor Rolando Chuaqui, Universidad Católica de Chile (chairman), Professor David Kaplan, University of California at Los Angeles, Professor Antonio Monteiro, Bahía Blanca, Argentina, and Professor Abraham Robinson, Yale University, New Haven, Connecticut. It will be the general function of the committee to encourage the development of Symbolic Logic in Latin America, and will be one of its major task to assist in the organizations of the activities outlined in the present proposal.

Abraham Robinson, President
Association for Symbolic Logic

Department of Mathematics
Yale University
New Haven, Connecticut 06520.

The I SLALM was held at the Catholic University of Chile in Santiago, in July 1970, according to Robinson's proposals: Three weeks of short courses, the Seminar, and one week of invited lectures and contributed papers, the Colloquium.

The courses in the Seminar were three:

A. Lévy (Israel), *Foundations of Set Theory*,

J. R. Shoenfield (U.S.A.), *Degrees of Unsolvability*,

R. Sikorski (Poland), *Boolean Algebras*.

A summary of the Colloquium appeared in the Journal of Symbolic Logic. We reproduce the main features (J.S.L. vol. 36, 576-581):

A Latin American meeting of the Association for Symbolic Logic was held in Santiago, Chile, from July 27 to July 31. The meeting was co-sponsored by the Catholic University of Chile, The Organization of American States, and the International Union for History and Philosophy of Science. The meeting was preceded by a Seminar on Mathematical Logic at the Catholic University of Chile.

There were thirteen invited one hour lectures and thirteen short communications presented at the meeting. The lectures were the following:

M. Morley (U.S.A.), *Some applications of topology to model theory* I.
R. Sikorski (Poland), *On strictly positive measures.*
S. Kochen (U.S.A.), *Quantum logic.*
J. Shoenfield (U.S.A.), *Hilbert's tenth problem.*
N. C. A. da Costa (Brazil), *Inconsistent formal systems.*
R. Cignoli (Argentina), *Moisil and Post algebras.*
G. Stahl (Chile), *Questions with numerical and totality requests.*
M. Morley (U.S.A.), *Some applications of topology to model theory*, II.
A. Lévy (Israel), *Normal ultrafilters and supercompact cardinals.*
A. Robinson (U.S.A.), *Model theoretic aspects of algebra.*
F. Alvim (Brazil). *Logic of quantum mechanics.*
G. Reyes (Canada), *Pro-discrete spaces.*
A. Robinson (U.S.A.), *Finite and infinite forcing in model theory.*

Below are reproduced the abstracts of the short communications presented at the meeting (we reproduce only the author and title):

M. M. Fidel (Argentina), *On the calculi* $\mathcal{C}_n$, $1 \leq n \leq \omega$.
M. M. Fidel (Argentina), *An algebraic study of logics with constructible falsity.*
M. M. Fidel (Argentina), *Moisil algebras and modal logic.*
I. Bicudo (Brazil), *Dually equivalent structures: an example.*
L. de Moraes (Brazil), *On discoursive predicate calculus.*
R. Chuaqui (Chile), *A representation theorem for linearly ordered cardinal algebras.*
N. C. A. da Costa (Brazil), *On the systems* T *and* T^*.
A. I. Arruda (Brazil), *On Griss' propositional calculus.*
E. Nemesszeghy (Chile), *A calculus of "δ" elimination (eliminability).*
L. P. de Alcantara (Brazil), *On the relative consistency of two systems of set theory.*
A. J. Engler (Brazil), *Symmetrical lattices.*

Professor Robinson participated actively in the meeting even to the point of giving one of his lectures in his hotel because he fell ill.

From the list of Latin American participants, it can be seen that they came only from Brazil, Argentina and Chile. From the beginning it was a problem to contact the Latin American logicians. The A.S.L. committee, at first, tried to obtain information directly from the members appearing in the roster of the A.S.L. living in Latin American countries. Only a few responses were received. Most contacts were made through friends. The main groups represented were from Bahīa Blanca, Argentina, from the State of São Paulo (Universidade Estadual de Campinas (UNICAMP), and Universidade de São Paulo (USP)), Brazil, and from Santiago, Chile.

Most of the financial support was provided by the Catholic University of Chile, and the Organization of American States (O.A.S.). The O.A.S. treated the Symposium as two meetings and gave double their usual amount.

During the I SLALM logicians from the different countries represented met with A. Robinson. It was decided that the next Symposium would be in Brazil. The new advisory Committee for Logic in Latin America was to be composed of the A.S.L. President (ex-officio), Newton C. A. da Costa (Brazil), Antonio Monteiro (Argentina), and Rolando Chuaqui (Chairman, Chile). (The A.S.L. President disappeared from the Committee in the December 1972 issue of the J.S.L.)

After consultations among Brazilian universities, the Brazilian delegation decided that it was best to hold the II SLALM at the University of Brasīlia, in July 1972. The last act of A. Robinson as President of the A. S.L. was to send a letter to the Rector of the University of Brasīlia offering the sponsorship of the A.S.L. for the meeting.

In preparation for the symposium, in January 1972, there was a pre-symposium mainly for Brazilian students, at the Technological Institute of Aeronautics (ITA) in São José dos Campos, São Paulo.

The chairman of the Organizing Committee of the II SLALM was Fausto Alvim. The meeting had a different character than the first as it was not composed of two parts: a Seminar and a Colloquium. It lasted for three weeks and its main activities were short courses of about ten lectures each. There were a few research lectures. The J.S.L. received no account of this meeting, so recollections of its courses and lectures may not be complete.

The following short courses were given:

R. Chuaqui (Chile), *Consistency and independence of the axiom of choice in the impredicative theory of classes.*

A. Robinson (U.S.A.), *Non Standard analysis.*

A. Robinson (U.S.A.), *Forcing in model theory.*

M. Dickmann (Chile), *Back and forth arguments in model theory.*

R. Cignoli (Argentina), *Algebra of logic.*

O. Porchat (Brazil), *Elementary logic.*

There were individual lectures by M. Guillaume (France), L. Monteiro (Argentina), P. Suppes (U.S.A.), A. I. Arruda (Brazil), E. Farah (Brazil). N. C. A. da Costa, who was at this time visiting the University of California at Berkeley, presented a paper by title.

The main support for this meeting came from Brazilian sources, especially from the Conselho Nacional de Pesquisas (CNPq), Fundação de Âmparo à Pesquisa do Estado de São Paulo (FAPESP), and the University of Brasília. There were a large number of Brazilian participants (most of them attending the course: Elementary Logic) and, as in Santiago, logicians from Chile, Argentina, plus one from Caracas, Venezuela.

It was decided at this meeting that the III SLALM would take place in Bahía Blanca, Argentina, in July 1974. However, due to the difficult situation in that country this was not possible, and there was no Symposium in 1974.

The revival in interest in these symposia came in March 1975. After an extended visit to the Catholic University of Chile, in Santiago, Professor Alfred Tarski visited UNICAMP with R. Chuaqui, during the first two weeks of March. A small logic meeting was organized there by Ayda I. Arruda with the participation of the two Visiting Professors and logicians from USP, UNICAMP, and University of Pernambuco, Brasil. The Proceedings of this meeting were published by the Institute of Mathematics, Statistics, and Computing Science (IMECC) of UNICAMP. Support for this meeting came from UNICAMP.

Due to the success of this meeting and the growing number of Brazilian logicians it was decided that the III SLALM would be held at the IMECC, UNICAMP, in July 1976. The A.S.L., through its President, J. R. Shoenfield, agreed to sponsor the meeting and named a new Advisory Committee on Logic in Latin America, formed, this time, by Newton C. A. da Costa (Brazil), Francisco Miró Quesada (Peru), and Rolando Chuaqui (Chairman, Chile). The incorporation of the Peruvian member signaled the participation of a new group of logicians from the universities in Lima, besides the Chilean, Brazilian and Argentinian groups.

After receiving a letter of J. R. Shoenfield, the Rector of UNICAMP, Dr. Zeferino Vaz, accepting the suggestion of the Director of the IMECC, Dr. Ubiratan D'Ambrosio named the Organizing Committee, formed by: Rolando Chuaqui (Universidad Católica de Chile, Santiago), Newton C. A. da Costa (USP, Brazil), Francisco Miró Quesada (Universidad Cayetano Heredia, Lima, Peru), and Ayda I. Arruda (Chairman, UNICAMP, Brazil).

In anticipation of the meeting there was a logic semester in the Department of Mathematics of IMECC, and the Center of Logic, Epistemology, and History of Science of UNICAMP. Advanced courses and seminars were given by R. Chuaqui (Visiting Professor at UNICAMP), N. C. A. da Costa (USP) and A. R. Raggio (IMECC). There were also advanced seminars from June 28 to July 9, with the participation of J. Kotas (Poland), C. Pinter (U.S.A.), M. Benda (USP), R. Routley (Australia), and R. Fraïssé (France). These courses and seminars were attended by logicians from Campinas and São Paulo as well as by students from UNICAMP, USP, and the Catholic University of Chile.

The Third Latin-American Symposium on Mathematical Logic was divided into three sections: Model Theory, Non-Classical Logics, and Applied Logic; and lasted for a week (July 11 to July 17, 1976). The financial support was given by: UNICAMP, Fundação de Amparo à Pesquisa do Estado de São Paulo (FAPESP), The Organization of American States (O.A.S.), The International Union for History and Philosophy of Science, Division of Logic, Methodology and Philosophy of Science (IUHPS/DLMPS), and the Bank of the State of São Paulo.

The scientific program was the following:

JULY 12.

09:15-09:30 A. M. — Opening session.

09:30-10:20 A. M. — R. Chuaqui (Chile, Brazil), *A semantical definition of probability.*

10:40-11:30 A. M. — N. C. A. da Costa (Brazil), *On Jaškowski discussive logic.*

2:00 - 2:20 P. M. — A. Loparić (Brazil), *A semantical study of some propositional calculi.*

2:20 - 2:40 P. M. — L. H. Lopez dos Santos (Brazil), *Some remarks on discussive logic.*

2:40 - 3:00 P. M. — E. H. Alves (Brasil), *On paraconsistent logic.*

3:00 - 3:20 P. M. — C. Lungarzo (Brazil), *A paraconsistent infinitary propositional calculus.*

4:00 - 5:00 P. M. — M. Benda (U.S.A., Brazil), *Some directions in model theory.*

JULY 13.

9:00 - 9:40 A. M. — A. I. Arruda (Brazil), *On the imaginary logic of* N. A. *Vasil'ēv.*

9:40-10:10 A. M. — A. R. Raggio (Brazil), *Semi-formal Beth tableaux.*

10:40-11:30 A. M. — F. G. Asenjo (U.S.A.), *Formalizing multiple location.*

2:00 - 2:20 P. M. — L. P. de Alcantara (Brazil), *On the equivalence of some axioms of strong infinity.*

2:20 - 2:40 P. M. — P. A. S. Veloso (Brazil), *Two model theoretic properties of class of representable relation algebras.*

2:40 - 3:00 P. M. — R. Morais (Brazil), *Projective logic and projective Boolean algebras.*

3:00 - 3:20 P. M. — X. Caicedo Ferrer (Colombia), *Independent sets of axioms in* $L_{\kappa\alpha}$.

4:00 - 5:00 P. M. — M. Krasner (France), *Galois theory of relations.*

JULY 14.

9:00-10:00 A. M. — A. Tarski (U.S.A.), *Relation Algebras* (replay of a lecture recorded in videotape at UNICAMP in March, 1975).

10:30-11:20 A. M. — C. Pinter (U.S.A.), *Some theorems on omitting types with applications to model completeness, and related properties.*

2:00 - 2:20 P. M. — J. Simon (Brazil), *Polynomially bounded quantification over higher types and a new hierarchy of the elementary sets.*

2:20 - 2:40 P. M. — M. S. de Gallego (Brazil), *On the abstract family of languages of categorical types.*

2:40 - 3:00 P. M. — H. P. Sankappanavar (Brazil), *On the decision problem of the congruence lattice of pseudocomplented semilattices.*

3:00 - 3:20 P. M. — I. Mikenberg (Chile), *A logical system for partial algebras.*

4:00 - 5:00 P. M. — R. Solovay (U.S.A.), *On random r. e. sets.*

JULY 15.

9:00 - 9:40 A. M. — M. Dascal (Brazil, Israel), *Conversational relevance.*

9:40-10:20 A. M. – F. Miró Quesada (Peru), *Heterodox logics and the problem of the unity of logic.*

10:50-11:40 A. M. – R. Routley (Australia), *Ultramodal logic as universal.*

2:00 - 2:20 P. M. – L. Piscoya (Peru), *Probability and empirical scientific statements.*

2:20 - 2:40 P. M. – D. M. de Souza Filho (Brazil), *Some remarks on meta-language.*

2:40 - 3:00 P. M. – R. Lintz (Brasil), *Organic and inorganic logic.*

3:00 - 3:20 P. M. – A. M. Sette (Brazil), *Fraïssé's and Robinson's forcing.*

4:00 - 5:00 P. M. – R. Fraïssé (France), *Present problems about intervals in relation theory and in logic.*

JULY 16.

9:00-10:00 A. M. – E. G. K. López-Escobar (U.S.A.), *Infinite rules in finite systems.*

10:30-11:30 A. M. – J. Kotas (Poland), *On some modal systems defined in connexion with Jaśkowski's problem.*

2:00 - 3:00 P. M. – J. R. Shoenfield (U.S.A.), *Quantifier elimination in fields.*

3:00 - 4:00 P. M. – Closing session.

The following communications were presented by title:

A. A. Mullin (U.S.A.), *Applications of fixed point theory to number theory.*

L. F. Monteiro (Argentina), *Algèbres de Post et de Moisil trivalentes monadiques libres.*

I. M. L. D'Ottaviano (Brazil), *Fuzzy sets in da Costa's systems T and T*.*

R. Routley (Australia), *Dialectical set theory.*

R. Routley (Australia), *Choice of logical foundations: ultramodal logic and dialectical foundations.*

There were 69 participants distributed by countries as follows:

Brazil - 43,	U.S.A. - 6,
Chile - 6 ,	France - 2,
Argentina - 3,	Poland - 1,
Peru - 4,	Canada - 1,
Colombia - 2,	Australia - 1.

This is the first time that full Proceedings of a Latin American Symposium are appearing. For the first symposium, North-Holland, after the meeting, offered to publish the invited lectures in full in its series Studies in Logic. However, as the publication was not planned in advance, it was too late to gather enough articles for the volume. Shoenfield's course, *Degrees of Unsolvability*, however, was published by North-Holland. For the second symposium, Springer-Verlag in its series Lecture Notes, agreed to publish three of the courses in one volume. Due to various problems the manuscript was never sent to the publisher.

When we compare the list of papers presented in 1970 with those of this meeting, we have reason to be encouraged: the number and overall quality of the contributions of Latin American logicians have certainly increased. The picture is not uniformly bright, however. In Chile, due mainly to economic difficulties, there has been no marked increase. The group from Bahía Blanca, Argentina, has been scattered, as most of its members have been excluded from the university. This explains the scarcity of papers for this symposium from this group, which made significant contributions to the algebra of logic.

On the bright side, we have the incorporation of the Peruvian group and the significant development of logic in Brazil, where besides the group around da Costa (USP, and UNICAMP) in the State of São Paulo, there are now participants from other sectors of the country. It is worthwhile to note also the support that the growing group of UNICAMP is receiving from the university, specially from the Director of the IMECC.

In 1969, A. Robinson said of the Latin America Logic group, that it was a small but growing group. Perhaps, now it is not so small, but, we hope, still growing.

Ayda I. Arruda,
Newton C. A. da Costa,
Rolando Chuaqui.

PART I

NON CLASSICAL LOGICS

Non-Classical Logics, Model Theory and Computability,
A.I. Arruda, N.C.A. da Costa and R. Chuaqui (eds.)

ON THE IMAGINARY LOGIC OF N. A. VASIL'ĒV

by *AYDA I. ARRUDA*

1. INTRODUCTION.

Nikolaj Alexandrovic Vasil'ēv (1880-1940) studied medicine and became a Professor of Philosophy at the University of Kazan, Russia. He wanted to do for Aristotelian Logic what Lobachevsky had done for Euclidean Geometry. The logical views of Vasil'ēv were presented in a series of papers published between 1910 and 1913 (Vasil'ēv 1910, 1911, 1912, and 1913), and in an abstract which appeared in 1924 (Vasil'ēv 1924). Due possible to the fact that the ideas of Vasil'ēv were too advanced for his time or because the most important of his papers were published only in Russian, his work passed almost unnoticed until 1962. Nonetheless, some of his papers were reviewed (Hessen 1910, and K. Smirnov 1911) and mentionned (Church 1936, and Korcik 1955). Vasil'ēv's conceptions began to receive his due with the paper of V. A. Smirnov 1962 , its review by D. D. Comey 1965 , and the paper of G. Kline 1965 where the author considers Vasil'ēv as a forerunner of many-valued logic (this opinion is also supported in N. Rescher 1969 , and M. Jammer 1974).

A deductive theory T is said to be *inconsistent nontrivial* if there is a formula A such that both A and its negation, $\neg A$, are theorems of T , and there is at least one formula which is not a theorem of T. If the underlying logic of T is the classical logic (or one of most of the common logics) T is trivial if and only if it is inconsistent. Then, for the study of inconsistent nontrivial theories it is necessary to construct new systems of logic (see, for example, Arruda 1967, Arruda and Costa 1970, Asenjo and Tamburino 1975, Costa 1974, and Routley and Meyer 1976).

In this paper we would like to emphasize that Vasil'ēv can (perhaps with much stronger reasons than for the case of many-valued logic) be considered as a forerunner of nonclassical logics constructed for the study of inconsistent nontrivial theories. To show this we construct three propositional calculi(*V1*, *V2*, and *V3*) according to some of Vasil'ēv's insights, and relate them to the propositional calculi of Costa 1974, and of Routley and Meyer 1976. Nevertheless, the main objective of this paper is to develop and study the systems *V1*, *V2*, and *V3*. Some results already obtained about the corresponding predicate calculi and their extensions will be published later.

Since Vasil'ēv's logical views may be interpreted in many different ways, it is difficult, if not impossible, to say that a formal system of logic is actually a formalization of his opinions. What we can say is that a certain formal system is a formalization of a given interpretation of Vasil'ēv *Imaginary Logic*. It is in accordance with this point of view that our systems may be named *Vasil'ēv propositional calculi*.

As we do not want to make an exegesis of Vasil'ēv's work, we will present here only a summary of his ideas (as formulated in Comey 1965, Kline 1965, and V. Smirnov 1962) which were taken as motivations for the construction of *V1*, *V2*, and *V3*. In the final section of this paper we will try to interpret these opinions of the Russian logician in a way which justifies our systems as Vasil'ēv propositional calculi.

Vasil'ēv intended to construct a "non-Aristotelian" and "universal" logic, universal in the sense that it might cover an infinite number of logical systems (V. Smirnov 1962). For him a logical system is composed of two parts: that which he calls *Metalogic*, i. e., "an indispensable core of laws related to thought which are necessary for any thinking and which cannot be eliminated from logic without its losing its logical character" (see Comey 1965); and a second part which we call here *ontological basis of logic*, i. e., "a varying range of laws which are functions of the properties of the known objects " (Comey 1965).

"Vasil'ēv wanted to see which postulates of logic could be changed or eliminated from logic without its ceasing to be logic. Thus he was led to drop the (ontological) law of excluded middle, and also the LAW OF CONTRADICTION which he took in the Kantian form:'no object can have a predicate which contradicts it'. Vasil'ēv distinguished the law of contradiction from the LAW OF NON-SELF-CONTRADICTION: 'one and the same judgement cannot be simultaneously true and false'. Vasil'ēv took this to be different laws"

(Comey 1965). The latter belongs to metalogic, and the first, if retained, would belong to ontological basis of logic.

Vasil'ēv considered worlds in connection to which there are only three sorts of different basic (predicative) judgements: *affirmative*, "S is P"; *negative*, "S is not P"; and *indifferent* (or *contradictory*) "S is P and not P", such that only one of these judgements can be true for a given object and predicate. From these hypotheses he delineated an Imaginary Logic with an ontological *law of excluded fourth* substituting the ontological law of excluded middle (Vasil'ēv 1910). Latter he generalized these ideas to a logic with an ontological *law of excluded* $(n+1)$*st*, $n \geq 1$. He also tried to show that his Imaginary Logic with his law of excluded fourth has a classical interpretation, as is the case with the *Imaginary Geometry* of Lobachevsky.

Vasil'ēv did not believe that contradictions in our *real world* exist, but that these may obtain only in an *imaginary world*. Perhaps, this belief might have been accepted as natural at the begining of this century, but after the last development of science and mathematics it seems not to be reliable. On the other hand, some ideas advanced by Vasil'ēv are the same as those supported nowadays by some quantum logicians who propose the use of many-valued logic for a quantum logical approach (Jammer 1974, chapter eight). Hence, perhaps, the Imaginary Logic of Vasil'ēv may become as *real* as the Imaginary Geometry of Lobachevsky.

2. THE PROPOSITIONAL CALCULUS *V1*.

Let *V1* be the propositional calculus whose language has the following symbols:

(1) A denumerably infinite set of *classical propositional letters*.

(2) A denumerable (possibly finite, but not empty) set of *propositional letters of Vasil'ēv*, which will be denoted by *S*.

(3) The connectives: $\supset$, &, V, and $\neg$.

(4) Parentheses: (,).

We define *atomic formula* and *formula* as usual, and employ capital Latin letters A, B, C,... as syntactical variables for formulas. The convention to omit parentheses is the one of Kleene 1952. To abbreviate "A is not a propositional letter of Vasil'ēv" we write: "$A \notin S$"

Axiom schemata of *V1*:

1) $A \supset (B \supset A)$.
2) $(A \supset B) \supset ((A \supset (B \supset C)) \supset (A \supset C))$.
3) $A \supset (B \supset A \& B)$.
4) $A \& B \supset A$.
5) $A \& B \supset B$.
6) $A \supset A \vee B$.
7) $B \supset A \vee B$.
8) $(A \supset C) \supset ((B \supset C) \supset (A \vee B \supset C))$.
9) $A \vee \neg A$.
10) $B \supset (\neg B \supset A)$, *if* $B \notin S$.

Rule of inference:

A , $A \supset B$ / B.

The notions of *proof*, and *deduction*, as well as the *assertion symbol*, $\vdash$, and the *equivalence symbol*, $\equiv$, are defined as in Kleene 1952. To indicate the use, say, of Theorem 2.1, item 3, we write: Theorem 2.1.3;the same procedure applies to Lemmas and Definitions. We omit the proofs of Theorems and Lemmas when they are immediate or similar to the corresponding classical ones (these definitions and conventions will be used in the remainder of the paper). "*F is a theorem of V1*" is abbreviated by "$\vdash_1 F$".

THEOREM 2.1. *In V1 we prove the following derived rules:*

(1) *If* $\Gamma, A \vdash B$, *then* $\Gamma \vdash A \supset B$;

(2) *If* $\Gamma, A \vdash C$ *and* $\Gamma, B \vdash C$, *then* $\Gamma, A \vee B \vdash C$;

(3) *If* $B \notin S$, *then* $B, \neg B \vdash A$;

(4) *Suppose that* $B \notin S$; *if* $\Gamma, A \vdash B$ *and* $\Gamma, A \vdash \neg B$, *then* $\Gamma \vdash \neg A$;

(5) *If* $B \notin S$, *then* $A \supset B \vdash \neg B \supset \neg A$ *and* $\neg A \supset \neg B \vdash B \supset A$.

THEOREM 2.2. *All formulas of the following forms are theorems of V1:*

(1) $((A \supset B) \supset ((A \supset \neg B) \supset \neg A))$, *if* $B \notin S$;

(2) $A \supset \neg\neg A$, *if* $A \notin S$;

(3) $\neg\neg A \supset A$;

(4) $\neg A \equiv \neg\neg\neg A$;

(5) $A \equiv \neg\neg A$, *if* $A \notin S$;

(6) $\neg(A \& B) \supset (\neg A \vee \neg B)$;

(7) $\neg(A \vee B) \supset (\neg A \& \neg B)$;

(8) $\neg(A \& \neg B) \supset (A \supset B)$;

(9) $(A \supset B) \supset (\neg A \vee B)$;

(10) $\neg(A \& B) \supset (A \supset \neg B)$;

(11) $\neg(\neg A \& \neg B) \supset (A \vee B)$;

(12) $\neg(\neg A \vee \neg B) \supset (A \& B)$;

(13) $\neg(\neg A \vee B) \supset \neg(A \supset B)$;

(14) $\neg(A \supset B) \supset (A \& \neg B)$;

(15) $\neg A \& \neg B \equiv \neg(\neg\neg A \vee \neg\neg B)$;

(16) $(A \supset \neg B) \supset \neg(A \& \neg\neg B)$;

(17) $(\neg A \vee \neg B) \equiv \neg(\neg\neg A \& \neg\neg B)$;

(18) $\neg A \vee B \equiv \neg\neg A \supset B$;

(19) $((A \supset B) \supset A) \supset A$;

(20) $(A \supset B) \vee (B \supset A)$;

(21) $(A \supset (B \vee C)) \equiv (A \supset B) \vee (A \supset C)$;

(22) $(A \supset \neg\neg A) \vee (B \supset \neg\neg B) \vee (A \supset B)$.

PROOF: We only give the proofs of some items.

(1) The proof is immediate after Theorem 2.1.4 and 2.1.1.

(2) If $A \notin S$, then $A, \neg A \vdash \neg\neg A$, and $A, \neg\neg A \vdash \neg\neg A$. Therefore, $A \vdash \neg\neg A$ and $\vdash A \supset \neg\neg A$.

(3) As $\neg A \notin S$, then $\neg\neg A, \neg A \vdash A$. But as $\neg\neg A, A \vdash A$, then $\neg\neg A \vdash A$.

(8) It is obvious that $\neg(A \,\&\, \neg B), A, \neg B \vdash (A \,\&\, \neg B) \,\&\, \neg(A \,\&\, \neg B)$. But as $A \,\&\, \neg B \notin S$, then $\neg(A \,\&\, \neg B), A, \neg B \vdash B$. As we have also $\neg(A \,\&\, \neg B), A, B \vdash B$, then $\vdash \neg(A \,\&\, \neg B) \supset (A \supset B)$.

(14) It is a consequence of (8), (5) and Theorem 2.1.3, since $(A \supset B)$, $\neg(A \,\&\, \neg B) \notin S$.

(19) By (14) we have $(A \supset B) \supset A, \neg(A \supset B) \vdash A$; but we have also $(A \supset B) \supset A, A \supset B \vdash A$; hence $\vdash ((A \supset B) \supset A) \supset A$.

Erasing all the propositional letters of Vasil'ēv from the language of *V1* we obtain a classical propositional language. So, restricting the axioms 1 - 10 and the detachement rule to formulas of the resulting language, we obtain the classical propositional calculus *C* .

THEOREM 2.3. *Let* B *be a non-atomic formula of* C *and* $B^{\S}$ *a formula obtained from* B *replacing non-atomic formulas of* $V1$ *for atomic components of* B. *If* $\vdash_C B$, *then* $\vdash_1 B^{\S}$.

PROOF: Similar to the proof of Theorem of Replacement of Kleene 1952, observing that substituting non-atomic formulas of *V1* for atomic components in the axioms of *C* we obtain axioms of *V1*.

THEOREM 2.4. *V1 is a conservative extension of C* .

PROOF: It is obvious that *V1* is an extension of *C*. To prove that this extension is a conservative one we proceed by induction on the length of the proofs in *V1*, observing that substituting classical propositional letters for the propositional letters of Vasil'ēv in an axiom of *V1* we obtain an axiom of *C*.

Let $G(A)$ be a formula where A may occur as a subformula, and $G(B)$ the formula obtained from $G(A)$, replacing specified occurrences of A by B.

THEOREM 2.5. *If* $A, B \notin S$, *then* $\vdash_1 (A \equiv B) \supset (G(A) \equiv G(B))$.

DEFINITION 2.1. Let P be a Vasil'ēv propositional letter occurring in a given formula F. Each not directly preceeded by negation occurrence of P in F is called a *normal occurrence of* P *in* F. A *negated occurrence of* P *in* F is an occurrence of $\neg P$ in F (for example, if F is $P \vee \neg P$ or $P \supset \neg\neg P$, the first occurrence of P is a normal occurrence, and the second is a negated occurrence).

Let us denote by $G(P,\neg P)$ a formula which have normal and negated occurrences of P. Then, $G(A,B)$ is the formula obtained from $G(P,\neg P)$ *replacing* specified normal and negated occurrences of P respectively by A and B. (In the above example, $G(A,B)$ will be $A \vee B$ or $A \supset \neg B$, if we replace both normal and negated occurrences of P.)

THEOREM 2.6. *If* $A, B \notin S$, *then* $\vdash_{I} ((P \equiv A) \& (\neg P \equiv B)) \supset (G(P,\neg P) \equiv G(A,B))$.

To prove the decidability of $V1$ we will follow two different methods. First, we prove the decidability of $V1$ through an interpretation of $V1$ in C; this method is interesting not only because Vasil'ēv wanted his Imaginary Logic to be interpretable in the classical logic, but also because it will be very useful in order to prove the completeness of the predicate calculus corresponding to $V1$. Second, we shall prove the decidability of $V1$ through the method of valuations, as in Costa and Alves 1976. By this way we obtain a semantics for $V1$, and from this a mechanical method to decide $V1$. (This remark is also valid for the systems $V2$ and $V3$ to be studied bellow.)

DEFINITION 2.2. Let $q_1, q_2, \ldots$ and $P_1, P_2, \ldots$ be respectively fixed enumerations of classical propositional letters and of Vasil'ēv propositional letters of $V1$. Now let $p_1, p_2, \ldots$ and $r_1, r_2, \ldots$ be respectively the prime numbers greater than 2 and the non-prime odd numbers, both series taken in the natural ordering. Then F^* is the formula obtained from F performing the following substitutions:

q_{p_i}	for	q_i ,
$q_{2i} \supset q_{r_i}$	for	normal occurrences of P_i ,
$q_{r_i} \supset q_{2i}$	for	negated occurrences of P_i (i. e., for $\neg P_i$).

THEOREM 2.7. *If* $\vdash_{I} F$, *then* $\vdash_{C} F^*$.

PROOF: By induction on the length of proofs in $V1$, observing that if A is an axiom of $V1$, then A^* is an axiom of C.

DEFINITION 2.3. Let $(F^*)^{\S}$ be the formula obtained from F^*, performing the following substitutions:

$$(\neg P_i \supset P_i) \supset (P_i \supset \neg P_i) \quad \text{for} \quad q_{2i},$$
$$\neg P_i \supset P_i \quad \text{for} \quad q_{ri},$$
$$q_i \quad \text{for} \quad q_{pi}.$$

LEMMA 2.1. *In V1 we prove the following schemata:*

$(A \supset B) \supset A \equiv A$, $(\neg A \supset A) \supset (A \supset \neg A) \equiv \neg A$,

$A \supset (A \supset B) \equiv A \supset B$, $\neg A \supset A \equiv A$.

LEMMA 2.2. *In V1 we prove:*

$(q_{2i} \supset q_{ri})^{\S} \equiv P_i$, *and* $(q_{ri} \supset q_{2i})^{\S} \equiv \neg P_i$.

PROOF: Immediate consequence of Lemma 2.1.

THEOREM 2.8. *If* $\vdash_C F^*$, *then* $\vdash_1 F$.

PROOF: If $\vdash_C F^*$, then by Theorem 2.3 we obtain $\vdash_1 (F^*)^{\S}$. Consequently, by Theorem 2.6 and Lemma 2.2, it follows that $\vdash_1 F$.

THEOREM 2.9. $\vdash_1 F$ *if and only if* $\vdash_C F^*$.

COROLLARY. $\vdash_1 F$ *if and only if* F^* *is a tautology.*

We pass now to prove the decidability of V1 by the method of valuations.

DEFINITION 2.4. Let F_1 be the set of formulas of V1. A *valuation* for V1 is a function $v\colon F_1 \to \{0,1\}$ such that:

$v(A \supset B) = 1$ *iff* $v(A) = 0$ *or* $v(B) = 1$,

$v(A \,\&\, B) = 1$ *iff* $v(A) = v(B) = 1$,

$v(A \vee B) = 1$ *iff* $v(A) = 1$ *or* $v(B) = 1$,

If $A \notin S$, *then* $v(A) = 1$ *iff* $v(\neg A) = 0$,

If $A \in S$ *and* $v(A) = 0$, *then* $v(\neg A) = 1$.

For the sake of simplicity we will use here the following abbreviations given in Costa 1974: $A^{o} =_{def} \neg(A \,\&\, \neg A)$ *and* $\neg^{*}A =_{def} \neg A \,\&\, A^{o}$.

LEMMA 2.3. *Let P be a syntactical variable for Vasil'év propositional letters. Then, a valuation* v *for V1 has the following properties:*

If $A \notin S$, *then* $v(A) = 0$ *iff* $v(\neg A) = 1$,

If $A \notin S$, *then* $v(A^{o}) = 1$,

If $v(\neg P) = 0$, *then* $v(P) = 1$,
If $v(\neg\neg P) = 1$, *then* $v(P) = 1$,
$v(A) = 0$ *iff* $v(\neg^* A) = 1$,
$v(P^o) = 0$ *iff* $v(P) = v(\neg P) = 1$.

DEFINITION 2.5. As in Costa and Alves 1976: *a formula A is valid if for every valuation* v, $v(A) = 1$. If Γ is a set of formulas of *V1*, then a valuation v is a *model* of Γ if $v(A) = 1$ for every $A \in \Gamma$ (if $v(A) = 1$ for every valuation v which is a model of Γ, we write $\Gamma \models A$; in particular, $\models A$ means that A is valid. The same definitions and notations will be used in the next two sections).

THEOREM 2.10. *If* $\Gamma \vdash_1 A$, *then* $\Gamma \models A$.

COROLLARY. *If* $\vdash_1 A$, *then* $\models A$.

DEFINITION 2.6. A set of formulas Γ is *nontrivial* if there exists a formula A such that A is not a consequence of Γ, i. e., $\Gamma \nvdash A$. A set of formulas Γ is *maximal nontrivial* if: 1) it is nontrivial, and 2) if $\Gamma \cup \{A\}$ is nontrivial, then $A \in \Gamma$.

LEMMA 2.4. *Let* Γ *be a maximal nontrivial set of formulas of V1, then we have:*

(1) $A \in \Gamma$ *iff* $\Gamma \vdash_1 A$,
(2) *if* $\vdash_1 A$, *then* $A \in \Gamma$,
(3) $A \in \Gamma$ *iff* $\neg^* A \notin \Gamma$,
(4) *if* $A, A^o \in \Gamma$, *then* $\neg A \notin \Gamma$,
(5) *if* $A \notin S$, *then* $A^o \in \Gamma$,
(6) $\neg^* A \in \Gamma$ *iff* $A \notin \Gamma$,
(7) *if* $\neg A$, $A^o \in \Gamma$, $A \notin \Gamma$,
(8) *if* $A, A \supset B \in \Gamma$ *then* $B \in \Gamma$,
(9) $A \in \Gamma$ *or* $\neg^* A \in \Gamma$,
(10) *if* $\Gamma \vdash_1 B$, *then* $\Gamma \vdash_1 A \supset B$.

PROOF: We give only the proof of item 3; for the others the proofs are immediate.

(3) Suppose that $A \in \Gamma$ and $\neg^* A \in \Gamma$. Then, $\Gamma \vdash A$ and $\Gamma \vdash \neg A \,\&\, A^o$, and so, $\Gamma \vdash (A \,\&\, \neg A) \,\&\, \neg(A \,\&\, \neg A)$. But as $A \,\&\, \neg A \notin S$, then by axiom 10, $\Gamma \vdash B$, for every formula B. Consequently, Γ would be trivial, which is impossible by the hypotheses of the Theorem. From this, if $A \in \Gamma$, then $\neg^* A \notin \Gamma$. On the other hand, if $\neg^* A \notin \Gamma$, then $\neg^* A \notin \Gamma \cup \{A\}$, and $\Gamma \cup \{A\}$ is nontrivial; but as Γ is maximal, then $A \in \Gamma$.

LEMMA 2.5. *Every nontrivial set is a subset of a maximal nontrivial set.*

PROOF: Similar to the proof of Theorem of Lindembaum.

LEMMA 2.6. *Every maximal nontrivial set of formulas of V1 has a model.*

PROOF: Let Γ be a maximal nontrivial set of formulas of *V1*, and $\upsilon: F_1 \to \{0,1\}$ a function such that if $A \in \Gamma$, then $\upsilon(A) = 1$, and if $A \notin \Gamma$, then $\upsilon(A) = 0$. It is easy to prove that υ is a valuation for *V1*, and then that υ is a model for Γ.

COROLLARY. *Every nontrivial set has a model.*

THEOREM 2.11. *Let Γ be a set of formulas of V1. If $\Gamma \models A$, then $\Gamma \vdash_{I} A$.*

PROOF: If Γ is trivial the proof is immediate. If Γ is nontrivial, supposing that $\Gamma \nvdash A$, then $\Gamma \cup \{\neg^* A\}$ is nontrivial, and has a model which is a model of Γ and such that $\upsilon(\neg^* A) = 1$. But, by the hypotheses $\Gamma \models A$, then $\upsilon(A) = 1$. Then, $\upsilon(A) = \upsilon(\neg^* A) = 1$, which is impossible. Consequently, $\Gamma \vdash A$.

THEOREM 2.12. $\vdash A$ *if and only if* $\models A$.

In order to decide *V1* by a three-element truth tables (Theorem 2.14 bellow) we begin rewritting Definition 2.4 in the following tables M, whose meanings are obvious (if the value of *F* according to M is always 1, then we write: $\models_{M} F$; in the remainder of this paper this notation will be used in an obvious sense):

A	B	$A \supset B$	$A \& B$	$A \vee B$
0	0	1	0	0
1	0	0	0	1
0	1	1	0	1
1	1	1	1	1

$A \notin S$.

A	$\neg A$
0	1
1	0

$A \in S$.

A	$\neg A$
0	1
1	0
1	1

Let *m* be the number of Vasil'ĕv propositional letters which have both normal and negated occurrences in a given formula *F*, and let *n* be the number of the other atomic components of *F* (i. e., the classical propositional letters, and the Vasil'ĕv propositional letters which have either only normal occurrences or only negated occurrences in *F*). Then, the table of *F* according to M has $3^m.2^n$ lines.

LEMMA 2.7. $\models F$ *if and only if* $\models_{M} F$.

THEOREM 2.13. $\vdash_{I} F$ *if and only if* $\models_{M} F$.

To obtain a mechanical method to decide if a formula *F* is a theorem of

V1, we consider the matrix M1 = <{0,1,2}, {1,2} , ⊃ , & , V , ˥> such that:

A	B	A ⊃ B	A & B	A V B
0	0	1	0	0
1	0	0	0	1
2	0	0	0	1
0	1	1	0	1
1	1	1	1	1
2	1	1	1	1
0	2	1	0	1
1	2	1	1	1
2	2	1	1	1

A ∉ S, A	˥A
0	1
1	0
2	0

A ∈ S A	˥A
0	1
1	0
2	1

LEMMA 2.8. *If* $\vdash_1 F$, *then* $\models_{M1} F$.

PROOF: By induction on the length of the proof of *F*. Let us note first that the value 2 is assumed only by atomic formulas, and that, by Theorem 2.12, atomic formulas cannot be theorens of *V1*; hence, if *F* is a theorem of *V1*, then *F* is not an atomic formula and never takes the value 2. Now it is easy to verify that the axioms of *V1* take always the value 1, and if *F* is a theorem of *V1*, and the value of A, and of A ⊃ *F*, is 1, then the value of *F* is also 1.

LEMMA 2.9. *If* $\models_{M1} F$, *then* $\models_{M} F$.

PROOF: In effect, substituting 1 for 2 in M1, and then erasing all repeated lines we obtain the tables M.

THEOREM 2.14. $\vdash_1 F$ *if and only if* $\models_{M1} F$.

PROOF: By Lemmas 2.7-2.9, and Theorem 2.12.

THEOREM 2.15. *If* *B* ∈ *S*, *then the following formulas are not valid in V1:*

˥A V ˥*B* ⊃ ˥(A & *B*), ˥A & ˥*B* ⊃ ˥(A V *B*), *B* ⊃ ˥˥*B*,
(A ⊃ *B*) ⊃ ˥(A & ˥*B*), ˥*B* V A ⊃ (*B* ⊃ A) *B* ⊃ (˥*B* ⊃ A),
A & *B* ⊃ ˥(˥A V ˥*B*), A V *B* ⊃ ˥(˥A & ˥*B*), *B* & ˥*B* ⊃ A,
A & ˥*B* ⊃ ˥(A ⊃ *B*), ˥(*B* ⊃ A) ⊃ ˥(˥*B* V A),
(A ⊃ *B*) ⊃ (˥*B* ⊃ ˥A), (A ⊃ *B*) ⊃ ((A ⊃ ˥*B*) ⊃ ˥A),
(˥A ⊃ ˥*B*) ⊃ (*B* ⊃ A), ˥(*B* & ˥*B*).

THEOREM 2.16. *The systems* $\mathcal{C}_n$, $1 \leq n \leq \omega$ (Costa 1974) *are proper subsystems of V1.*

PROOF: *The systems* $\mathcal{C}_n$, $1 \leq n \leq \omega$, are proper subsystems of *C*, which is a proper subsystem of *V1*.

3. THE PROPOSITIONAL CALCULUS *V2*.

We consider now the propositional calculus *V2*, which is obtained from *V1* introducing a new axiom schema (in this section P is taken as a syntactical variable for Vasil'ěv propositional letters):

11) $P \mathbin{\&} \neg P$.

Therefore the propositional calculus *V2* is inconsistent, but as we shall prove later, it is nontrivial.

THEOREM 3.1. *All the rules and schemata of* Theorem 2.1, *and* 2.2 *are provable in V2, as well as the following schemata:*

(1) $\neg\neg P \supset \neg P$, (2) $P \equiv \neg P$, (3) $\neg(P \mathbin{\&} \neg P) \supset B$,
(4) $\neg\neg P \supset B$, (5) P, (6) $\neg P$.

PROOF: The proofs of rules and schemata of Theorema 2.1 and 2.2 are the same, and (1)-(6) are immediate consequnces of Axiom 11.

DEFINITION 3.1. Let $q_1, q_2, \ldots$ and $P_1, P_2, \ldots$ be respectively fixed enumerations of the classical propositional letters, and Vasil'ěv propositional letters of *V2*. Then F^* is the formula obtained from F performing the following substitutions:

q_{2i-1} for q_i,
$q_{2i} \vee \neg q_{2i}$ for normal occurrences of P_i,
$\neg(q_{2i} \mathbin{\&} \neg q_{2i})$ for negated occurrences of P_i (i. e., for $\neg P_i$).

LEMMA 3.1. *If* $\vdash_2 F$, *then* $\vdash_C F^*$.

PROOF: Similar to the proof of Theorem 2.7.

REMARK: The Theorems 2.3, 2.4, 2.5, and 2.6 are valid for *V2*.

DEFINITION 3.2. $(F^*)^{\S}$ is the formula obtained from F^* performing the following substitutions:

q_i for q_{2i-1}
$\neg P_i$ for q_{2i}.

LEMMA 3.2. *In V2 we prove:*
$(q_{2i} \vee \neg q_{2i})^{\S} \equiv P_i$, and $\neg(q_{2i} \,\&\, \neg q_{2i})^{\S} \equiv \neg P_i$.

PROOF: It is immediate from the fact that both sides of the two equivalences are theorems of $V2$.

LEMMA 3.3. *If* $\vdash_C F^*$, *then* $\vdash_2 F$.

PROOF: $\vdash_C F^*$, then, by Theorem 2.3, we obtain $\vdash_2 (F^*)^{\S}$; but by Lemma 3.2 and Theorem 2.6, it follows that $\vdash_2 F$.

THEOREM 3.2. $\vdash_2 F$ *if and only if* $\vdash_C F^*$.

PROOF: By Lemmas 3.1 and 3.3.

COROLLARY. $\vdash_2 F$ *if and only if* F^* *is a tautology.*

Now, we shall use the decision method already obtained for $V2$, to obtain a semantics and a mechanical decision method for $V2$.

DEFINITION 3.3. Let F_2 be the set of formulas of $V2$. A *valuation* for $V2$ is a function $v: F_2 \longrightarrow \{0,1\}$ such that:
$v(A \supset B) = 1$ *iff* $v(A) = 0$ *or* $v(B) = 1$,
$v(A \,\&\, B) = 1$ *iff* $v(A) = v(B) = 1$,
$v(A \vee B) = 1$ *iff* $v(A) = 1$ *or* $v(B) = 1$,
If $A \notin S$, $v(A) = 1$ *iff* $v(\neg A) = 0$,
If $A \in S$, $v(A) = v(\neg A) = 1$.

Now we rewrite the Definition 3.2 in the following truth-tables M2, where MI constitute the usual truth-tables for the classical propositional calculus:

MI) A	B	$A \supset B$	$A \,\&\, B$	$A \vee B$	$A \notin S$, A	$\neg A$
0	0	1	0	0	0	1
1	0	0	0	1	1	0
0	1	1	0	1		
1	1	1	1	1		

MII) $A \in S$ A	$\neg A$
1	1

LEMMA 3.4. $\models F$ *if and only if* $\models_{M2} F$.

PROOF: It is immediate, from the fact that in M2 we have retained all the items of Definition 3.2, and considered all the different ways of assigning the values 0 and 1 to the atomic components of a given formula.

LEMMA 3.5. $\models_{M2} F$ *if and only if* F^* *is a tautology.*

PROOF: It is immediate from the fact that MI is the usual matrix for the classical propositional calculus, and that, if $A \in S$, A^* and $(\neg A)^*$ are tautologies.

THEOREM 3.3. $\vdash_2 F$ *if and only if* $\models_{M2} F$.

PROOF: By Theorem 3.2 and Lemma 3.5.

THEOREM 3.4. Theorem 2.15 *is valid for V2. Therefore, V2 is nontrivial.*

It is worth observing that in *V2*, Vasil'ēv propositional letters as well as its successive negations operate as constants, as p_0 in the system DM (or DL) of Routley and Meyer 1976. Another common feature between *V2* and DM (or DL) is that both are inconsistent but nontrivial. Nonetheless, *V2* is much stronger than DM and DL.

4. THE PROPOSITIONAL CALCULUS *V3*.

The language of *V3* is composed of:

(1) A denumerably infinite set of propositional letters.

(2) The connectives: $\supset$, &, V, $\neg$, $\cdot$, and $\bar{\ }$ (where $\cdot$ and $\bar{\ }$ are respectively a *non-classical conjunction* and a *non-classical negation*).

(3) Parentheses: (,).

In the following we will use Q as a syntactical variable for *atomic formulas* (defined as usual) and the other capital Latin letters as syntactical variables for formulas. The symbols $\vdash$, $\models$ and $\models_M$ have clear neanings.

DEFINITION 4.1. 1) *An atomic formula is a formula;* 2) *if Q is an atomic formula, then $\bar{Q}$ and $Q \cdot \bar{Q}$ are formulas;* 3) *if* A *and* B *are formulas, then* $A \supset B$, $A \& B$, $A \vee B$ *and* $\neg A$ *are formulas;* 4) *The only formulas are those given by the clauses* 1-3 .

Axiom schemata of *V3*:

1) $A \supset (B \supset A)$,
2) $(A \supset B) \supset ((A \supset (B \supset C)) \supset (A \supset C))$,
3) $A \supset (B \supset A \& B)$,
4) $A \& B \supset A$,
5) $A \& B \supset B$,
6) $A \supset A \vee B$,
7) $B \supset A \vee B$,
8) $(A \supset C) \supset ((B \supset C) \supset (A \vee B \supset C))$,
9) $A \vee \neg A$,
10) $A \supset (\neg A \supset B)$,
11) $\neg Q \equiv \overline{Q} \vee (Q.\overline{Q})$,
12) $\neg \overline{Q} \equiv Q \vee (Q.\overline{Q})$,
13) $\neg(Q.\overline{Q}) \equiv Q \vee \overline{Q}$.

Rule of inference:
$A , A \supset B / B$.

REMARK: Without changing the characteristics of $V3$, we can, by the introduction of new axioms (for example, $\overline{Q}.Q \equiv Q.\overline{Q}$) reinforce the properties of the connectives . and $\overline{}$. However, we leave this question to be studied together with the problem of giving many-valued interpretations to the connectives of the systems developed in this paper.

THEOREM 4.1. *The classical propositional calculus* C *is a proper subsystam of* $V3$.

THEOREM 4.2. *The following schemata are theorems of* $V3$:
(1) $\overline{Q} \vee Q \vee (Q.\overline{Q})$,
(2) $(Q \& \overline{Q}) \vee (Q \& (Q.\overline{Q})) \vee (\overline{Q} \& (Q.\overline{Q})) \supset B$,
(3) $\neg(Q \& \overline{Q})$,
(4) $\neg(Q \& (Q.\overline{Q}))$,
(5) $\neg(\overline{Q} \& (Q.\overline{Q}))$,
(6) $Q.\overline{Q} \supset \neg Q \& \neg \overline{Q}$,
(7) $\overline{Q} \supset \neg Q$.

PROOF: (1) is a consequence of Axioms 9 and 11; (3), (4), and (5) are consequences of (1); (6), and (7) are immediate. Then we prove:
(2): From Axioms 10 and 11 we obtain $Q \& (\overline{Q} \vee (Q.\overline{Q})) \supset B$,and from axioms 10 and 12 we obtain $\overline{Q} \& (Q \vee (Q.\overline{Q})) \supset B$. Hence, $(Q \& \overline{Q}) \vee (Q \& (Q.\overline{Q})) \supset B$, and $(\overline{Q} \& Q) \vee (\overline{Q} \& (Q.\overline{Q})) \supset B$. Therefore, $(Q \& \overline{Q}) \vee (Q \& (Q.\overline{Q})) \vee (\overline{Q} \& (Q.\overline{Q})) \supset B$.

THEOREM 4.3. *Let* F *be a formula of* C *and* $F^{\S}$ *a formula obtained from* F *replacing non-atomic formulas of* $V3$ *for atomic components of* F. *Under this conditions, if* $\vdash_C F$, *then* $\vdash_3 F^{\S}$.

DEFINITION 4.2. When Q occurs in a formula F not in a subformula of the form $\overline{Q}$ and $Q.\overline{Q}$, we have a *normal occurrence of* Q *in* F. When $\overline{Q}$ occurs in F not in a subformula of the form $Q.\overline{Q}$, we have a *simple occurrence of* Q

in F. Each occurrence of $Q.\overline{Q}$ as a subformula of F is called a *composite occurrence of* Q *in* F. (For example, in the formula $(Q \,\&\, \neg Q) \vee (\overline{Q} \supset Q.\overline{Q})$, the first two occurrences of Q are normal, the third is a simple occurrence of Q, and the last two (i.e., $Q.\overline{Q}$) together, constitute a composite occurrence of Q .)

Let $G(\overline{Q}, (Q.\overline{Q}))$ be a formula which may have simple and composite occurrences of Q. Then $G(A,B)$ is the formula obtained from $G(\overline{Q},(Q.\overline{Q}))$ replacing specified simple and composite occurrences of Q respectively by A and B. (For example, if $G(\overline{Q},(Q.\overline{Q}))$ is $Q \,\&\, \neg\overline{Q} \supset (Q \vee \overline{Q} \vee (Q.\overline{Q}))$, then $G(A,B)$ is $Q \,\&\, \neg A \supset (Q \vee A \vee B)$.)

THEOREM 4.4. *If* A *and* B *are non-atomic formulas of* $V3$, *then*
$\vdash_3 ((\overline{Q} \equiv A) \,\&\, (Q.\overline{Q} \equiv B)) \supset (G(\overline{Q},(Q.\overline{Q})) \equiv G(A,B))$.

DEFINITION 4.3. Let $q_1, q_2, \ldots$ be a fixed enumeration of the propositional letters of $V3$. Then F^* is the formula obtained from F performing the following substitutions:

$$
\begin{array}{rll}
q_{2i} & \text{for} & \text{normal occurrences of } q_i\,, \\
\neg q_{2i} \,\&\, q_{2i-1} & \text{for} & \text{composite occurrences of } q_i \text{ (i.e., for } q_i.\overline{q}_i\text{)}, \\
\neg q_{2i} \,\&\, \neg q_{2i-1} & \text{for} & \text{simple occurrences of } q_i \text{ (i.e., for } \overline{q}_i\text{)}.
\end{array}
$$

Now, let $(F^*)^{\S}$ be the formula obtained from F^* performing the following substitutions:

$$
\begin{array}{rll}
q_i & \text{for} & q_{2i}\,, \\
q_i.\overline{q}_i & \text{for} & q_{2i-1}\,.
\end{array}
$$

LEMMA 4.1. *If* $\vdash_3 F$, *then* $\vdash_C F^*$.

LEMMA 4.2. *The following schemata are theorems of* $V3$:
$(\neg q_{2i} \,\&\, q_{2i-1})^{\S} \equiv q_i.\overline{q}_i$, *and* $(\neg q_{2i} \,\&\, \neg q_{2i-1})^{\S} \equiv \overline{q}_i$.

PROOF: It is a consequence of the schemata $\neg Q \,\&\, (Q.\overline{Q}) \equiv Q.\overline{Q}$ and $\neg Q \,\&\, \neg(Q.\overline{Q}) \equiv \overline{Q}$, which are easily proved in $V3$.

LEMMA 4.3. *If* $\vdash_C F^*$, *then* $\vdash_3 F$.

PROOF: If $\vdash_C F^*$ then by Theorem 4.3 we obtain $\vdash_3 (F^*)^{\S}$. Then, by Lemma 4.2 and Theorem 4.4 it follows that $\vdash_3 F$.

THEOREM 4.5. $\vdash_3 F$ *if and only if* $\vdash_C F^*$.

COROLLARY. $\vdash_3 F$ *if and only if* F^* *is a tautology.*

In the following we give a semantics and a mechanical decision method for *V3*.

DEFINITION 4.4. Let F_3 be the set of formulas of *V3* . A *valuation for V3* is a function $v: F_3 \longrightarrow \{0,1\}$ such that:

$$
\begin{aligned}
v(A \supset B) &= 1 \quad \textit{iff} \quad v(A) = 0 \ \textit{or} \ v(B) = 1,\\
v(A \,\&\, B) &= 1 \quad \textit{iff} \quad v(A) = v(B) = 1,\\
v(A \vee B) &= 1 \quad \textit{iff} \quad v(A) = 1 \ \textit{or} \ v(B) = 1,\\
v(\neg A) &= 1 \quad \textit{iff} \quad v(A) = 0\ ,\\
v(Q) &= 1 \quad \textit{iff} \quad v(Q) = v(Q\cdot\overline{Q}\,) = 0,\\
v(\,\overline{Q}\,) &= 1 \quad \textit{iff} \quad v(Q) = v(Q\cdot\overline{Q}\,) = 0,\\
v(Q\cdot\overline{Q}\,) &= 1 \quad \textit{iff} \quad v(Q) = v(\,\overline{Q}\,) = 0\ .
\end{aligned}
$$

THEOREM 4.6. *If* $\vdash_3 F$, *then* $\models F$.

In a way similar to the usual one we prove that every maximal non-trivial set of formulas of *V3* has a model, and that every nontrivial set of formulas of *V3* has a model. Consequently, we prove, as is usual, the two following theorems:

THEOREM 4.7. *Let* Γ *be a set of formulas of V3. If* $\Gamma \models F$, *then* $\Gamma \vdash_3 F$, *and conversely.*

THEOREM 4.8. $\vdash_3 F$ *if and only if* $\models F$.

In order to prove that *V3* is decidable by a three-values truth-tables we begin rewritting Definition 4.3 in the following tables M, where we retained all the items of this definition and considered all the possibilities of assigning the values 0 and 1 to the atomic components of a given formula:

A	B	$A \supset B$	$A \,\&\, B$	$A \vee B$	$\neg A$
0	0	1	0	0	1
1	0	0	0	1	0
0	1	1	0	1	
1	1	1	1	1	

Q	$\overline{Q}$	$Q\cdot\overline{Q}$
0	0	1
1	0	0
0	1	0

If m be the number of propositional letters $\mathcal{Q}$ such that $\overline{\mathcal{Q}}$ occurs in F, and n is the number of the other atomic components of F, then it is easy to prove that the tables of F, according to M, have $3^m.2^n$ lines. Now, taking into account the relation between the Definition 4.3 and the tables M, it is immediate that:

LEMMA 4.4. $\models_{M} F$ *if and only if* $\models F$.

Let M3 = $\langle\{0,1,2\}, \{1\}, \supset, \&, \vee, \neg, \cdot, {}^{-}\rangle$ be such that:

A	B	$A \supset B$	$A \& B$	$A \vee B$	$\neg A$
0	0	1	0	0	1
1	0	0	0	1	0
2	1	1	0	0	1
0	1	1	0	1	
1	1	1	1	1	
2	1	1	0	1	
0	2	1	0	0	
1	2	0	0	1	
2	2	1	0	0	

$\mathcal{Q}$	$\overline{\mathcal{Q}}$	$\mathcal{Q}\cdot\overline{\mathcal{Q}}$
0	0	1
1	0	0
2	1	0

LEMMA 4.5. $\models_{M3} F$ *if and only if* $\models_{M} F$.

THEOREM 4.9. $\vdash_{3} F$ *if and only if* $\models_{M3} F$.

THEOREM 4.10. *The following schemata are not valid in V3. Consequently, V3 is nontrivial.*

$\mathcal{Q} \vee \overline{\mathcal{Q}}$, $\mathcal{Q} \vee (\mathcal{Q}\cdot\overline{\mathcal{Q}})$, $\neg(\mathcal{Q}\cdot\overline{\mathcal{Q}}) \supset \mathcal{Q}$, $\neg\overline{\mathcal{Q}} \supset (\mathcal{Q}\cdot\overline{\mathcal{Q}})$, $\mathcal{Q}\cdot\overline{\mathcal{Q}}$,
$\neg\mathcal{Q} \supset \overline{\mathcal{Q}}$, $\overline{\mathcal{Q}} \vee (\mathcal{Q}\cdot\overline{\mathcal{Q}})$, $\neg(\mathcal{Q}\cdot\overline{\mathcal{Q}}) \supset \overline{\mathcal{Q}}$, $\neg\mathcal{Q} \supset (\mathcal{Q}\cdot\overline{\mathcal{Q}})$, $\neg\overline{\mathcal{Q}} \supset \mathcal{Q}$.

PROOF: Employing the truth-tables M3.

5. Conclusion.

According to our interpretation, the main characteristic of Vasil'ēv's logical views is to accept the possibility of a logic in which the law of contradiction is not valid in general. Since in Imaginary Logic the law of non-self-contradiction is valid, but not the (ontological) law of contradiction, we may interpret it as a logic in which it is possible to handle two sorts of negation (*logical* and *ontological*). We say that a logic is adequate to handle two sorts of negation if: 1) in its language there are two different symbols of negation, or 2) even if there is only one symbol of negation it is possible to characterize another negation, as in the systems $\mathscr{C}_n$, $1 \leq n \leq \omega$, of Costa 1974, or in the systems *V1* and *V2* .

Moreover, anlyzing the conceptions of Vasil'ēv from the point of view of present day formal logic, it seems that the laws of metalogic should be formulated in the propositional level, and the laws of ontological basis of logic in the predicate level. However, in accordance with the same point of view, the laws concerning negation belong mostly to the propositional level. Then, as we do not want here to make an exegesis of Vasil'ēv's work, we will not talk of the metalogic and ontological basis of logic, but of propositional and predicate calculi of Vasil'ēv.

Therefore, we call Vasil'ēv propositional calculus either an extension of the classical propositional calculus or a propositional calculus in which it is possible to obtain the classical propositional calculus, with the requirement, in both cases, that the calculus be adequate (as defined above) to handle two sorts of negation. One of these negations must be the *classical* one (the *logical negation*) and for the other, the *non-classical negation*, the law of contradiction must not be valid. In the predicate level the non-classical negation will be actually an *ontological negation*.

The systems $\mathscr{C}_n$, $1 < n \leq \omega$, are examples of Vasil'ēv propositional calculi in which it is possible to obtain the classical propositional calculus.

Subsequently we point out some of the characteristics of the systems *V1*, *V2*, and *V3* which allow us to name them Vasil'ēv propositional calculi. The first, common to all, is that they are extensions of the classical propositional calculus in which there is (or it is possible to characterize) a negation for which the law of contradiction is not valid.

V1 is constructed with only one symbol of negation, but employing the alternative device of two sorts of propositional letters. Consequently, *V1* is able to handle two sorts of negation because the symbol $\neg$ when applied to classical propositional letters or to non-atomic formulas has classical properties, and when applied to Vasil'ēv propositional letters, has non-classical properties; for example, the schema $\neg(A \ \& \ \neg A)$, when A is a propositional letter of Vasil'ēv, is not valid (as well as the other schemata of Theorem 2.15).

V2 has all the characteristics just mentioned for *V1*, but its most interesting feature is its being a propositional calculus which is inconsistent nontrivial. In this aspect it is similar to DM and DL of Routley and Meyer 1976, but much stronger than they are, and also decidable by a two-elements truth-table.

V1 and *V2* are particularly interesting because they seem to be adequate for the study of Meinong's theory of objects.

V3 is a propositional calculus with two symbols of negation: $\neg$ is the classical negation, and $\overline{\ }$ the non-classical negation. For the symbol $\overline{\ }$ the syntactical law of excluded middle is not valid, i. e., the schema $Q \vee \overline{Q}$ is not valid. But it is valid a syntactical law of excluded fourth: $Q \vee \overline{Q} \vee (Q \cdot \overline{Q})$ (it is easy to verify that in this formula the disjunction is exclusive). Nonetheless, the law of contradiction cannot even be formulated for this negation; in effect $\overline{Q \cdot \overline{Q}}$ is not a well formed formula of *V3*, since the symbol $\overline{\ }$ applies only to propositional letters. Employing the truth-table which decides *V3*, we can see that the truth-values of Q, $\overline{Q}$, and $Q \cdot \overline{Q}$ satisfy the following laws: *If one is true, then the other two are false; if one is false, then one, but not both of the others, is true; if two are false, then the third is true.* These laws are the ones, according to V. A. Smirnov 1962, proposed by Vasil'ēv to govern the truth-values of the affirmative, negative, and indifferent propositions in his logic with a law of *excluded fourth*. Thus, *V3* is a propositional calculus which appears adequate to be used in the development of Imaginary Logic based on affirmative, negative, and indifferent propositions.

With the above said, we reach the first objective of this paper: to emphasize that Vasil'ēv is a forerunner of the logics constructed for the study of inconsistent nontrivial theories, since the main characteristic of these logics is the non-validity in general of the law of contradiction. Furthermore, we can say that Vasil'ēv is more a forerunner of such logics than of many-valued ones, because we can construct a great variety of propositional calculi which may be called Vasil'ēv's, and which are not many-valued. For example, the $\mathcal{C}_n$, $1 \leq n \leq \omega$, and some denumerable hierarchies of propositional calculi that exist between $\mathcal{C}_n$ and $\mathcal{C}_{n+1}$. Moreover, the calculi presented in this paper (specially *V3* which fits best with some of the ideas of Vasil'ēv), although decidable by finite truth-tables, cannot really said to be many-valued, since we did not give any intuitive many-valued interpretations to their connectives, specially to non-classical negation and non-classical conjunction.

The second objective of the present paper was only partially obtained. Although we developed to some extent the calculi *V1*, *V2*, and *V3*, their study was not completed, since we left the development of the corresponding predicate calculi and their applications for future papers. Then, one can argue that in present stage of development of these Vasil'ēv propositional calculi, their value is only that of formal games. Nevertheless, the answer to this kind of question was already given by C. S. Peirce, when he wrote to William James, in 1909:

> "I have long felt that it is a serious defect in the existing logic that it takes no heed of the *limit* between two realms. I do not say that the Principle of Excluded Middle is downright *false*; but I *do* say that in every field of thought whatsoever there is an intermediate ground between *positive assertion* and *positive negation* which is just as Real as they. Mathematicians always recognize this, and seek for that limit as the presumable lair of powerful concepts; while metaphysicians and oldfashioned logicians, - the sheep & goat separators,- never recognize this. The recognition does not involve any denial of existing logic, but it involves a great addition to it." (Quoted in M. Fisch and A. R. Turquette, *Peirce's Triadic Logic*, The Transactions of the Charles S. Peirce Society, vol. 2 (1966), p. 81.)

REFERENCES.

Arruda, A. I.

1967, *Sur certaines hiérarchies de calculs propositionnels*, C. R. Acad. Sc. Paris, 265, 641-644.

Arruda, A. I. and N. C. A. da Costa

1970, *Sur le schéma de la séparation*, Nagoya Math. Journal, 38, 71-84.

Asenjo, F. G. and J. Tamburino

1975, *Logic of Antinomies*, Notre Dame Journal of Formal Logic, XV, 17-43.

Church, A.

1936, *A Bibliography of Symbolic Logic*, The Journal of Symbolic Logic, 1, 121-216, and 3 (1938), 178-192.

Comey, D. D.

1965, *Review of V. A. Smirnov 1962*, The Journal of Symbolic Logic, 30, 368-370.

Costa, N. C. A. da

1974, *On the theory of inconsistent formal systems*, Notre Dame Journal of Formal Logic, XV, nº 4, 479-510.

Costa, N. C. A. da and E. H. Alves

1976, *Une sémantique pour le calcul* $\mathcal{C}_1$, C. R. Acad. Sc. Paris, 383 A, 729-731.

Hessen, S. I.

1910, *Review of* N. A. *Vasil'ēv 1910*, Logos 2, 287-288.

Jammer, M.

1974, The Philosophy of Quantum Mechanics, John Wiley & Sons, New York.

Kleene, S. C.

1952, Introduction to Metamathematics, Nan Nostrand, Princeton.

Kline, G.

1965, *N. A. Vasil'ēv and the development of many-valued logic*, Contributions to Logic and Methodology in Honor of J. M. Bochenski, edited by A. T. Tymieniecka, North-Holland, Amsterdam, 315-326.

Korcik, A.

1965, *Przyczynek do historii klasycznej teorii opozycji zdań asertorycznych (A contribution to the history of the classical theory of opposition of assertoric propositions)*, Roczniki Filozoficzne, 4, 33-49.

Rescher, N.

1969, Many-valued Logic, McGraw-Hill, New York.

Routley, R. and R. K. Meyer

1976, *Dialectical logic, classical logic, and the consistency of the world*, to appear in Studies in Soviet Thought.

Smirnov, K. A.

1911, *Review of* N. A. *Vasil'ēv 1910*, Žurnal Ministērstva Narodnogo Prosveščeniā, XXXII, 144-154.

Smirnov, V. A.

1962, *Logičēskiē vzglādy* N. A. *Vasil'ēva* (N. A. *Vasil'ēv's logical views*), Očerki po istorii logiki v Rossii (Essays in the

History of logic in Russia), Izdatel'stvo, MGU, 242-257.

Vasil'ēv, N. A.

1910, *O častnyh suždēniāh, o trēugol'nikē protivopoložnostēj, o zakonē isklūčēnnogo čētvertogo* (*On particular propositions, the triangle of oppositions, and the law of excluded fourth*), Učēniē zapiski Kanzan'skogo Universitēta, 42 p.

1911, *Voobražaēmaā logika: Konspēkt lektsii* (*Imaginary logic, Abstract of a lecture*), 6 pp.

1912, *Voobražaēmaā (nēariatotēlēva) logika* (*Imaginary (non-Aristotelian) logic*) Žurnal Ministērstva Narodnogo Prosvēščeniā, vol.40, 207-246.

1913, *Logika i mētalogika* (*Logic and metalogic*), Logos, vols. 2-3, 53-81.

1925, *Imaginary (Non-Aristoteliam) Logic*, Atti del V Congresso Internazionale di Filosofia (Napoli, 5-9 maggio, 1924), Naples, 107-109.

Departamento de Matemática
Universidade Estadual de Campinas
Campinas, São Paulo, Brazil.

Non-Classical Logics, Model Theory and Computability,
A.I. Arruda, N.C.A. da Costa and R. Chuaqui (eds.)

FORMALIZING MULTIPLE LOCATION

by F. G. ASENJO

0. PURPOSE.

Whitehead criticized the concept of simple location as insufficiently descriptive of the spacial relationships of actual entities in the physical world (see Whitehead 1967, chapters III and IV, specially page 65). The concept of field of forces provides a good example of the kind of formal ubiquity which reality exhibits and which simple location cannot convey. Entities in a field exert their dynamic influence throughout that field and are in turn influenced by all the field's other entities as well as by its general distribution of forces. Because space is thought to be a single and uniform depository of physical entities, this plurality of dynamic effects is usually described as being merely part of the general phenomenon of action at a distance. Actually, entities are not placed in an indifferent space; space is a changing property of the dynamic conditions of a field, which is specially evident at the microphysical level. Every region in a field has what elsewhere we have called *multiple location* (see Asenjo 1962), i. e., a real and efficient extension of each region into and throughout other regions of the field. Such an idea of multiple location provides us with a more concrete conceptual approach to the realities of the physical world. No longer must one take refuge in the mysterious action at a distance; instead, one can think of entities as acting upon one another by virtue of their mutual coextensionality. This idea is what we shall formalize here. The problems in realizing this project are as follows. Classical set theory lends itself too naturally to interpretation in terms of simple location. Elements in a set are clearly distinguished from one another and all are related externally to the set that gathers them together through membership - a relationship that unavoidably performs a radical selection from the elements' many properties. It is appropriate to collect numbers into a set, say, but one cannot collect the entities of a physical field into a set without doing violence to the wealth of concrete relationships that those entities have between one another. Current point set topology is of no help, either, because it is a form of applied set theory. We need to be able to formally place one simple location into another; that is, we want to formalize the fact that simple location V_1 has location V_2 as well, that a point located at V_1 also moves into the location V_2 — although not necessarily vice versa—just as forces may be either exclusively outgoing or exclusively ingoing in a given region (the so-called *sources* and *sinks*). We shall use the concept of directed graph for this purpose, vertices representing prima facie simple locations, and directed edges, symbolized by the notation $\overrightarrow{V_iV_j}$, representing multiple locations — explicitely, V_i is the inicial vertex and V_j the terminal one in the location of V_i into V_j. $\overrightarrow{V_iV_j}$ will be called an outgoing edge at V_i, and an ingoing one at V_j. Extreme cases will be (i) those in which every bilocation $\overrightarrow{V_iV_j}$ is symmetric (that is, the existence of $\overrightarrow{V_iV_j}$ implies that of $\overrightarrow{V_jV_i}$ in the graph), and (ii) those in which edges are totally absent, multiple location then being reduced to simple location (points becoming mere vertices or sets of vertices without directed links). In general, points here will be graphs, finite or infinite, and the pattern of directed edges of a given point will represent the network of multiple locations intrinsic to that point. A point, then, will have structure, its vertices and directed edges making up its internal constitution. Further, it is essential to the notion of multiple location that this internal constitution not be closed, but open to enlargement and additional structural articulation. This requires that points not be sealed elements to be collected; rather; they must be sets of some kind that can be included, embedded in other larger points. Points are not to be taken as irreducible members of a set, immodifiable topological atoms in a neighborhood, but as entities at the same logical level as that of any set or neighborhood which contains them. Indeed, here points will themselves be sets of a special kind, and in turn a topological space will sometimes be one or more point among others.

From formalizing multiple location it follows that no figure has a single geometric structure — an "absolute appearence", to use a paradoxical expression that describes our ordinary, naive idea of form. The topological aspect of a figure is relative to various points of view, a matter of topological perspective; that is, considered from different vertices within a point in a figure (thought of as an assemblage of vertices), different topological configurations describe that same figure. Further, there are vertices from which the figure cannot be described by any topological configuration; also, there are vertices without neighborhoods, as well as vertices from which no topology at all can be built (see examples in Section 4). A torus is not a torus from all viewpoints. If this appears bewildering at first, let us stop to think: Why should a figure have a unique topological configuration? In the physical world the singleness of a figure's topology is a macroscopic preconception, a matter of choosing from a wealth of appearences whose primary or secondary character depends on viewpoint. Indeed, it is the conclusions of current ultramicroscopic physics that force us to acknowledge this basic perspectivism of space as a routine property of matter.

I. A Set Theory Based on the Notion of Inclusion.

1. Inclusion.

The primitive ideas are those of set, inclusion, and binary relation. Capital letters stand for sets and the inclusion relation is denoted by $\subseteq$. In addition we have an unlimited number of symbols for binary relations (R, F, f, g, ...). Also, let us assume the formal arithmetic of non-negative integers, including ordinary induction.

DEFINITION 1. (*Equality.*) $X = Y$ stands for $(Z)(Z \subseteq X \equiv Z \subseteq Y)$.

AXIOM 1. (*Extensionality.*) $(X)(Y)(X = Y \rightarrow (Z)(X \subseteq Z \equiv Y \subseteq Z))$.

Def. 1 establishes that equal sets are those having the same subsets and no others, whereas Ax. 1 determines further that equal sets are subsets of the same sets. Obviously, $X = X$.

DEFINITION 2. (*Proper inclusion.*) $X \subset Y$ stands for $X \subseteq Y$ & $X \neq Y$.

AXIOM 2. (*Nul set.*) $(\exists X)(Y)(Y \nsubseteq X \ \& \ (Z)(Z \neq X \equiv X \subseteq Z))$.

There exists a set without subsets which is included in every other set except itself. By Def. 1 and Ax. 1, this set is unique . It will be represented by $\emptyset$.

Since the axiom of extensionality guarantees that a set is uniquely determined by its subsets, the notation $S = \{X, Y, Z, \ldots\}$ is then in order (where $X, Y, Z, \ldots$ are all the subsets of S in finite or infinite number). Of course, for every set $S \neq \emptyset$, S itself and $\emptyset$ are to be listed between brackets, and because $\{\emptyset\}$ is meaningless, so is the equation $S = \{\emptyset\}$ Further, it is never the case that $X = \{X\}$.

AXIOM 3. (*Reflexivity, Antisymmetry,* and *Transitivity of Inclusion.*)
$(X)(X \neq \emptyset \longrightarrow X \subseteq X)$ & $(X)(Y)(X \subseteq Y$ & $Y \subseteq X \longrightarrow X = Y)$ &
$(X)(Y)(Z)(X \subseteq Y$ & $Y \subseteq Z \longrightarrow X \subseteq Z)$.

AXIOM 4. (*Separation.*) $(X)(\exists Y)$ $[Y \subseteq X$ & $(Z)(Z \subseteq X$ & $\phi(Z) \longrightarrow Z \subseteq Y)$ &
$(V)(Z)((Z \subseteq X$ & $\phi(Z) \longrightarrow Z \subseteq V) \longrightarrow Y \subseteq V)]$,

where $\phi(Z)$ is any wff with one free variable. Notice that Y may also contain sets U such that $\neg\phi(U)$. Obviously, if $\phi(X)$, then X itself satisfies Ax. 4. If $\neg\phi(X)$, Ax. 4 guarantees the existence of a least set included in X that contains all subsets of X with the property ϕ (plus any other subsets of X without such property but not separable from Y because of their being included in some subset Z of X with the property ϕ).

The notation $Y = \{Z: Z \subseteq X$ & $\phi(Z)\}$ is now justified: Y is the least subset of X that contains all the subsets of X that satisfy $\phi(Z)$.

AXIOM 5. (*Expansion.*) $(X)(\exists Y)(\exists Z)(X \subseteq Y$ & $Z \subseteq Y$ & $Z \nsubseteq X$ & $X \nsubseteq Z)$.

As a consequence of this axiom there is no class of all sets. Also, there exists at least a countable infinity of sets. In fact, there is an infinity of countably infinite sequences of sets. Ax. 5 can be applied successively to assert the existence of nested sequences of distinct sets, each properly included in the following ones (chains),as well as to asset the existence of sequences of sets that are pairwise incomparable with respect to inclusion (antichains).These are the two extreme poles in the spectrum of all the possible arbitrary sequences of sets whose existence derives from this axiom. Let $Exp_1(X,U)$ indicate that the set U is obtained by applying Ax. 5 to X once, U being either a set that properly contains X or a set incompable to X . Let $Exp_k(X,U)$ indicate that U is obtained from X after k applications of Ax.5 (where k is a non-negative integer and $Exp_0(X,Y)$ denotes X itself), and where the k successive choices are made either by following some recursive schema or at random.

AXIOM 6. (*Infinity*). $(X)(\exists Y)(Z)(U)(X \subseteq Y \ \&\ (Exp_k(X,Z) \subseteq Y \longrightarrow Exp_{k+1}(X,U) \subseteq Y))$:

The infinite set Y (denoted $Exp\ (X)$), whose existence is asserted by this axiom, collects all the sets obtainable by a finite number of sequential applications of Ax. 5 to a given set X. Notice that $Exp\ (X)$ does not collect all the supersets of X, but at most only a countable sequence of them (plus all the subsets of each term of such sequence).

Let $Seq\ (X)$ denote a particular infinite sequence $X, X_1, X_2, \ldots X_k, \ldots$ of sets obtained by successive applications of Ax.5 (∅ could be the initial set X of any such sequence, but it could not occupy any other place in the sequence). Let $Seq_k(X,U)$ indicate the k-th term of such sequence with $Seq_o(X,U) = X$.

AXIOM 7. (*Union of a sequence*) $(X)(\exists Y)(Z)(U)(X \subseteq Y \ \&\ (Seq_k(X,Z) \subseteq Y \longrightarrow Seq_{k+1}(X,U) \subseteq Y))$.

Obviously the union of the terms of a sequence (denoted $\cup Seq\ (X)$) is a subset of $Exp\ (X)$.

AXIOM 8. (*Union*). $(X)(Y)(\exists Z)(U)(U \subseteq X \vee U \subseteq Y \equiv U \subseteq Z)$.

AXIOM 9. (*Intersection*) $(X)(Y)(\exists Z)(U)(U \subseteq X \ \&\ U \subseteq Y \equiv U \subseteq Z)$.

Union and intersection, which are uniquely determined, will be denoted by $X \cup Y$ and $X \cap Y$, respectively. It is clear that both operations are associative and satisfy the distributive laws.

2. ELEMENTHOOD AND MEMBERSHIP.

DEFINITION 3. (*Element*) $E\ (X)$ stands for $(X \neq \emptyset)\ \&\ (Y)(Y \neq \emptyset \longrightarrow Y \nsubseteq X \vee Y = X)$.

Elements are nonempty sets without nonempty proper subsets. The null set is not an element.

AXIOM 10. (*Regularity*) $(X)(X \neq \emptyset \Rightarrow (\exists Y)(Y \subseteq X \;\&\; E(Y)))$.

Every nonempty set contains at least one element (eventually itself only).

AXIOM 11. (*Element expansion*) $(X)(\exists Y)(\exists Z)(X \subseteq Y \;\&\; Z \subseteq Y \;\&\; Z \nsubseteq X \;\&\; E(Z))$:

Hence, there is no set of all elements, and there is at least a countable infinity of them.

AXIOM 12. (*Pairing*) $(X)(Y)(\exists Z)(E(X) \;\&\; E(Y) \Rightarrow (U)(U \subseteq Z \equiv U = X \vee U = Y \vee U = \emptyset))$.

There exists the set that contains exclusively a given pair of elements (plus $\emptyset$). The bracket notation $\{x,y\}$ is now in order; small letters indicate elements, and $\{x,y\}$ is the unique set that contains x, y, and $\emptyset$. $\{x\}$ is $\{x,x\}$.

DEFINITION 4. (*Successor*) $S(X)$ stands for $X \cup Y$ where Y is any set such that $E(Y)$ and $Y \nsubseteq X$.

The successor of a set is not uniquely determined, but by Ax. 11 and Ax. 8, countably infinite sequences of sets can be assumed to exist such that, begining with a given set, every set that follows is the successor of the preceding one.

DEFINITION 5. (*Membership*) $X \in Y$ stands for $X \subseteq Y \;\&\; E(X)$

Only elements are members.

THEOREM 1. $(X)(E(X) \Rightarrow X \in X)$.

PROOF: Since every set is a subset of itself, every element is a member of itself.

THEOREM 2. $E(X) \;\&\; E(Y) \;\&\; X \neq Y \Rightarrow X \notin Y \;\&\; Y \notin X$.

PROOF: Antisymmetry of inclusion and Def. 3.

AXIOM 13. (*Comprehension*) $(\exists X)(Y)(Y \in X \equiv \phi(Y))$,
where ϕ is any wff with one free variable. The axiom asserts the existence of a set containing all the elements that have the property ϕ.

THEOREM 3. *The set of all elements X which are not members of themselves is empty.*

PROOF: Theo. 1 for $X \neq \emptyset$, and the fact that $\neg E(\emptyset)$ by Def. 3.

3. CARTESIAN PRODUCT, FUNCTIONS, CARDINALITY, ORDER.

DEFINITION 6. (*Cartesian product*) Given any two sets A and B, their *Cartesian product* is the binary relation defined as follows:

$$(A \times B)(X,Y) \equiv X \subseteq A \ \& \ Y \subseteq B).$$

(Cartesian products are not sets.) $\emptyset \times \emptyset$ holds for no pair of sets, $(\emptyset,\emptyset)$ included.

DEFINITION 7. (*Correspondences*) Given a Cartesian product $(A \times B)$, a *correspondence* between A and B is any binary relation R that satisfies

$$R(X,Y) \longrightarrow (A \times B)(X,Y).$$

DEFINITION 8. (*Functions*) A *function* on A into B is a correspondence F between A and B such that for each $X \subseteq A$ there is one and only one $Y \subseteq B$ such that $F(X,Y)$. In symbols:

$$(X)(X \subseteq A \longrightarrow ((\exists Y)\, F(X,Y) \ \& \ (Y)(Z)(F(X,Y) \ \& \ F(X,Z) \longrightarrow Y = Z))).$$

If for each $Y \subseteq B$ for which there is an $X \subseteq A$ such that $F(X,Y)$ there is only one such X, the function is called monomorphic. If for every $Y \subseteq B$ there is at least one $X \subseteq A$ such that $F(X,Y)$, the function F is called surjective.

DEFINITION 9. (*Cardinality*) Two sets A and B have the same cardinality (denoted $|A| = |B|$) if there exists a monomorphic and surjective function on A into B. If there exists a monomorphic and surjective function on A into a subset of B, but not one on B into a subset of A, then A is said to have lesser cardinality than B (denoted $|A| < |B|$). Obviously, for all X, $|X| \leq |USeq\ (X)| \leq |Exp\ (X)|$.

AXIOM 14. (*Total order*) $(X)(\exists R)((Y)(Y \subseteq X \longrightarrow R(Y,Y)$ &
$(Y)(Z)(Y \subseteq X \ \& \ Z \subseteq X \longrightarrow (R(Y,Z) \ \& \ R(Z,Y) \longrightarrow Y = Z))$ &
$(Y)(Z)(U)(Y \subseteq X \ \& \ Z \subseteq X \ \& \ U \subseteq X \longrightarrow (R(Y,Z) \ \& \ R(Z,U) \longrightarrow R(Y,U))$ &
$(Y)(Z)(Y \subseteq X \ \& \ Z \subseteq X \longrightarrow R(Y,Z) \vee R(Z,Y))).$

Every set can be totally ordered.

II. The Topology of Multiple Location.

4. A graph topology.

Henceforward, the notions of set, element, inclusion, and sequence are those presented in Part I. A topological space *relative to a vertex* V shall be a directed graph X_V , the product graph of all the graphs labeled points relative to V (not every subgraph of X_V is a V-point), and a sequence T_V of subsets of X_V, called neighborhoods, that satisfy the definition and axioms given below. Graphs are arrays of vertices (elements) and directed edges (introduced in the usual way, though not as a Cantorian ordered pair of elements, but as elements themselves that are symbolized $\overrightarrow{V_iV_j}$). It is understood that every graph that contains $\overrightarrow{V_iV_j}$ as an element also contains V_i and V_j, although not necessarily $\overrightarrow{V_jV_i}$. Given X_V, it is determined which of its subsets are V-points. It is assumed, further, that it is always possible to ascertain for a given vertex in a given subgraph whether the number of outgoings edges is greater, equal, or less then the number of ingoing edges (or whether those two numbers are incomparable). Note that, in accordance with Part I, a set of graphs is their own product graph, which includes all the new graphs that can be formed with the assorted vertices and edges of the given graphs. The topological space X_V shall be, then, both a directed graph and a set, true also of points and neighborhoods.

DEFINITION 1. Given a vertex W in X_V, a *neighborhood* of W (denoted $N_V(W)$) is any point or product graph of points of X_V that (i) contains W, and such that (ii) the number of outgoing edges of W in $N_V(W)$ is greater than or equal to the number of its ingoing edges.

(Note that just as neither the subgraph nor the supergraph of a point are necessarily points, neither are the subgraph nor the supergraph of a neighborhood necessarily neighborhoods.)

Let T_V be a sequence whose terms are all neighborhoods and such that every neighborhoods of X_V is a subset of a term of T_V . Since the predominance of outgoing edges is preserved by finite or infinite unions, the existence of T_V follows. T_V is called a *topology* relative to V if the following are satisfied.

AXIOM 1. X_V is a term of T_V .

AXIOM 2. Given two neighborhoods $N_V(W)$ and $N'_V(W)$, their intersection is also a V-neighborhood of W.

Let us look at two very simple examples given here merely to add some intuitive interpretation to the previous concepts. Consider the graph X, composed of the vertices V_1, V_2, V_3, and the edges $\overrightarrow{V_1V_2}$, $\overrightarrow{V_3V_2}$, and $\overrightarrow{V_3V_1}$. Let the points of X_{V_1} be V_1; V_2; $\overrightarrow{V_1V_2}$ (i.e., the subgraph composed of three elements V_1, V_2, $\overrightarrow{V_1V_2}$); and $\overrightarrow{V_3V_2}$. Let the points of X_{V_2} be V_3; $\overrightarrow{V_1V_2}$; and $\overrightarrow{V_3V_1}$. Finally, let the points of X_{V_3} be V_1; V_2; $\overrightarrow{V_1V_2}$; $\overrightarrow{V_3V_2}$; and $\overrightarrow{V_3V_1}$. The neighborhoods in X_{V_1} (the product graph of its points) are the following. Neighborhoods of V_1: V_1; V_1, V_2; and $\overrightarrow{V_1V_2}$. Neighborhoods of V_2: V_2; V_2, V_1. Neighborhoods of V_3: $\overrightarrow{V_3V_2}$; V_1, $\overrightarrow{V_3V_2}$; V_3, V_2, $\overrightarrow{V_1V_2}$. These eight neighborhoods constitute a topology for the space X_{V_1}. The space X_{V_2} also has a a topology, although V_2 has no neighborhoods. X_{V_3} is not a topological space.

Let us now consider the graph X, composed of all positive integers as vertices and of all the edges $\overrightarrow{k, k+m}$ for all positive integers k, $m \geq 1$. For each vertex k, let the points of X_k be the edges $\overrightarrow{k,k+m}$ for every $m > 1$. Only k has neighborhoods in X_k, but these form a topology for X_k.

5. Closure, Derived Set, Boundary.

DEFINITION 2. Given a V-point P and a neighborhood $N_V(W)$, $N_V(W)$ is called a V-neighborhood of P iff $P \subseteq N_V(W)$.

DEFINITION 3. Let X_V be a topological space and S a subset of X_V, a V-point P is said to be a *limit point* of S iff every V-neighborhood of P contains at least one vertex W of S not in P.

DEFINITION 4. The product graph of all limit points of a set $S \subseteq X_V$ is called the *derived set* of S, denoted S'.

DEFINITION 5. A set $S \subseteq X_V$ that contains all its limit points is called *closed*. We shall call $S \cup S' = \overline{S}$ the *closure* of S.

DEFINITION 6. The *boundary* of a set $S \subseteq X_V$ (denoted $Bd(S)$) is the product graph of those points common to the closure of S and the closure of

$X_V - S$, the latter being the graph spanned by all vertices and edges not in S (note that although S and $X_V - S$ have no edges in common, they can have some vertices in common).

DEFINITION 7. The *interior* of a set $S \subseteq X_V$ is $S - Bd(S)$.

Obviously, if $S_1 \subseteq S_2$, then $S'_1 \subseteq S'_2$. Hence, if S_1 and S_2 are both closed (and therefore $S'_1 \subseteq S_1$ and $S'_2 \subseteq S_2$), then $(S_1 \cap S_2)' \subseteq S'_1$ and $(S_1 \cap S_2)' \subseteq S'_2$. But every limit point of $S_1 \cap S_2$ is also a limit point of S_1 as well as a limit point of S_2, therefore $(S_1 \cap S_2)' \subseteq S'_1 \cap S'_2$. We then have the following.

THEOREM 1. *The intersection of a finite sequence of closed sets is closed (therefore the boundary of a set is closed).*

6. Homeomorphism, connectedness, compactness.

DEFINITION 8. Given two topological spaces X_V and X_W from the same graph X (or X_V and Y_W from different graphs X and Y), a function $f: X_V \longrightarrow X_W$ (or $f: X_V \longrightarrow Y_W$ respectively) which maps vertices into vertices, and edges into edges of corresponding vertices in a direction-preserving manner (i.e., with each edge $\overrightarrow{V_1 V_2}$ mapped into the edge $\overrightarrow{f(V_1) f(V_2)}$) is said to be *continuous* at a vertex $V_0 \subseteq X_V$ iff for every W-neighborhood $N_W(f(V_0))$ of $f(V_0)$, $f^{-1}(N_W(f(V_0)))$ is a V-neighborhood of V_0, where $f^{-1}(N_W(f(V_0)))$ is the graph of all vertices and edges whose image under f is $N_W(f(V_0))$. f is said to be continuous if f is continuous at each vertex of X_V.

DEFINITION 9. Two topological spaces are said to be *homeomorphic* iff there exist inverse functions $f: X_V \longrightarrow X_W$ (or $f: X_V \longrightarrow Y_W$) and $g: X_W \longrightarrow X_V$ (or $g: Y_W \longrightarrow X_V$ respectively) such that f and g are continuous; f and g are then called *homeomorphisms*.

DEFINITION 10. Two nonempty sets S_1 and S_2 of a topological space X_V are said to be *separated* iff $S_1 \cap S_2 = S_1 \cap S'_2 = S'_1 \cap S_2 = \emptyset$.

DEFINITION 11. A set which is not the union of two separated sets is called *connected*.

THEOREM 2. *A closed set S is connected iff it is not the union of two nonempty disjoint closed sets.*

PROOF: Assume $S = A \cup B$ where A and B are nonempty and disjoint closed sets. We have $A' \cap S \subseteq A$ and $B' \cap S \subseteq B$. Then $A' \cap B = A' \cap S \cap B \subseteq A \cap B = \emptyset$ and $A \cap B' = A \cap S \cap B' \subseteq A \cap B = \emptyset$. Therefore, A and B are separated and S is not connected.

If S is not connected, then $S = A \cup B$ where A and B are separated sets. Then $A' \cap S = A' \cap (A \cup B) = (A' \cap A) \cup (A' \cap B) = A' \cap A \subseteq A$, which means that A is closed. Similarly, so is B.

THEOREM 3. *The closure of a connected set S is connected.*

PROOF: By contradiction: assume S not connected. Since nonempty subsets of separated sets are correspondingly separated, it is easy to show that then S could not be connected (against the premise of the theorem).

COROLLARY 4. *A connected set that is contained in the union of two separated sets is contained in one of the two sets.*

DEFINITION 12. A set S is *compact* iff it is finite (composed of a finite number of vertices and edges) or if every infinite subset of S has a nonempty derived set.

THEOREM 5. *The subset of a compact set is compact. Further, the union of a finite number of compact sets is compact.*

PROOF: Let $S = S_1 \cup S_2 \ldots \cup S_n$ be the union of n compact sets. If T is an infinite subset of S, then at least one of the sets $T \cap S_i$ is infinite. Since $T \cap S_i \subseteq S_i$, and S_i is compact, T possesses at least one limit point.

(NOTE: As these definitions and theorems demonstrate, general topology can be constructed without excessive differences on a set theory based on inclusion; this shows that membership is not in itself an indispensable relationship for topology. Furthermore, to base topology on inclusion also indicates a way in which to make topological notions play a basic role in graph theory — general topology and graph theory being disciplines that so far have very little to do with one another.)

REFERENCES.

Asenjo, F. G.

1962, El Todo y las Partes, Editorial Tecnos, Madrid.

Whitehead, A. N.

1967, Science and the Modern World, The Free Press, New York.

Department of Mathematics
University of Pittsburgh
Pittsburgh, Pennsylvania, U.S.A.

Non-Classical Logics, Model Theory and Computability,
A.I. Arruda, N.C.A. da Costa and R. Chuaqui (eds.)

ON JAŚKOWSKI'S DISCUSSIVE LOGIC (*)

by *N.C.A. da COSTA and L.DUBIKAJTIS*

We call a deductive theory (or system) T consistent if for no formula α of T, both α and $\neg\alpha$ are theses of T ($\neg\alpha$ is the negation of α); otherwise, T is called inconsistent. A system T is said to be nontrivial (or not over-complete) if not every formula of T is a thesis; otherwise, T is said to be trivial (or over-complete). If the underlying logic of T is classical logic (or one of most of existent logical systems), then T is inconsistent if, and only if, it is trivial. Therefore, if we want to construct and study inconsistent but nontrivial theories, we have to develop new kinds of logic.

Jaśkowski (in 1948) was the first to develop a propositional calculus which could be employed in the *direct* study of nontrivial and inconsistent deductive theories. He named his calculus *discussive* (or *discoursive*) *propositional calculus*.

Several are the problems that motivated the development of discussive calculus. For example, Jaśkowski makes reference to the following (all citations are from Jaśkowski 1948):

1) To systematize conceptions which contain contradictions (i. e., in which a proposition α and its negation, $\neg\alpha$, can both be true), as is the case for instance with dialectics: "'The principle that two contradictory statements are not both true is the most certain of all'. This is how Aristotle ... formulates his opinion known as the logical principle

(*) This work was partially supported by a research grant of the Fundação de Amparo à Pesquisa do Estado de São Paulo (FAPESP), Brazil.

of contradiction. Examples of convincing reasonings which nevertheless yield contradictory conclusions were the reason why others sometimes disagreed with the Stagirite's firm stand. That was why Aristotle's opinion was not in the least universally shared in antiquity. His opponents included Heraclitus of Ephesus, Antisthenes the Cynic, and others, ... In the early 19th century Heraclitus' idea was taken up by Hegel, who opposed to classical logic a new logic, termed by him dialectics, in which co-existence of two contradictory statements is possible. That opinion remains to this day as one of the theoretical foundations of Marxist philosophy. Chwistek...voices his doubts as to whether dialectics is necessary for that *Weltanschauung*, and Ossowski... holds that people whose views differ widely from Marxism accept obvious contradictions".

2) To handle theories in which there are contradictions caused by certain kinds of vagueness: "The contemporary formal approach to logic increases the precision of research in many fields, but it would not be correct to formulate Aristotle's principle of contradiction as: 'Two contradictory sentences are not both true'. We have namely to add: 'in the same language' or 'if the words occurring in those sentences have the same meaning'. This restriction is not always observed in every day usage, and in science too we often use terms that are more or less vague (in the sense explained by Kotarbiński...), as was noticed by Chwistek... Any vagueness of the term a can result in a contradiction of sentences, because with reference to the same object X we may say that 'X is a' and also 'X is not a', according to the meaning of the term a adopted for the moment."

3) To treat directly some empirical disciplines whose main postulates are not consistent: "... it is known that the evolution of empirical disciplines is marked by periods in which the theorists are unable to explain the results of experiments by a homogeneous and consistent theory, but use different hypotheses, which are not always consistent with one another, to explain the various groups of phenomena. This applies, for instance, to physics in its present-day stage. Some hypotheses are even termed working hypotheses when they result in certain correct predictions, but have no chance of being accepted for good, since they fail in some other cases. A hypothesis which is known to be false is sometimes termed a fiction. In the opinion of Vaihinger... fictions are characteristic of contenporary science and are indispensable instruments of scientific research. Whether we accept that extremist and doubtful opinion or not, we have to take into account the fact

that in some cases we have to do with a system of hypotheses which, if subjected to a too precise analysis, would show a contradiction among them or with a certain accepted law,but which we use in a way that is restricted so as not to yield a self-evident falsehood."

Moreover the discussive propositional calculus has a logical and mathematical importance by itself, and has originated several interesting problems (see Kotas 1975).

Our paper is divided in two parts. In the first, we extend the discussive propositional calculus to a higher-order discussive logic; this extension is a generalization of some previous results of ours (Costa and Dubikajtis 1968, and Costa 1975). In the second part, we present a new axiomatization of discussive propositional calculus that constitutes a solution of Problems I and II of Costa 1975.

This paper is not self-contained,presupposing on the part of the reader some familiarity with higher-order logic, modal logic and the contents of Costa 1975.

PART I

THE HIGHER-ORDER DISCUSSIVE LOGIC BASED ON S5

1 - One can introduce a discussive predicate logic of order ω, as Jaśkowski defined his discussive propositional calculus (Jaśkowski 1948 and 1949) and as we constructed a discussive predicate calculus of first order (Costa and Dubikajtis 1968 and Costa 1975). To begin with we shall describe the language of modal higher-order predicate logic: S5ω.

DEFINITION 1. 1) ι is a *type;* 2) If $t_1, t_2, \ldots, t_m$, $1 \leq m < \omega$, are *types*, then $< t_1, t_2, \ldots, t_m >$ is a *type;* 3) The only *types* are those given by clauses 1 and 2.

Primitive symbols of S5ω: 1) For each type t, a denumerably infinite set of variables of this type; 2) For each type t, an arbitrary set of constants of type t (which may be empty); 3) The connectives: $\neg$ (not) and $\vee$ (or); 4) the modal operator of necessity: $\Box$; 5) The general quantifier: $\forall$ (for all); 6) Parentheses: (,).

A *term* of type t is any variable or constant of type t. The terms of type t are supposed to be well ordered; in particular,the ordinal number of the set of variables of type t is ω.

DEFINITION 2. 1) An expression consisting of a term of type $<t_1, t_2, \dots, t_m>$, followed by m terms of types respectively $t_1, t_2, \dots, t_m$ is a *formula* (*an atomic formula*); 2) If α is a *formula*, so are $\Box\alpha$ and $\neg\alpha$; 3) If α and β are *formulas*, so is $(\alpha \vee \beta)$; 4) If α is a *formula* and X is a variable, then $(\forall X)\alpha$ is a *formula;* 5) The only *formulas* are those given by the preceding clauses.

The notions of free occurrence of a variable in a formula, of a variable free for another in a formula, etc. are defined as usual. In general, our terminology is that of Hughes and Cresswell 1968, with clear adaptations.

DEFINITION 3. (The implicit restrictions are obvious.)

$(\alpha \supset \beta) =_{\text{def}} (\neg\alpha \vee \beta)$; $(\alpha \cdot \beta) =_{\text{def}} \neg(\neg\alpha \vee \neg\beta)$;

$(\alpha \equiv \beta) =_{\text{def}} ((\alpha \supset \beta) \cdot (\beta \supset \alpha))$; $(\exists X)\alpha =_{\text{def}} \neg(\forall X)\neg\alpha$;

$\Diamond\alpha =_{\text{def}} \neg\Box\neg\alpha$;

$(\alpha \longrightarrow \beta) =_{\text{def}} (\Diamond\alpha \supset \beta)$ (*discussive implication*);

$(\alpha \wedge \beta) =_{\text{def}} (\Diamond\alpha \cdot \beta)$ (*left discussive conjunction*);

$(\alpha \,\&\, \beta) =_{\text{def}} (\alpha \cdot \Diamond\beta)$ (*right discussive conjunction*);

$(\alpha \longleftrightarrow \beta) =_{\text{def}} ((\alpha \longrightarrow \beta) \wedge (\beta \longrightarrow \alpha))$ (*discussive equivalence*).

DEFINITION 4. If X and Y are terms of type t and Z is the first variable of type $<t>$, then:

$$X = Y =_{\text{def}} (\forall Z)(ZX \equiv ZY)$$

2 - We now proceed to the semantical study of $S5\omega$.

Let $\mathcal{T}$ denote the set of all types and W and D two disjoint nonempty sets. The *W-D-frame*, whose set of worlds and whose domain of individuals are respectively W and D, is the function $\mathcal{G}$ on $\mathcal{T}$ such that: 1) $\mathcal{G}(\iota) = D$; 2) if $n \geq 1$ and $t_1, t_2, \dots t_n \in \mathcal{T}$, then $\mathcal{G}(<t_1, t_2, \dots, t_n>) = \mathbb{P}(\mathcal{G}(t_1) \times \mathcal{G}(t_2) \times \dots \times \mathcal{G}(t_n) \times W)$, where $\mathbb{P}(A)$ denotes the set of all subsets of A.

Generalizing the exposition of Hughes and Cresswell 1968, Chapter 8, one may define an $S5\omega$-model, $<W, D, V>$, where D and W are the domain of individuals and the set of worlds of a W-D-frame, and V is a value-assignment which enables us to calculate the value (*0* or *1*) for any formula of $S5\omega$ in any world ω_i of W. V assigns to each term of type t an element of $\mathcal{G}(t)$;

given $V(\tau)$ for any given term τ, we can calculate the value $V(\alpha,\omega_i)$ for every formula α and every $\omega_i \in W$, by the following rules:

1) If ϕ is a term of type $<t_1,t_2,\ldots,t_n>$ and $\tau_1, \tau_2, \ldots, \tau_n$ are terms of types respectively $t_1,t_2,\ldots,t_n$, then: $V(\phi\tau_1\tau_2\ldots\tau_n, \omega_i) = 1$ if $<V(\tau_1), V(\tau_2),\ldots,V(\tau_n), \omega_i> \in V(\phi)$; otherwise $V(\phi\tau_1\tau_2\ldots\tau_n, \omega_i) = 0$.

2) If α is formula and $\omega_i \in W$, then $V(\neg\alpha, \omega_i) = 1$ if $V(\alpha, \omega_i) \doteq 0$, and $V(\neg\alpha, \omega_i) = 0$ if $V(\alpha, \omega_i) = 1$.

3) If α and β are formulas and $\omega_i \in W$, then $V((\alpha \vee \beta), \omega_i) = 1$ if either $V(\alpha, \omega_i) = 1$ or $V(\beta, \omega_i) = 1$; otherwise $V((\alpha \vee \beta), \omega_i) = 0$.

4) If α is a formula, X a variable, and ω_i a world, then $V((\forall X)\alpha, \omega_i) = 1$ if for every value-assignment V' which coincides with V on all terms other than X, $V'(\alpha, \omega_i) = 1$. Otherwise, $V((\forall X)\alpha, \omega_i) = 0$.

5) If α is a formula and $\omega_i \in W$, then $V(\Box\alpha, \omega_i) = 1$ in the case that for every $\omega_j \in W$ we have $V(\alpha, \omega_j) = 1$. In all other cases $V(\Box\alpha, \omega_i) = 0$.

A formula α is said to be valid if for every model $<W, D, V>$, $V(\alpha, \omega_i) = 1$ for every $\omega_i \in W$. We write $\models \alpha$ to express the fact that α is valid.

3 - The higher - order discussive predicate logic (or the discussive predicate logic of order ω), $\mathbb{J}\omega$, is defined as follows: 1) $\mathbb{J}\omega$ and $S5\omega$ have the same language; 2) A formula α is valid in $\mathbb{J}\omega$ if and only if $\Diamond\alpha$ is valid in $S5\omega$.

In other words, α is valid in $\mathbb{J}\omega$ if and only if for every model $<W, D, V>$, $V(\alpha, \omega_i) = 1$ for some $\omega_i \in W$. $\models_{\mathbb{J}\omega} \alpha$ means that the formula α is valid in $\mathbb{J}\omega$.

It is immediate that : $\models \alpha$ implies $\models_{\mathbb{J}\omega} \Box\alpha$, and conversely.

By Gödel's incompleteness theorem, the usual higher-order predicate calculus is not axiomatizable. As a consequence, $S5\omega$ and $\mathbb{J}\omega$ are not axiomatizable either. Nonetheless, we shall present axiom systems $S5\omega^*$ and $\mathbb{J}\omega^*$ for $S5\omega$ and $\mathbb{J}\omega$, which are sufficient for our purposes.

The postulates (axiom schemata and primitive rules of inference) of $S5\omega^*$ are the following:

Axiom schemata (α and β are formulas):

1) If α is a substitution instance of a tautology, then α is an axiom;

2) $\Box(\alpha \supset \beta) \supset \Box(\Box\alpha \supset \Box\beta)$;

3) $\Box\alpha \supset \alpha$;

4) $\alpha \supset \Box\Diamond\alpha$;

5) $(\forall X)\alpha(X) \supset \alpha(Y)$, where $\alpha(X)$ is a formula, X is a variable, Y is a term of the same type as X, Y is free for X in $\alpha(X)$ and $\alpha(Y)$ is the result of replacing the free occurrences of X in $\alpha(X)$ by Y.

6) $(\exists P)(\forall X_1)(\forall X_2)\ldots(\forall X_n)(PX_1X_2\ldots X_n \equiv \alpha(X_1, X_2, \ldots, X_n))$, where $n \geq 1$, $\alpha(X_1, X_2, \ldots, X_n)$ is a formula in which the variables $X_1, X_2, \ldots, X_n$ may occur free, P is a variable of type $<t_1, t_2, \ldots, t_n>$, where $t_1, t_2, \ldots t_n$ are respectively the types of $X_1, X_2, \ldots, X_n$, and P does not appear free in $\alpha(X_1, X_2, \ldots, X_n)$.

Derivation rules:

I) $\alpha, \alpha \supset \beta / \beta$,

II) $\alpha / \Box\alpha$,

III) $\alpha \supset \beta(X) / \alpha \supset (\forall X)\beta(X)$, when the variable X does not occur free in α.

The notions of proof, of theorem (or thesis) and the symbol $\vdash$ are introduced as usual.

To the postulates of S5ω* one may add the postulates of extensionality, of choice and of infinite (cf. Church 1956, and Hilbert and Ackermann 1950).

THEOREM 1. *If* $\vdash \alpha$, *then* $\models \alpha$.

THEOREM 2. *In* S5ω* *we have:*

$\vdash (\forall X)\Box\alpha \supset \Box(\forall X)\alpha$ (the generalized Barcan formula),

$\vdash X = Y \supset \Box(X = Y)$ (the law of necessary identity),

$\vdash X \neq Y \supset \Box(X \neq Y)$.

In general, all theorems of first-order modal predicate calculus based on S5 (cf. Hughes and Cresswell 1968), with or without equality, can be extended to S5ω*.

The next theorem shows that an axiom system for Jω, Jω*, exists, such that α is a thesis of Jω* if and only if $\vdash \Diamond\alpha$ in S5ω*.

THEOREM 3. *Let* A *be the set* $\{\alpha : \Diamond\alpha$ *is a thesis of* S5ω*$\}$. *Then* A *may be axiomatized by means of the following axiom system* Jω* (the notations have clear meanings and are subject to the standard restrictions):

Axiom schemata:

1) $\Box\alpha$, *whenever* α *is a substitution - instance of tautology* :

2) $\Box(\Box(\alpha \supset \beta) \supset \Box(\Box\alpha \supset \Box\beta))$;

3) $\Box(\Box\alpha \supset \alpha)$;

4) $\Box(\alpha \supset \Box\Diamond\alpha)$;

5) $\square((\forall X)\,\alpha(X) \supset \alpha(Y))$;

6) $\square((\exists P)(\forall X_1)(\forall X_2)\ldots(\forall X_m)(PX_1X_2\ldots X_m \equiv \alpha(X_1, X_2, \ldots, X_m)))$.

Derivation rules:

I) $\alpha,\ \square(\alpha \supset \beta)\ /\ \beta$

II) $\Diamond\alpha / \alpha$,

III) $\square(\alpha \supset \beta(X))\ /\ \square(\alpha \supset (\forall X)\,\alpha(X))$.

PROOF: First of all, we note that it is easy to prove the following lemmas:

LEMMA 1. $\square(\square\alpha \supset \square\square\alpha)$ *is a derivable schema in* $\mathbb{J}\omega^*$.

LEMMA 2. *If* α *is a thesis of* $\mathbb{J}\omega^*$ *, then* $\Diamond\alpha$ *is a thesis of* $S5\omega^*$.

LEMMA 3. *If* α *is a thesis of* $S5\omega^*$ *,then* $\square\alpha$ *is a thesis of* $\mathbb{J}\omega^*$.

Taking into account the above results, we reason as in Theorem 4 of Costa 1975 and complete the proof of Theorem 3.

THEOREM 4. *The following schemata are derivable in* $\mathbb{J}\omega^*$.

$\alpha \rightarrow (\beta \rightarrow \alpha)$, $\quad (\alpha \rightarrow \beta) \rightarrow ((\alpha \rightarrow (\beta \rightarrow \gamma)) \rightarrow (\alpha \rightarrow \gamma))$,

$(\alpha \wedge \beta) \rightarrow \alpha$, $\quad (\alpha \wedge \beta) \rightarrow \beta$, $\quad (\alpha \,\&\, \beta) \rightarrow \alpha$,

$(\alpha \,\&\, \beta) \rightarrow \beta$, $\quad (\alpha \rightarrow (\beta \rightarrow (\alpha \wedge \beta)))$, $\quad \alpha \rightarrow (\beta \rightarrow (\alpha \,\&\, \beta))$,

$\alpha \rightarrow (\alpha \vee \beta)$, $\quad \beta \rightarrow (\alpha \vee \beta)$, $\quad (\alpha \leftrightarrow \beta) \rightarrow (\alpha \rightarrow \beta)$,

$(\alpha \longleftrightarrow \beta) \rightarrow (\beta \rightarrow \alpha)$, $\quad (\alpha \rightarrow \gamma) \rightarrow ((\beta \rightarrow \gamma) \rightarrow ((\alpha \vee \beta) \rightarrow \gamma))$,

$((\alpha \rightarrow \beta) \rightarrow \alpha) \rightarrow \alpha$, $\quad (\alpha \rightarrow \beta) \rightarrow ((\beta \rightarrow \alpha) \rightarrow (\alpha \longleftrightarrow \beta))$,

$\alpha \vee \neg\alpha$, $\quad \neg(\alpha \wedge \neg\alpha)$, $\quad \neg(\alpha \,\&\, \neg\alpha)$,

$\alpha \longleftrightarrow \neg\neg\alpha$, $\quad (\neg\alpha \rightarrow \alpha) \rightarrow \alpha$, $\quad (\alpha \rightarrow \neg\alpha) \rightarrow \neg\alpha$,

$(\alpha \cdot \neg\alpha) \rightarrow \beta$, $\quad (\alpha \cdot \beta) \rightarrow \alpha$, $\quad (\alpha \cdot \beta) \rightarrow \beta$,

$(\alpha \longleftrightarrow \neg\alpha) \rightarrow \neg\alpha$, $\quad (\alpha \longleftrightarrow \neg\alpha) \rightarrow \alpha$, $\quad \neg(\alpha \longleftrightarrow \neg\alpha)$,

$\neg(\alpha \rightarrow \beta) \rightarrow \alpha$, $\quad ((\alpha \supset \beta) \cdot (\alpha \supset \neg\beta)) \rightarrow \neg\alpha$, $\quad \neg(\alpha \rightarrow \beta) \rightarrow \neg\beta$;

$\square(\alpha \supset \beta) \rightarrow (\alpha \rightarrow \beta)$, $\quad \square(\alpha \supset \beta) \supset (\alpha \rightarrow \beta)$, $\quad (\alpha \rightarrow \beta) \supset \Diamond(\alpha \supset \beta)$,

$\neg\Diamond\alpha \rightarrow (\alpha \rightarrow \beta)$, $\quad (\alpha \rightarrow \beta) \rightarrow ((\alpha \rightarrow \neg\Diamond\beta) \rightarrow \neg\Diamond\alpha)$,

$\neg\Diamond\alpha \vee \alpha$, $\quad \square(\alpha \equiv \beta) \rightarrow (\alpha \rightarrow \beta)$, $\quad \square(\alpha \equiv \beta) \rightarrow (\beta \rightarrow \alpha)$,

$\square((\alpha \wedge \beta) \equiv (\beta \,\&\, \alpha))$, $\quad \square(\square\alpha \equiv (\neg\alpha \rightarrow (\alpha \cdot \neg\alpha)))$,

$\Diamond\alpha \longleftrightarrow \neg(\alpha \rightarrow (\alpha \cdot \neg\alpha))$, $\quad \square(\Diamond\alpha \equiv \neg(\alpha \rightarrow \alpha \cdot \neg\alpha))$, $\quad \square\alpha \longleftrightarrow (\neg\alpha \rightarrow (\alpha \cdot \neg\alpha))$,

$$(\forall X)\,\alpha(X) \rightarrow \alpha(Y), \qquad \Box(\forall X)\ \alpha(X) \longleftrightarrow (\forall X)\Box\alpha(X),$$
$$\alpha(Y) \rightarrow (\exists X)\,\alpha(X), \qquad \Diamond(\exists X)\ \alpha(X) \longleftrightarrow (\exists X)\Diamond\alpha(X).$$

PROOF: If γ is one of the above schemata, then $\Diamond\gamma$ is derivable in $S5\omega^*$. Therefore, γ is also derivable in $\mathbb{J}\omega^*$.

THEOREM 5. *If* α *is a thesis of* $\mathbb{J}$ * *(in symbols:* $\vdash_{\mathbb{J}\omega^*}$ *), then* $\models_{\mathbb{J}\omega}$ α.

THEOREM 6. *The following schemata, among others, are not theses of* $\mathbb{J}\omega^*$.

$$\alpha \rightarrow (\neg\alpha \rightarrow \beta) \qquad \alpha \rightarrow (\neg\alpha \rightarrow \neg\beta), \qquad \neg\alpha \rightarrow (\alpha \rightarrow \beta),$$
$$\neg\alpha \rightarrow (\alpha \rightarrow \neg\beta), \qquad (\alpha \rightarrow \beta) \rightarrow ((\alpha \rightarrow \neg\beta) \rightarrow \neg\alpha), \qquad (\alpha \wedge \neg\alpha) \rightarrow \beta,$$
$$(\alpha \longleftrightarrow \neg\alpha) \rightarrow \beta, \qquad (\alpha \rightarrow \beta) \rightarrow (\neg\beta \rightarrow \neg\alpha), \qquad \neg(\alpha \rightarrow \alpha) \rightarrow \beta,$$
$$(\neg\alpha \rightarrow \neg\beta) \rightarrow (\beta \rightarrow \alpha), \qquad (\neg\alpha \rightarrow \beta) \rightarrow ((\neg\alpha \rightarrow \neg\beta) \rightarrow \alpha), \qquad (\alpha \wedge \neg\alpha) \rightarrow \neg\beta,$$
$$\alpha \rightarrow (\beta \rightarrow (\alpha \cdot \beta)) \qquad (\alpha \longleftrightarrow \neg\alpha) \rightarrow (\alpha \cdot \neg\alpha), \qquad (\alpha \mathbin{\&} \neg\alpha) \rightarrow \beta,$$
$$(\alpha \rightarrow \beta) \rightarrow \Box(\alpha \supset \beta), \qquad \Box(\Box(\alpha \supset \beta) \supset (\alpha \rightarrow \beta)), \qquad (\alpha \mathbin{\&} \neg\alpha) \rightarrow \neg\beta,$$
$$\Box((\alpha \rightarrow \beta) \supset \Box(\alpha \supset \beta)), \qquad (\forall X)(\alpha \wedge \beta) \rightarrow ((\forall X)\alpha \wedge (\forall X)\beta),$$
$$((\exists X)\alpha \wedge \neg(\exists X)\alpha) \rightarrow \beta, \qquad ((\forall X)\alpha \wedge \neg(\forall X)\alpha) \rightarrow \beta,$$
$$((\exists X)\alpha \mathbin{\&} \neg(\exists X)\alpha) \rightarrow \neg\beta, \qquad ((\forall X)\alpha \mathbin{\&} \neg(\forall X)\alpha) \rightarrow \neg\beta,$$
$$((\forall X)(\alpha \mathbin{\&} \beta) \rightarrow ((\forall X)\alpha \wedge (\forall X)\beta).$$

PROOF: Let γ denote any of the above schemata. Hence, if $\vdash_{\mathbb{J}\omega^*} \gamma$, then $\models_{\mathbb{J}\omega} \gamma$ and, consequently, $\models \Diamond\gamma$. But one can prove, without difficulty, with the help of the semantics of $S5\omega$, that $\models \Diamond\gamma$ is false.

The proofs of the next theorems are immediate.

THEOREM 7. *The rule* $\alpha, \alpha \rightarrow \beta / \beta$ *is valid in* $\mathbb{J}\omega^*$, *but* $\alpha, \alpha \supset \beta / \beta$ *is not.*

THEOREM 8. *The rules* $\alpha \rightarrow \beta(X) / \alpha \rightarrow (\forall X)\beta(X)$ *and* $\beta(X) \rightarrow \alpha / (\exists X)\beta(X) \rightarrow \alpha$, *where the symbols have clear meanings and are subject to the usual restrictions, are valid in* $S5\omega^*$. *Nonetheless, the second is true in* $\mathbb{J}\omega^*$, *though the first is not.*

4 - $S5\omega^*$ is not complete. Nevertheless, extending the notion of $S5\omega$ - model, it is possible to establish a weak completeness theorem for $S5\omega^*$ (and for $\mathbb{J}\omega^*$). Usual higher-order predicate logic and $S5\omega$ ($\mathbb{J}\omega$) are similar in this respect. In fact, we can define a Henkin generalized kind of model, $\langle W, D, V \rangle$, the same way as we have introduced the concept of $S5\omega$-model, except for the fact that the second condition in the definition of frame is

replaced by the following: 2') If $m \geq 1$ and $t_1, t_2, \ldots, t_m \in \mathcal{T}$, then $\emptyset \neq \mathcal{G}(<t_1, t_2, \ldots, t_m>) \subseteq \mathbb{P}(\mathcal{G}(t_1) \times \mathcal{G}(t_2) \times \ldots \times \mathcal{G}(t_m) \times W)$. The notion of value-assignment formally remains the same. A normal model is a model which satisfies, in an obvious sense, the schema of separation (the sixth axiom schema of S5ω*). A formula α is called S5ω-weakly valid (𝕁ω-weakly valid) if for every normal model $<W,D,V>$, $V(\alpha, \omega_i) = 1$ for every $\omega_i \in W$ (for some $\omega_i \in W$). Modifying the methods of Hughes and Cresswell 1968, it is a simple matter to prove:

THEOREM 9. *For every formula* α *of* S5ω, α *is* S5ω-*weakly valid* (𝕁ω-*weakly valid*) *if and only if it is a thesis of* S5ω* (𝕁ω*).

PART II

A NEW AXIOMATIZATION FOR THE DISCUSSIVE PROPOSITIONAL CALCULUS

In this part of our paper we present an axiomatization for the discussive propositional calculus 𝕁 (cf. Jaśkowski 1948), in which the primitive connectives are $\rightarrow$ (discussive implication), $\wedge$ (left discussive conjunction), V (disjunction), and $\neg$ (negation). From this axiomatization it is an easy task to derive another in which the primitive connectives are $\rightarrow$, & (right discussive conjunction), V and $\neg$. The formulas of 𝕁 are constructed with propositional variables, the connectives and parentheses. In order to facilitate the writing of formulas, in the remainder of the paper, small Latin letters will be employed as syntactical variables (instead of small Greek letters).

LEMMA A. *If* $\supset$, V *and* $\neg$ *are chosen as primitive, then the classical propositional calculus may be axiomatized by means of the following axiom schemata and derivation rule:*

$(p \supset q) \supset ((q \supset r) \supset (p \supset r))$, $\quad p \supset (\neg p \supset q)$, $\quad (\neg p \supset p) \supset p$,

$(p \supset r) \supset ((q \supset r) \supset ((p \vee q) \supset r)$, $\quad p \supset (p \vee q)$, $\quad q \supset (p \vee q)$,

$p, p \supset q / q$.

PROOF: In effect, the first three schemata and rule constitute a complete axiomatic system for classical propositional calculus in which $\supset$ and $\neg$ are the sole primitive connectives (Łukasiewicz)

LEMMA B. 𝕁 *may be axiomatized as follows* ($\supset$, V, $\neg$ *and* $\Box$ *are the sole primitive connectives,* $p \cdot q$ *is an abbreviation of* $\neg(\neg p \vee \neg q)$, *and* $\equiv$, $\Diamond$, $\longrightarrow$, $\wedge$, &, *and* $\longleftrightarrow$ *are defined as in Definition* 3).

Axiom schemata :

I) $\Box((p \supset q) \supset ((q \supset r) \supset (p \supset r)))$,

II) $\Box(p \supset (\neg p \supset q))$, III) $\Box((\neg p \supset p) \supset p)$,

IV) $\Box(p \supset (p \vee q))$, V) $\Box(q \supset (p \vee q))$,

VI) $\Box((p \supset r) \supset ((q \supset r) \supset ((p \vee q) \supset r)))$,

VII) $\Box(\Box(p \supset q) \supset \Box(\Box p \supset \Box q))$,

VIII) $\Box(\Box p \supset p)$, IX) $\Box(p \supset \Box\Diamond p)$;

Derivation rules:

p, $\Box(p \supset q)$ / q

$\Diamond p$ / p;

PROOF: Consequence of the preceding Lemma and of Theorem 4 of Costa 1975.

Let us consider now the propositional calculus $\tilde{\mathbb{J}}$ characterized as follows: 1) The primitive symbols of $\tilde{\mathbb{J}}$ are $\longrightarrow$, $\wedge$, $\vee$, $\neg$, (,), and the propositional variables; 2) $\tilde{\mathbb{J}}$ has the following postulates:

Axiom schemata:

A01) $p \longrightarrow (q \longrightarrow p)$,

A02) $(p \longrightarrow (q \longrightarrow r)) \longrightarrow ((p \longrightarrow q) \longrightarrow (p \longrightarrow r))$,

A03) $((p \longrightarrow q) \longrightarrow p) \longrightarrow p$,

A04) $(p \wedge q) \longrightarrow p$,

A05) $(p \wedge q) \longrightarrow q$,

A06) $p \longrightarrow (q \longrightarrow (p \wedge q))$,

A07) $p \longrightarrow (p \vee q)$,

A08) $q \longrightarrow (p \vee q)$,

A09) $(p \longrightarrow r) \longrightarrow ((q \longrightarrow r) \longrightarrow ((p \vee q) \longrightarrow r))$,

A1) $p \longrightarrow \neg\neg p$,

A2) $\neg\neg p \longrightarrow p$,

A3) $\neg(p \vee \neg p) \longrightarrow q$,

A4) $\neg(p \vee q) \longrightarrow \neg(q \vee p)$,

A5) $\neg(p \vee q) \longrightarrow (\neg p \wedge \neg q)$,

A6) $\neg(\neg\neg p \vee q) \longrightarrow \neg(p \vee q)$,

A7) $(\neg(p \vee q) \longrightarrow r) \longrightarrow ((\neg p \longrightarrow q) \vee r)$,

A8) $\neg((p \vee q) \vee r) \longrightarrow \neg(p \vee (q \vee r))$,

A9) $\neg((p \longrightarrow q) \vee r) \longrightarrow (p \wedge \neg(q \vee r))$,

A10) $\neg((p \wedge q) \vee r) \longrightarrow (p \longrightarrow \neg(q \vee r))$,

A11) $\neg(\neg(p \vee q) \vee r) \longrightarrow (\neg(\neg p \vee r) \vee \neg(\neg q \vee r))$,
A12) $\neg(\neg(p \rightarrow q) \vee r) \longrightarrow (p \longrightarrow \neg(\neg q \vee r)$,
A13) $\neg(\neg(p \wedge q) \vee r) \longrightarrow (p \wedge \neg(\neg q \vee r))$.

Derivation rule:
R0) $p, p \rightarrow q / q$.

Definitions:

Df1) $(p \supset q) =_{def} (\neg p \vee q)$,

Df2) $Op =_{def} \neg(p \vee \neg p)$,

Df3) $\Box p =_{def} (\neg p \rightarrow Op)$,

Df4) $\Diamond p =_{def} \neg \Box \neg p$,

Df5) $(p \cdot q) =_{def} \neg(\neg p \vee \neg q)$,

Df6) $(p \equiv q) =_{def} ((p \supset q) \cdot (q \supset p))$.

REMARK. When $\longrightarrow$, $\wedge$, $\vee$ and $\neg$ are interpreted respectively as discussive implication, left discussive conjuction, disjunction, and negation, A01-A13 are theses of $\mathbb{J}$ and R0 is a valid rule of this calculus.

Several theses and three derived rules of $\tilde{\mathbb{J}}$ will be estabished in the sequel. The method of setting out the (formal) proofs is self explamatory, and from now on we shall abbreviate "*p is a thesis of* $\tilde{\mathbb{J}}$" by $\vdash p$. The symbol $\Longrightarrow$ denotes the metalinguistic relation of implication.

THEOREM 0.

T01) $\vdash (p \longrightarrow q) \longrightarrow ((q \longrightarrow s) \longrightarrow (p \longrightarrow s))$;
T02) $\vdash (p \vee p) \longrightarrow p$;
T03) $\vdash (p \vee q) \longrightarrow (q \vee p)$;
T04a) $\vdash (p' \longrightarrow p'') \longrightarrow ((p' \vee q) \longrightarrow (p'' \vee q))$;
T04b) $\vdash (q' \longrightarrow q'') \longrightarrow ((p \vee q')) \longrightarrow (p \vee q'')$;
T05) $\vdash (p \wedge q) \longrightarrow (q \wedge p)$;
T06a) $\vdash (s \longrightarrow p) \longrightarrow ((s \longrightarrow q) \longrightarrow (s \longrightarrow (p \wedge q)))$;
T06b) $\vdash (s \longrightarrow q) \longrightarrow ((s \longrightarrow p) \longrightarrow (s \longrightarrow (p \wedge q)))$;
T07a) $\vdash (p' \longrightarrow p'') \longrightarrow ((p' \wedge q) \longrightarrow (p'' \wedge q))$;
T07b) $\vdash (q' \longrightarrow q'') \longrightarrow ((p \wedge q') \longrightarrow (p \wedge q''))$;
T08) $\vdash (p \longrightarrow q) \wedge p) \longrightarrow q$;
T09) $\vdash (p \wedge (q \wedge s)) \longrightarrow (p \wedge q)$;

T010) $\vdash (s \rightarrow (p' \wedge q')) \rightarrow ((p' \rightarrow p'') \rightarrow (s \rightarrow (p'' \wedge q)))$;
T011) $\vdash (s' \rightarrow s'') \rightarrow ((p \rightarrow (q \rightarrow s')) \rightarrow (p \rightarrow (q \rightarrow s'')))$;
T012) $\vdash (q \rightarrow s) \rightarrow ((p \rightarrow q) \rightarrow (p \rightarrow s))$;
T013) $\vdash p \rightarrow p$.

PROOF: T01 - T013 are known to be consequences of the postulates of classical positive logic, that is to say, of A01 - A09 and R0.

THEOREM 1. (T1) $\vdash 0p \rightarrow 0q$.

PROOF: A3: $(0q / q)$; Df2 $\Longrightarrow$ T1 .

THEOREM 2. (T2) $\vdash (p \vee 0q) \rightarrow p$.

PROOF: $(p \vee 0q) \xrightarrow{Df2} (p \vee \neg(q \vee \neg q)) \xrightarrow{A3,T04b,T01} (p \vee p) \xrightarrow{T02} p \xRightarrow{T01}$ T2 .

THEOREM 3. (T3) $\vdash \neg(p \vee q) \rightarrow \neg p$.

PROOF: $\neg(p \vee q) \xrightarrow{A5} (\neg p \wedge \neg q) \xrightarrow{A04} \neg p \xRightarrow{T01}$ T3.

In the next proofs applications of T01 will not be made explicit.

THEOREM 4. (T4) $\vdash (\Box p \wedge \neg p) \rightarrow 0q$.

PROOF: $(\Box p \wedge \neg p) \xrightarrow{Df3} ((\neg p \rightarrow 0p) \wedge \neg p) \xrightarrow{T08} 0p \xrightarrow{T1} 0q \Longrightarrow$ T4.

THEOREM 5. (T5) $\vdash (\neg 0 p \wedge p) \rightarrow 0q$.

PROOF: $(\neg 0p \wedge p) \xrightarrow{Df4} (\neg\neg\Box\neg p \wedge p) \xrightarrow{A2,T07a} (\Box\neg p \wedge p) \xrightarrow{A1,T07b} (\Box\neg p \wedge \neg\neg p) \xrightarrow{T4} 0p \Longrightarrow$ T5.

THEOREM 6. (T6) $\vdash \neg(p \supset q) \rightarrow (p \wedge \neg q)$.

PROOF: $\neg(p \supset q) \xrightarrow{Df1} \neg(\neg p \vee q) \xrightarrow{A5} (\neg\neg p \wedge \neg q) \xrightarrow{A2,T07a} (p \wedge \neg q) \Longrightarrow$ T6.

THEOREM 7. (T7) $\vdash \Box(\Box p \supset p)$.

PROOF: $\neg(\Box p \supset p) \xrightarrow{Df1} \neg(\neg\Box p \vee p) \xrightarrow{Df3} \neg(\neg(\neg p \rightarrow 0p) \vee p) \xrightarrow{A5} (\neg\neg(\neg p \rightarrow 0p) \wedge \neg p) \xrightarrow{A2,T07a} ((\neg p \rightarrow 0p) \wedge \neg p) \xrightarrow{T08} 0p \xrightarrow{T1} 0(\Box p \supset p) \xRightarrow{Df3}$ T7.

THEOREM 8. (T8) $\vdash \neg(p \vee (q \vee s)) \rightarrow \neg((p \vee q) \vee s)$.

PROOF: $\neg(p \vee (q \vee s)) \xrightarrow{A4} \neg((q \vee s) \vee p) \xrightarrow{A8} \neg(q \vee (s \vee p)) \xrightarrow{A4} \neg((s \vee p) \vee q) \xrightarrow{A8} \neg(s \vee (p \vee q)) \xrightarrow{A4} \neg((p \vee q) \vee s) \Longrightarrow$ T8.

THEOREM 9. (T9) $\vdash \Box(p \supset (\neg p \supset q))$.

PROOF: (Abbreviation: $\alpha = p \supset (\neg p \supset q)$.) $\neg\alpha \xrightarrow{Df1} \neg(\neg p \vee (\neg\neg p \vee q)) \xrightarrow{T8} \neg((\neg p \vee \neg\neg p) \vee q) \xrightarrow{T3} \neg(\neg p \vee \neg\neg p) \xrightarrow{Df2} 0(\neg p) \xrightarrow{T1} 0\alpha \xRightarrow{Df3}$ T9.

THEOREM 10. (T10) $\vdash \Box(p \supset (p \vee q))$.

PROOF: (Abbreviation: $\alpha = p \supset (p \vee q)$.) $\neg\alpha \xrightarrow{Df1} \neg(\neg p \vee (p \vee q)) \xrightarrow{T8} \neg((\neg p \vee p) \vee q) \xrightarrow{T3} \neg(\neg p \vee p) \xrightarrow{A4} \neg(p \vee \neg p) \xrightarrow{Df2} 0p \xrightarrow{T1} 0\alpha \xRightarrow{Df3}$ T10.

THEOREM 11. (T11) $\vdash \Box(q \supset (p \vee q))$.

PROOF: (Abbreviation: $\alpha = q \supset (p \vee q)$.) $\neg\alpha \xrightarrow{Df1} \neg(\neg q \vee (p \vee q)) \xrightarrow{A4} \neg((p \vee q) \vee \neg q) \xrightarrow{A8} \neg(p \vee (q \vee \neg q)) \xrightarrow{A4} \neg((q \vee \neg q) \vee p) \xrightarrow{T3} \neg(q \vee \neg q) \xrightarrow{Df2} 0q \xrightarrow{T1} 0\alpha \xRightarrow{Df3}$ T11.

THEOREM 12. (T12) $\vdash \Box(p \supset q) \rightarrow (p \rightarrow q)$.

PROOF: $\Box(p \supset q) \xrightarrow{Df1} \Box(\neg p \vee q) \xrightarrow{Df3} (\neg(\neg p \vee q) \rightarrow 0(\neg p \vee q)) \xrightarrow{A7} ((\neg\neg p \rightarrow q) \vee 0(\neg p \vee q)) \xrightarrow{T2} (\neg\neg p \rightarrow q) \xrightarrow{A1} (p \rightarrow q) \Longrightarrow$ T12.

THEOREM 13. (T13) $\vdash p$, $\vdash \Box(p \supset q) / \vdash q$.

PROOF: If $\vdash p$ and $\vdash \Box(p \supset q)$, then, by T12, and R0, $\vdash p \rightarrow q$; therefore, by R0, $\vdash q$.

THEOREM 14. (T14) $\vdash \Box(p \supset q) \rightarrow (\Box p \rightarrow \Box q)$.

PROOF: $\Box(p \supset q) \xrightarrow{Df3} (\neg(p \supset q) \rightarrow 0(p \supset q)) \xrightarrow{Df1} (\neg(\neg p \vee q) \rightarrow 0(p \supset q)) \xrightarrow{A4} (\neg(q \vee \neg p) \rightarrow 0(p \supset q)) \xrightarrow{A7} ((\neg q \rightarrow \neg p) \vee 0(p \supset q)) \xrightarrow{T2} (\neg q \rightarrow \neg p) \xrightarrow{T01} ((\neg p \rightarrow 0p) \rightarrow (\neg q \vee 0p)) \xrightarrow{T011,\ T1} ((\neg p \vee 0p) \rightarrow (\neg q \rightarrow 0q)) \xrightarrow{Df3} (\Box p \rightarrow \Box q) \Longrightarrow$ T14 .

THEOREM 15. (T15) $\vdash \Box p$, $\vdash \Box(p \supset q)$ / $\vdash \Box q$.

PROOF: Consequence of T14 and R0.

THEOREM 16. (T16) $\vdash \Box p \vee \neg(p \vee \circ p)$.

PROOF: By T013, $\vdash \neg(p \vee \circ p) \rightarrow \neg(p \vee \circ p)$. By A7, $\vdash (\neg p \rightarrow \circ p) \vee \neg(p \vee \circ p)$; hence, by Df3, $\vdash \Box p \vee \neg(p \vee \circ p)$.

THEOREM 17. (T17) $\vdash \neg p \rightarrow \neg(p \vee \circ p)$.

PROOF: Since $((p \rightarrow q) \vee s) \rightarrow (p \rightarrow (q \vee s))$ is a valid schema of positive logic, we have: $\vdash ((\neg p \vee \circ p) \vee \neg(p \vee \circ p)) \rightarrow (\neg p \rightarrow (\circ p \vee \neg(p \vee \circ p))$; by Df3 and T16, $\vdash \neg p \rightarrow (\circ p \vee \neg(p \vee \circ p))$. By T2, $\vdash \neg p \rightarrow \neg(p \vee \circ p)$.

THEOREM 18. (T18) $\vdash \neg(p \rightarrow q) \rightarrow (p \vee \neg q)$.

PROOF: $\neg(p \rightarrow q) \xrightarrow{T17} \neg((p \rightarrow q) \vee \circ(p \rightarrow q)) \xrightarrow{A9} (p \wedge \neg(q \vee \circ(p \rightarrow q))$; hence: $\vdash \neg(p \rightarrow q) \rightarrow (p \wedge \neg(q \vee \circ(p \rightarrow q))$. By T3 and positive logic, $\vdash \neg(p \rightarrow q) \rightarrow (p \wedge \neg q)$.

THEOREM 19. (T19) $\vdash \neg((p \rightarrow q) \vee s) \rightarrow (p \wedge \neg s)$.

PROOF: $\neg((p \rightarrow q) \vee s) \xrightarrow{A9} (p \wedge \neg(q \vee s)) \xrightarrow{A4} (p \wedge \neg(s \vee q) \xrightarrow{T3,T07b} (p \wedge \neg s) \Longrightarrow$ T19.

THEOREM 20. (T20) $\vdash \Box(p \supset \Box \circ p)$.

PROOF: (Abbreviation: $\alpha = p \supset \Box \circ p$.) $\neg \alpha \xrightarrow{Df1} (\neg p \vee \Box \circ p) \xrightarrow{A4}$ $\neg(\Box \circ p \vee \neg p) \xrightarrow{Df3} \neg((\neg \circ p \rightarrow \circ(\circ p)) \vee \neg p) \xrightarrow{T19} (\neg \circ p \wedge \neg\neg p) \xrightarrow{A2,T07b}$ $(\neg \circ p \wedge p) \xrightarrow{T5} \circ q \xrightarrow{T1} \circ\alpha \xRightarrow{Df3}$ T20.

THEOREM 21. (T21) $\vdash \neg \Box p \rightarrow \neg p$.

PROOF: $\neg \Box p \xrightarrow{Df3} \neg(\neg p \rightarrow \circ p) \xrightarrow{T18} (\neg p \wedge \neg \circ p) \xrightarrow{A04} \neg p \Longrightarrow$ T21.

THEOREM 22. (T22) $\vdash \Diamond p \rightarrow p$.

PROOF: $\Diamond p \xrightarrow{Df4} \neg \Box \neg p \xrightarrow{T21} \neg\neg p \xrightarrow{A2} p \Longrightarrow$ T22.

THEOREM 23. (T23) $\vdash \Diamond p$ / $\vdash p$.

PROOF: Immediate consequence of T22.

THEOREM 24. (T24) $\vdash (p \longrightarrow \Box q) \longrightarrow \Box(p \supset \Box q)$.

PROOF: $\neg(p \supset \Box q) \xrightarrow{T6} (p \wedge \neg\Box q) \xrightarrow{T21,T07b} p \wedge \neg q$; hence:
(1) $\vdash \neg(p \supset \Box q) \longrightarrow (p \wedge \neg q)$.
But $(p \longrightarrow \Box q) \xrightarrow{(1),T010} (\neg(p \supset \Box q) \longrightarrow (\Box q \wedge \neg q)) \xrightarrow{T4,T012} (\neg(p \supset \Box q) \longrightarrow O(p \supset \Box q)) \xrightarrow{Df3} \Box(p \supset \Box q) \Longrightarrow$ T24.

THEOREM 25. (T25) $\vdash \Box(p \supset q) \longrightarrow \Box(\Box p \supset \Box q)$.

PROOF: $\Box(p \supset q) \xrightarrow{T14} (\Box p \longrightarrow \Box q) \xrightarrow{T24} \Box(\Box p \supset \Box q) \Longrightarrow$ T25.

THEOREM 26. (T26) $\vdash \Box(\Box(p \supset q) \supset \Box(\Box p \supset \Box q))$.

PROOF: By T24, $\vdash (\Box(p \supset q) \longrightarrow \Box(\Box p \supset \Box q)) \longrightarrow \Box(\Box(p \supset q) \supset \Box(\Box p \supset \Box q))$; therefore, by T25 and R0, T26.

THEOREM 27. (T27) $\vdash \Box((p \supset q) \supset ((q \supset s) \supset (p \supset s)))$.

PROOF: (Abbreviations: $\alpha = (p \supset q) \supset ((q \supset s) \supset (p \supset s))$;
$\beta = \neg(\neg q \vee s) \vee (\neg p \vee s)$; $\gamma = (\neg p \vee s) \vee q$.)

I) $\neg(\neg\neg q \vee \gamma) \xrightarrow{T8} \neg((\neg\neg q \vee (\neg p \vee s)) \vee \neg q) \xrightarrow{A4} \neg(\neg q \vee (\neg\neg q \vee (\neg p \vee s))) \xrightarrow{T8} \neg((\neg q \vee \neg\neg q) \vee (\neg p \vee s)) \xrightarrow{T3} \neg(\neg q \vee \neg\neg q) \xrightarrow{Df2} O(\neg q) \xRightarrow{T1}$

[1] $\vdash \neg(\neg\neg q \vee \gamma) \longrightarrow Os$;

II) $\neg(\neg s \vee \gamma) \xrightarrow{T8} \neg((\neg s \vee (\neg p \vee s)) \vee \neg q) \xrightarrow{T3} \neg(\neg s \vee (\neg p \vee s)) \xrightarrow{A4} \neg((\neg p \vee s) \vee \neg s)) \xrightarrow{A8} \neg(\neg p \vee (s \vee \neg s)) \xrightarrow{A4} \neg((s \vee \neg s) \vee \neg p) \xrightarrow{T3} \neg(s \vee \neg s) \xrightarrow{Df2} Os \Longrightarrow$

[2] $\vdash \neg(\neg s \vee \gamma) \longrightarrow Os$;

III) $\neg(\neg\neg p \vee \beta) \xrightarrow{T8} \neg((\neg\neg p \vee \neg(\neg q \vee s)) \vee (\neg p \vee s)) \xrightarrow{T8} \neg(((\neg\neg p \vee \neg(\neg q \vee s)) \vee \neg p) \vee s) \xrightarrow{T3} \neg((\neg\neg p \vee \neg(\neg q \vee s)) \vee \neg p) \xrightarrow{A4} \neg(\neg p \vee (\neg\neg p \vee \neg(\neg q \vee s))) \xrightarrow{T8} ((\neg p \vee \neg\neg p) \vee \neg(\neg q \vee s)) \xrightarrow{T3} \neg(\neg p \vee \neg\neg p) \xrightarrow{Df2} O(\neg p) \xrightarrow{T1} Os \Longrightarrow$

[3] $\vdash \neg(\neg\neg p \vee \beta) \longrightarrow Os$;

IV) $\neg(\neg q \vee \beta) \xrightarrow{A4} \neg(\beta \vee \neg q) \xrightarrow{A8} \neg(\neg(\neg q \vee s) \vee ((\neg p \vee s) \vee \neg q))$

$\longrightarrow \neg(\neg(\neg q \vee s) \vee \gamma) \xrightarrow{A11} (\neg(\neg\neg q \vee \gamma) \vee \neg(\neg s \vee \gamma)) \xrightarrow{[1],[2];T04}$
$0s \vee 0s \xrightarrow{T02} 0s \Longrightarrow$

[4] $\vdash \neg(\neg q \vee \beta) \longrightarrow 0s$;

V) $\neg\alpha \xrightarrow{Df1} \neg(\neg(\neg p \vee q) \vee \beta) \xrightarrow{A11} (\neg(\neg\neg p \vee \beta) \vee \neg(\neg q \vee \beta)$
$\xrightarrow{[3],[4],A09} 0s \Longrightarrow$ T27 .

THEOREM 28. (T28) $\vdash \Box((p \supset s) \supset ((q \supset s) \supset ((p \vee q) \supset s)))$.

PROOF: (Abbreviations: $\alpha = (p \supset s) \supset ((q \supset s) \supset (p \vee q) \supset s))$;
$\beta = s \vee (\neg(\neg p \vee s) \vee \neg(\neg q \vee s))$.)

I) $\neg(\neg p \vee \beta) \xrightarrow{T8} \neg((\neg p \vee s) \vee (\neg(\neg p \vee s) \vee \neg(\neg q \vee s))) \xrightarrow{T8}$
$\neg(((\neg p \vee s) \vee \neg(\neg p \vee s)) \vee \neg(\neg q \vee s)) \xrightarrow{T3} \neg((\neg p \vee s) \vee \neg(\neg p \vee s))$
$\xrightarrow{Df2} 0(\neg p \vee s) \xrightarrow{T1} 0t \Longrightarrow$

[1] $\vdash \neg(\neg p \vee \beta) \longrightarrow 0t$;

II) $\neg(\neg q \vee \beta) \xrightarrow{T8} \neg(((\neg q \vee s) \vee (\neg(\neg p \vee s) \vee \neg(\neg q \vee s))) \xrightarrow{T8}$
$\neg(((\neg q \vee s) \vee \neg(\neg p \vee s)) \vee \neg(\neg q \vee s)) \xrightarrow{A4} \neg(\neg(\neg q \vee s) \vee ((\neg q \vee s)$
$\vee \neg(\neg p \vee s))) \xrightarrow{T8} \neg((\neg(\neg q \vee s) \vee (\neg q \vee s)) \vee \neg(\neg p \vee s)) \xrightarrow{T3}$
$\neg(\neg(\neg q \vee s) \vee (\neg q \vee s)) \xrightarrow{A4} \neg((\neg q \vee s) \vee \neg(\neg q \vee s)) \xrightarrow{Df2}$
$0(\neg q \vee s) \xrightarrow{T1} 0t \Longrightarrow$

[2] $\vdash \neg(\neg q \vee \beta) \longrightarrow 0t$;

III) $\neg\alpha \xrightarrow{Df1} \neg(\neg(\neg p \vee s) \vee (\neg(\neg q \vee s) \vee (\neg(p \vee s) \vee s)) \xrightarrow{T8}$
$\neg((\neg(\neg p \vee s) \vee \neg(\neg q \vee s)) \vee (\neg(p \vee q) \vee s)) \xrightarrow{A4} \neg((\neg(p \vee q) \vee s) \vee$
$(\neg(\neg p \vee s) \vee \neg(\neg q \vee s))) \xrightarrow{A8} \neg(\neg(p \vee q) \vee \beta) \xrightarrow{A11} (\neg(\neg p \vee \beta) \vee$
$\neg(\neg q \vee \beta)) \xrightarrow{[1],[2],A09} 0t \Longrightarrow$ T28 .

THEOREM 29. (T29) $\vdash \Box((\neg p \supset p) \supset p)$.

PROOF: (Abbreviation: $\alpha = (\neg p \supset p) \supset p$.) $\neg\alpha \xrightarrow{Df1} \neg(\neg(\neg\neg p \vee p) \vee p)$
$\xrightarrow{A11} (\neg(\neg\neg\neg p \vee p) \vee \neg(\neg p \vee p)) \xrightarrow{A6,T04} (\neg(\neg p \vee p) \vee \neg(\neg p \vee p))$
$\xrightarrow{T02} \neg(\neg p \vee p) \xrightarrow{A4} \neg(p \vee \neg p) \xrightarrow{Df2} 0p \xrightarrow{T1} 0\alpha \Longrightarrow$ T29.

THEOREM 30. (T30) $\vdash \Box((\Diamond p \supset q) \supset (p \longrightarrow q))$.

PROOF: (Abbreviations: $\alpha = (\Diamond p \supset q) \supset (p \longrightarrow q)$,

$\beta_1 = \lnot(\lnot\lnot\Diamond p \lor (p \longrightarrow q))$ and $\beta_2 = \lnot(\lnot q \lor (p \longrightarrow q))$.)

I) $\beta_1 \xrightarrow{\text{A5}} (\lnot\lnot\lnot\Diamond p \land \lnot(p \longrightarrow q)) \xrightarrow{\text{A2,T07a}} (\lnot\Diamond p \land \lnot(p \longrightarrow q))$

$\xrightarrow{\text{T18,T07b}} (\lnot\Diamond p \land (p \land \lnot q)) \xrightarrow{\text{T09}} (\lnot\Diamond p \land p) \xrightarrow{\text{T5}} \Diamond s \Longrightarrow$

$[1] \vdash \beta_1 \longrightarrow \Diamond\alpha$;

II) $\beta_2 \xrightarrow{\text{A4}} \lnot((p \longrightarrow q) \lor \lnot q) \xrightarrow{\text{A9}} (p \land \lnot(q \lor \lnot q)) \xrightarrow{\text{A05,Df2}} \Diamond q \xrightarrow{\text{T1}} \Diamond\alpha =>$

$[2] \vdash \beta_2 \longrightarrow \Diamond\alpha$;

III) $\lnot\alpha \xrightarrow{\text{Df1}} \lnot(\lnot(\lnot\Diamond p \lor q) \lor (p \longrightarrow q)) \xrightarrow{\text{A11}} (\beta_1 \lor \beta_2) \xrightarrow{[1],[2],\text{T04}}$

$(\Diamond\alpha \lor \Diamond\alpha) \xrightarrow{\text{T02}} \Diamond\alpha \xrightarrow{\text{Df3}} \text{T30}$.

THEOREM 31. (T31) $\vdash \Box((p \longrightarrow q) \supset (\Diamond p \supset q))$.

PROOF: (Abbreviation: $\alpha = (p \longrightarrow q) \supset (\Diamond p \supset q)$.)

I) $\lnot\alpha \xrightarrow{\text{Df1}} \lnot(\lnot(p \longrightarrow q) \lor (\lnot\Diamond p \lor q)) \xrightarrow{\text{A12}} (p \longrightarrow \lnot(\lnot q \lor (\lnot\Diamond p \lor q)))$

$\xrightarrow{\text{A4,T012}} (p \longrightarrow \lnot((\lnot\Diamond p \lor q) \lor \lnot q)) \xrightarrow{\text{A8,T012}} (p \longrightarrow \lnot(\lnot\Diamond p \lor$

$(q \lor \lnot q))) \xrightarrow{\text{A4,T012}} (p \longrightarrow \lnot((q \lor \lnot q) \lor \lnot\Diamond p)) \xrightarrow{\text{T3,T012}} (p \longrightarrow (q \lor \lnot q))$

$\xrightarrow{\text{Df2,T012}} (p \longrightarrow \Diamond q) \xrightarrow{\text{T1}}$

$[1] \vdash \lnot\alpha \longrightarrow (p \longrightarrow \Diamond\alpha)$;

II) $\lnot\alpha \xrightarrow{\text{Df1,A4}} \lnot((\lnot\Diamond p \lor q) \lor \lnot(p \longrightarrow q)) \xrightarrow{\text{A8}} \lnot(\lnot\Diamond p \lor (q \lor \lnot(p \longrightarrow q)))$

$\xrightarrow{\text{T3}} \lnot\lnot\Diamond p \xrightarrow{\text{A2}} \Diamond p \xrightarrow{\text{T22}} p \Longrightarrow$

$[2] \vdash \lnot\alpha \longrightarrow p$;

III) By T06, $[1]$ and $[2]$: $\vdash \lnot\alpha \longrightarrow ((p \longrightarrow \Diamond\alpha) \land p)$; hence, by T08, $\vdash \lnot\alpha \longrightarrow \Diamond\alpha$, and, by Df3, $\vdash \Box((p \longrightarrow q) \supset (\Diamond p \supset q))$.

THEOREM 32. (T32) $\vdash \Box(\lnot(\lnot\Diamond p \lor \lnot q) \supset (p \land q))$.

PROOF: (Abbreviation: $\alpha = \lnot(\lnot\Diamond p \lor \lnot q) \supset (p \land q)$.)

I) $\lnot\alpha \xrightarrow{\text{Df1}} \lnot(\lnot\lnot(\lnot\Diamond p \lor \lnot q) \lor (p \land q)) \xrightarrow{\text{A6}} \lnot((\lnot\Diamond p \lor \lnot q) \lor (p \land q))$

$\xrightarrow{\text{A4}} \lnot((p \land q) \lor (\lnot\Diamond p \lor \lnot q)) \xrightarrow{\text{A10}} (p \longrightarrow \lnot(q \lor (\lnot\Diamond p \lor \lnot q)))$

$\xrightarrow{A4,T012} (p \to \neg((\neg \circ p \vee \neg q) \vee q))) \xrightarrow{A8,T012} (p \to \neg(\neg \circ p \vee$
$(\neg q \vee q))) \xrightarrow{A4,T012} (p \to \neg((\neg q \vee q) \vee \neg \circ p))) \xrightarrow{T3,T012} (p \to (\neg q \vee q))$
$\xrightarrow{A4,T012} (p \to \neg(q \vee \neg q)) \xrightarrow{Df2,T012} (p \to 0q) \xrightarrow{T01} (p \to 0\alpha) \Longrightarrow$

[1] $\vdash \neg\alpha \to (p \to 0\alpha)$;

II) $\neg\alpha \xrightarrow{Df1,A6,A8} \neg(\neg \circ p \vee (\neg q \vee (p \wedge q))) \xrightarrow{T3} \neg\neg \circ p \xrightarrow{A2} \circ p$
$\xrightarrow{T22} p \Longrightarrow$

[2] $\vdash \neg\alpha \to p$;

III) From T06, [1] and [2] it follows that $\vdash \neg\alpha \to (\neg \circ p \wedge p)$;hence by T5, $\vdash \neg\alpha \to 0\alpha$, and $\vdash \Box(\neg(\neg \circ p \vee \neg q) \supset (p \wedge q))$.

THEOREM 33. (T33) $\vdash \Box((p \wedge q) \supset \neg(\neg \circ p \vee \neg q))$.

PROOF: (Abbreviation: $\alpha = (p \wedge q) \supset \neg(\neg \circ p \vee \neg q)$.)

I) $\neg\alpha \xrightarrow{Df1} \neg(\neg(p \wedge q) \vee \neg(\neg \circ p \vee \neg q)) \xrightarrow{A13,A05}$
$\neg(\neg q \vee \neg(\neg \circ p \vee \neg q)) \xrightarrow{A4} \neg(\neg(\neg \circ p \vee \neg q) \vee \neg q) \xrightarrow{A11} (\neg(\neg\neg \circ p \vee \neg q) \vee$
$(\neg\neg q \vee \neg q)) \xrightarrow{T3,A4,T04} (\neg\neg\neg \circ p \vee (\neg q \vee \neg\neg q)) \xrightarrow{A2,Df2,T04}$
$(\neg \circ p \vee 0(\neg q)) \xRightarrow{T2}$

[1] $\vdash \neg\alpha \to \neg \circ p$;

II) $\neg\alpha \xrightarrow{Df1,T3} \neg\neg(p \wedge q) \xrightarrow{A2} (p \wedge q) \xrightarrow{A04} p \Longrightarrow$

[2] $\vdash \neg\alpha \to p$;

III) From T06, [1] and [2] it follows that $\vdash \neg\alpha \to (\neg \circ p \wedge p)$;hence, by T5, $\vdash \neg\alpha \to 0\alpha$, and $\vdash \Box((p \wedge q) \supset \neg(\neg \circ p \vee \neg q))$.

THEOREM 34. $\tilde{\mathbb{J}}$ *is a complete axiomatization for the discussive proprositional calculus.*

PROOF: In effect, this is an immediate consequence of T7, T9, T10, T11, T13, T20, T26, T27, T28, T29, T30, T31, T32, and T33.

It deserves to be noted that the postulates of $\tilde{\mathbb{J}}$ are not independent.

From the above considerations emerges the following question:

PROBLEM: To modify the axiom system $\tilde{\mathbb{J}}$ for the discussive propositional calculus in order to get an axiom system composed of independent postulates.

Finally it should also be emphasized that the usual rules of the positive propositional logic are valid in $\tilde{\mathbb{J}}$; in particular,the deduction theorem is true, as Furmanowski has shown employing very different methods (cf. Kotas 1975).

REFERENCES.

Church, A.

1956, An Introduction to Mathematical Logic, I, Princeton University Press, Princeton.

Costa, N. C. A. da and L. Dubikajtis

1968, *Sur la logique discoursive de Jaśkowski*, Bulletin Acad. Polonaise des Sciences Math. Astr. et Phys., 16, 551 - 557.

Costa, N. C. A. da

1975, *Remarks on Jaśkowski's discussive logic*, Reports on Math. Logic, 4, 7 - 16.

Hilbert, D. and W. Ackermann

1950, Principles of Mathematical Logic, Chelsea, New York.

Hughes, G. E. and M. J. Cresswell

1968, An introduction to Modal Logic, Methuen, London.

Jaśkowski, S.

1948, *Rachunek zdań dla systemów dedukcyjnych sprzecznych*, Studia Societatis Scientiarun Torunensis, Sectio A, I, nº 5, 55 - 77. (An English translation of this paper appeared in Studia Logica, XXIV (1969), 143 - 157.)

1949, *O konjunkcji dyskusyjnej w rachunku zdań dla systemów dedukcyjnych sprzecznych*, Studia Societatis Scientiarun Torunensis, Sectio A, I, nº 8, 171-172.

Kotas, J.
1975, *Discussive sentencial calculus of Jaśkowski*, Studia Logica, XXXIV, 149 - 168.

Instituto de Matemática
Universidade de São Paulo
São Paulo, S.P., Brazil

and

Instytut Matematyki
Uniwersytetu Ślaskiego
Kotowice, Poland.

Non-Classical Logics, Model Theory and Computability,
A.I. Arruda, N.C.A. da Costa and R. Chuaqui (eds.)

On some Modal Logical Systems Defined in Connexion with Jaśkowski's Problem

by *J. KOTAS* and *N. C. A. da COSTA*.

In 1948 Stanislaw Jaśkowski formulated the problem of finding logical systems which could be employed as underlying logics of deductive systems not devoid of inconsistency (see Jaśkowski 1948). This is a very important problem, not only from the philosophical but also from the practical and the mathematical points of view. In the mentioned paper Jaśkowski presented one solution of the problem at the level of the propositional calculus. He defined by an interpretation in Lewis' S5 propositional calculus a new logical system D_2 which he called *discussive* (or *discoursive*) *propositional calculus*. This system has been the subject of many studies (cf. e.g.,Costa 1975, Costa and Dubikajtis 1968 and 1977, Furmanowski 1975, and Kotas 1971 and 1974). It can be immediately seen that we may identify D_2 with M-S5, where M-S5 is the set of all formulas that become theses of S5 when preceded by the possibility sign M. Analogously, if S is any modal system, then the symbol M^k-S denotes the set of all formulas which become theses of S when preceded k times by M. M^k-S is called (cf. Perzanowski 1975) the M^k-*counterpart* of S.Therefore D_2 is the M-counterpart of S5. It is obvious that if we replace S5 by any other modal system, then we can obtain various corresponding M-counterparts and, eo ipso, various discussive logics. Furmanowski investigated the M-counterparts of all normal systems placed between S4 and S5 , and proved that for any normal modal calculus S such that S4 $\subset$ S $\subset$ S5, we have M-S = M-S5 (see Furmanowski 1975). In connection with this result, the following question arises: Are S4 and S5 respectively the smallest and the largest normal systems whose M-counterparts are identical to D_2? It is well-known that Sobociński's $S4_1$ system is the same as S4; hence, by Furmanowski's result, all M^k-counterparts of $S4_1$ are equal to M-counterparts of this system. Similarly, all M^k-counterparts of Sobociński's $S4_n$ system, for $k \geq n$, coincide with the M^n-counterpart of $S4_n$. Then, it is natural to pose the question of characterizing the smallest and the largest normal modal systems whose M^k-counterparts are identical to $M^n - S4_n$. The same question may be formulated with reference to Thomas' or to some other modal calculi.

In this paper we limit ourselves to *normal modal systems* and their M^n-counterparts. By a normal modal system we mean a set of modal propositional formulas which is closed under substitution, detachement for material implication and the rule of Gödel, i. e., the rule: *If* A, *then* LA. Our aim is to survey certain modal systems obtained by some authors in the course of investigations of M^k-counterparts of well-known modal logics, and to present some new interesting properties of their M^k-counterparts.

Our terminology and symbolism are standard (cf. Feys 1965, and Hughes and Cresswell 1968).

1. NOTATIONS.

Throughout this paper the symbols $\rightarrow$, $\neg$, L, and M mean material implication, negation, necessity and possibility, respectively. Capitals A, B, C,... denote any modal formulas. We put $L^0A = A$ and $L^{n+1}A = LL^nA$. The abbreviation M^nA is defined similarly. Let us put:

$$A_K = \{(A \rightarrow B) \rightarrow ((B \rightarrow C) \rightarrow (A \rightarrow C)),\ \neg A \rightarrow (A \rightarrow B), (\neg A \rightarrow A) \rightarrow A, L(A \rightarrow B) \rightarrow (LA \rightarrow LB)\},$$

$$A_T = A_k \cup \{LA \rightarrow A\},$$

$$A_B = A_T \cup \{MLA \rightarrow A\},$$

$$A_{S4_n} = A_T \cup \{L^nA \rightarrow L^{n+1}A\},\ n \geq 1,$$

$$A_{T_n^+} = A_B \cup \{L^nA \rightarrow L^{n+1}A\},\ n \geq 1.$$

In this paper we shall employ the following derivation rules:

MP : *If* A *and* $A \rightarrow B$, *then* B;

M^n : *If* M^nA, *then* A, $n \geq 1$;

L^n : *If* L^nA, *then* A, $n \geq 1$;

M^nL^n : *If* M^nL^nA, *then* A, $n \geq 1$;

M_n^{n+1} : *If* $M^{n+1}A$, *then* M^nA, $n \geq 1$.

The symbol R_N denotes the set $\{MP, L_1\}$ where L_1 is Gödel's rule. For any set X of modal formulas, we put:

$L^nX = \{L^nA, A \in X\}$, $n \geq 0$,

$M^n\text{-}X = \{A, M^nA \in X\}$, $n \geq 0$.

If r is a derivation rule, then L^nr denotes the derivation rule obtained from r by preceding its premisses and conclusion n-times by L. For any set R of derivation rules we put:

$L^nR = \{L^nr, r \in R\}$, $n \geq 0$.

For any set X of formulas and any set R of derivation rules, $Cn(R,X)$ denotes the set of formulas derivable from X by R. Logical systems will be treated as sets of formulas. Let us observe that A_K, A_T, A_B, A_{S4_n}, and $A_{T_n^+}$, together with MP and L_1, constitute the well-known axiomatics for Kripke's, Feys', Brouwerian, Sobociński's and Thomas' modal systems respectively.

2. THE SYSTEMS $S5_n$, $S4_n$, T_n^*, C_n^*, $n \geq 1$,

Perzanowski defines a system T^* (see Perzanowski 1975) in the following way:

DEFINITION 1.

$T^* = Cn(R_N \cup \{M_1^2\}, A_T \cup \{ML(LA \Rightarrow L^2A), ML(MLA \Rightarrow LA)\})$.

He next proves the following proposition:

THEOREM 1. (i) $M\text{-}T^* = M\text{-}S4$,

(ii) *For any normal system* S, *if* $T \subset S$ *and* $M\text{-}S = M\text{-}S4$, *then* $T^* \subset S$.

It follows from Theorem 1 that T^* is the smallest in the class of all normal modal systems which contain **T** and whose M-counterparts are equal to M-S4 (eo ipso to M-S5). In Błaszczuk and Dziobiak 1975a it is proved that $ML(LA \Rightarrow L^2A)$ and $ML(MLA \Rightarrow LA)$, treated by Perzanowski as axioms, belong to $Cn(R_N \cup \{M_1^2\}, A_T)$. Then we can obtain an axiomatics for T^* by adding M_1^2

to the usual axiomatics for T. If in the above mentioned axiomatics for T* we replace the rule M_1^2 by the formula $LA \to ML^2A$ (or $LM^2A \to MA$), then we get another axiomatics for T*. Because M_1^2 is not permissible in T, $T \nsubseteq T^*$; because M_1^2 is permissible in S4 and there exists a T*-matrix according to which $LA \to L^2A$ is not valid, then $T^* \subsetneq S4$. It is also easy to prove that T* and $S4_n$, $n \geq 2$, are *independent*, i. e., it is not true that for some $n \geq 2$ we have either $T^* \subset S4_n$ or $S4_n \subset T^*$.

Since the system $S4_1$ of Sobociński and the system S4 of Lewis are equal, and for any $n \geq 1$ and any $n \geq 0$ we have M^{n+k}-$S4_n$ = M^n-$S4_k$, then the question arises, whether the result of Perzanowski can be generalized to all systems of Sobociński. A solution to this problem is given in Błaszczuk and Dziobiak 1975b.

DEFINITION 2. $T_n^* = Cn(R_N \cup \{M_n^{n+1}\}, A_T)$, $n \geq 1$.

THEOREM 2. (i) M^n-T_n^* = M^n-S4, $n \geq 1$,

(ii) *For any normal modal system* S, *if* $T \subset S$ *and* M^n-S = M^n-$S4_n$, *then* $T_n^* \subset S$, $n \geq 1$.

Let us notice that T_1^* is the same as T*. T_n^* can be defined in another way, namely in order to obtain T_n^* we may only add $L^nA \to M^nL^{n+1}A$ (or $L^nM^{n+1}A \to M^nA$) to the set of axioms for T. It is easy to see that T_n^* and $S4_k$ are independent for $k \geq n$, and that $T_n^* \subsetneq S4_n$ for any $n \geq 1$. It is also easy to prove that $T_1^* \supsetneq T_2^* \supsetneq T_3^* \supsetneq \ldots$, and $\bigcap_{n \geq 1} T_n^* \supset T$.

If S is a normal modal system such that $S \supset S5$ and M-S = M-$S4_1$, then S = S5. Therefore, S5 is the largest system among all normal systems whose M-counterparts are equal to M-$S4_1$. Accordingly, the following question arises, whether for any $S4_n$ there exists an analogous maximal modal system. This question is answered in Błaszczuk and Dziobiak 1975b.

DEFINITION 3. $S5_n = Cn(R_N, A_{S4_n} \cup \{M^nL^nA \to L^nA\})$, $n \geq 1$.

Let us note that $S5_1$ is identical to S5. For the above defined family of systems we can prove the following

THEOREM 3. (i) M^n-$S5_n$ = M^n-$S4_n$, $n \geq 1$,

(ii) *For any normal modal system* S, *if* $T \subset S$ *and* M^n-S = M^n-$S4_n$, *then* $S \subset S5_n$, $n \geq 1$.

All the systems T^*_n, $S4_n$, and $S5_n$ contains T. It is easy to see that no formula of the form MA is a thesis of the system K of Kripke; hence M-K (or the discussive logic based on K) is empty. In Perzanowski 1975 it is proved that $D = Cn(R_N, A_K \cup \{M(A \rightarrow A)\})$ (in Lemmon 1966 this system is denoted by T(D)) is the smallest normal modal system whose M-counterpart is not empty. Hence it is obvious that if S is a normal modal system and M^n-S $\neq \emptyset$, then $S \supset D$, but it is not necessary that $S \supset T$. If we omit the clause that a normal modal system contains T, then the $S5_n$ are still the largest systems in the class of all normal logics whose M^n-counterparts are equal to M^n-$S4_n$, but T^*_n are not now the smallest systems. We can define a new family of normal modal logics (see Dziobiak 197+) as follows:

DEFINITION 4. $C^*_n = Cn(R_N, A_K \cup \{M^nL^n(LA \rightarrow A), L^nA \rightarrow M^nL^nA, L^nA \rightarrow M^nL^{n+1}A\})$, $n \geq 1$.

About the systems of this family we can prove the following:

THEOREM 4. (i) M^n-C^*_n = M^n-$S4_n$, $n \geq 1$.

(ii) *For any normal modal system* S, *if* M^n-S = M^n-$S4_n$, *then* $C^*_n \subset S$.

Of course, $C^*_n \subset T^*_n$, since $L^nA \rightarrow M^nA$ and $L^nA \rightarrow M^nL^nA$ are theses of T and $LA \rightarrow A$ is not a thesis of C^*_n, $n \geq 1$. We can also prove that $C^*_1 \not\supseteq C^*_2 \not\supseteq C^*_3 \not\supseteq \ldots$, and $\bigcap_{n\geq 1} C^*_n \supset D$.

We can say about S4 and T that they are placed *very far*, because there exists between them an infinite sequence of intermediate axiomatizable normal modal systems. The calculi $S4_n$ are just such a kind of systems. T^*_n and $S4_n$, as well as C^*_n and T^*_n, are also placed very far. For any pair T^*_n and $S4_n$, as well as any pair C^*_n and T^*_n, we can define the corresponding intermediate systems as follows:

DEFINITION 5. $T^k_n = Cn(R_N, A_T \cup \{L^nA \rightarrow M^nL^{n+1}A, L^kA \rightarrow L^{k+1}A\})$, $k \geq 1$, $1 \leq n \leq k$.

DEFINITION 6. $C^k_n = Cn(R_N \cup \{M^{k+1}_k\}, A_k \cup \{M^nL^n(LA \rightarrow A), L^nA \rightarrow M^nL^nA, L^nA \rightarrow M^nL^{n+1}A\})$, $k \geq 1$, $1 \leq n \leq k$.

We can prove that the above defined normal modal systems have the following properties:

(i) $T_n^n = S4_n$, $n \geq 1$.

(ii) $T_1^k \supsetneq T_2^k \supsetneq T_3^k \supsetneq \ldots \supsetneq T_k^k$, $k \geq 1$.

(iii) $T_n^n \supsetneq T_n^{n+1} \supsetneq T_n^{n+2} \supsetneq \ldots$ and $\bigcap_{k \geq 0} T_n^{n+k} \supset T_n^*$, $n \geq 1$.

(iv) $M^n - T_n^k = M^n - S4_n$, $n \geq 1$, $k \geq n$.

(v) $C_n^n \supsetneq C_n^{n+1} \supsetneq C_n^{n+2} \supsetneq \ldots$ and $\bigcap_{k \geq 0} C_n^{n+k} \supset C_n^*$, $n \geq 1$.

(vi) $M^n - C_n^k = M^n - S4_n$, $n \geq 1$, $k \geq n$.

(vii) $C_n^{n+k} \supsetneq C_{n+1}^{n+k+1} \supsetneq C_{n+2}^{n+k+2} \supsetneq \ldots$ and $\bigcap_{n \geq 1} C_n^{n+k} \supset D$, $k \geq 0$.

With the above theorems as a basis we may describe the connections among C_n^* , C_n^k , T_n^* , T_n^k , $S4_n$, $S5_n$, T_n^+ , T, D, K, and B by the following diagram:

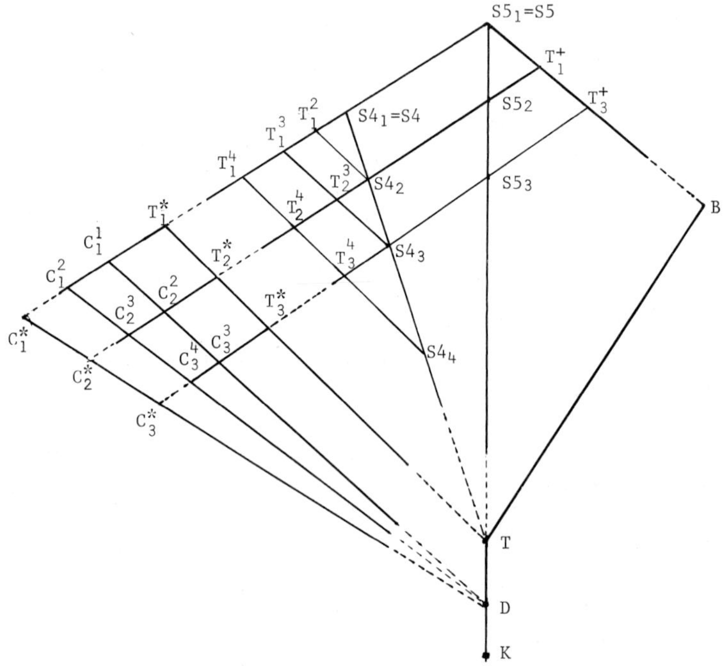

(In the diagram, if we have two systems such that one is placed in the lower part and the other in the higher part and they are connected by a straight line, then this means that the first is contained in the second.)

3. KRIPKE'S SEMANTICS FOR C_n^*, T_n^*, $S4_n$, $S5_n$, $n \geq 1$.

By a *frame* we understand any pair $<W,R>$, where W is a non-empty set and $R \subset W \times W$. If $R \subset W \times W$ and $A \subset W$, then we put:

(i) $R^{-1} = \{(w,w') \in W \times W,\ w'Rw\}$.

(ii) $R/_A = R \cap (A \times A)$.

(iii) $R^n = \begin{cases} R \text{ for } n=1, \\ \{(w,w') \in W \times W,\ \bigvee_{(w_i)_{i \leq n-1}} (wRw_1 \wedge \ldots \wedge w_{n-1}Rw')\} \text{ for } n>1. \end{cases}$

The set $\{(w,w),\ w \in W\}$ is denoted by Δ. For any frame $<W,R>$ and any $w \in W$, by a *tree with top* w we understand the smallest set T_w having the following two properties:

(i) $w \in T_w$,

(ii) *If* $w'Rw''$ *and* $w' \in T_w$, *then* $w'' \in T_w$.

We can prove (see Dziobiak 197+) the following four theorems about the semantical characterization of C_n^*, T_n^*, $S4_n$, $S5_n$, $n \geq 1$:

THEOREM 5. *For any* $n \geq 1$, $S5_n$ *is characterized by the class of all frames* $<W,R>$ *such that:*

(i) $\Delta \subset R$,

(ii) $R^{n+1} \subset R^n$,

(iii) $R^n \subset (R^n)^{-1}$.

THEOREM 6. *For any* $n \geq 1$, $S4_n$ *is characterized by the class of all frames* $<W,R>$ *such that:*

(i) $\Delta \subset R$,

(ii) $R^{n+1} \subset R^n$,

(iii) $\bigwedge_{w \in W} \bigvee_{w' \in W} (wR^n w' \wedge \bigwedge_{w'' \in W} (w'R^n w'' \rightarrow wR^n w'') \wedge (R^n/_{T_{w'}} \subset (R^n/_{T_{w'}})^{-1}))$.

THEOREM 7. *For any* $n \geq 1$, T_n^* *is characterized by the set of all frames* $<W,R>$ *such that:*

(i) $\Delta \subset R$,

(ii) $\bigwedge_{w \in W} \bigvee_{w' \in W} (wR^n w' \wedge \bigwedge_{w'' \in W} (w'R^n w'' \rightarrow wR^n w'') \wedge (R^n/_{T_{w'}} \subset (R^n/_{T_{w'}})^{-1})$

$$\wedge (R^{n+1}/_{T_{w'}} \subset R^n/_{T_{w'}})).$$

THEOREM 8. *For any* $n \geq 1$, C^*_n *is characterized by the class of all frames* $<W,R>$ *such that:*

$$\bigwedge_{w \in W} \bigvee_{w' \in W} (wR^n w' \wedge \bigwedge_{w'' \in W} (w'R^n w'' \to wR^n w'') \wedge \Delta \subset R \wedge$$
$$(R^{n+1}/_{T_{w'}} \subset R^n/_{T_{w'}}) \wedge (R^n/_{T_{w'}} \subset R^n/_{T_{w'}})^{-1})).$$

The condition (iii) of Theorem 6 is superfluous, of course, and the condition (ii) of Theorem 7 may be formulated in a simpler form. Nonetheless, from our formulations of Theorems 5-8, it easily follows a semantical substantiation of the following equalities:

$$M^n\text{-}C^*_n = M^n\text{-}T^*_n = M^n\text{-}S4_n = M^n\text{-}S5_n \quad , n \geq 1,$$

which we have earlier obtained in a syntactical way.

The above semantics for $S5_n$, $S4_n$, T^*_n , C^*_n suggest a general method of constructing normal modal systems whose M^k-counterparts are equal to M^k-counterparts of a given system S . (In order to simplify our considerations, we will only consider M-counterparts.)

Let S be a normal modal logic determined by frames $<W,R>$, where R is a relation characterized by the independent properties w_1 , w_2 ,..., w_n. Let us denote by P the set $\{w_1, w_2, \ldots, w_n\}$. If $P' \subset P$, then the symbol (P') will denote the following property:

$$(P') \qquad \bigwedge_{w \in W} \bigvee_{w' \in W} (wRw' \wedge \bigwedge_{w'' \in W} (w'Rw'' \to wRw'') \wedge (R/_{T_{w'}} \textit{ has all}$$
properties belonging to P')).

For any $P' \subset P$, let us put:

(i) $S^{P'}$ is the normal modal logic determined by frames whose relations have the properties belonging to $P-P'$.

(ii) $S_{P'}$ is the normal modal logic determined by frames whose relations are characterized by the properties belonging to $(P-P') \cup \{(P')\}$.

The systems $S^{P'}$ as well as the systems $S_{P'}$, $P' \subset P$, form two diagrams in which Kripke's system K is the smallest and S is the largest system. The following theorem is true:

THEOREM 9. (i) *If* $P' \subset P'' \subset P$, *then* $S^{P''} \subset S^{P'}$ *and* $S_{P'} \subset S_{P''}$.

(ii) *For any* P', *if* $P' \subset P$, *then* $M\text{-}S_{P'} = M\text{-}S$.

(iii) *If* $P' \subset P$, *then* (P') *and* $((P'))$ *are equivalent.*

Let us note, for example, that S5 is characterized by frames $<W, R>$, where R is reflexive (w_1), transitive (w_2) and symmetric (w_3). It is obvious that the property $(\{w_3\})$ holds in each finite frame with a reflexive and transitive relation. Because S4 and S5 have the finite model property, it follows from the above that $S4 = S5^{\{w_3\}}$ and M-S4 = M-S5. Błaszczuk and Dziobiak show that S5 is the smallest normal system which contains the Brouwerian system B and whose M-counterpart is equal to M-S5 (Błaszczuk and Dziobiak 1975b). Then we see that S5 can be characterized by frames in which the relations have the properties w_1, $(\{w_2\})$, and w_3.

The systems $S5^{P'}$, where $P' \subset \{w_1, w_2, w_3\}$, form a diagram in which the following occur: K, T, $S4^o$, B^o, B, S4, $S5^o$, and S5. If we consider any system $S5_n$, $n \geq 1$, we obtain an analogous diagram. We can say about the systems of the diagrams associated with the systems $S5_n$, that they correspond to the systems of the diagram associated with S5. In this way we can define new families of normal modal logics having properties very similar to the properties of the systems of the diagram associated with S5.

4. AXIOMATIZATIONS OF M^n-COUNTERPARTS OF SOME NORMAL MODAL SYSTEMS.

As we have already said, Jaśkowski defined discussive logic D_2 (or M-S5) by an interpretation. It is easy to see that D_2 is not a normal system. Moreover, the rule of detachement for material impliction is not permissible, though the rule of detachement for strict implication as well as the rule "*If* LA *and* L(A → B), *then* LB" are permissible. Of course, to axiomatize a given system is in general a very important achievement. The first axiomatics for M-S5 was formulated in Costa and Dubikajtis 1968. Other axiomatics for M-S5 were presented by Makinson (see Costa 1975) and Kotas (see Kotas 1974). Although Makinson's axiomatics is interesting, we will describe an axiomatics studied in Kotas 1974, because the method applied there can also be extended to the axiomatization of some M^n-counterparts of other modal systems.

In Kotas 1974 the following theorem is proved:

THEOREM 10. $M\text{-}S5 = Cn(LR_N \cup \{L^1, M^1\}, L(A_{S4} \cup \{MLA \to A\}))$.

Let us observe that the preceding axiomatics for M-S5 is obtained from the axiomatics for S5 in a simple way. We add new derivation rules to the set of rules of the axiomatics for S5 and simultaneously we write L before each axiom of S5 in order to restrict the ranges of the derivation rules. Błaszczuk and Dziobiak have applied the same method to axiomatize M^n-counterparts of some normal systems (see Błaszczuk and Dziobiak 1975b, and 197+).

A system S defined by a set X of axioms and a set R of derivation rules is called a normal extension of the Brouwerian system B, when $A_B \subset S$ and the rules of Gödel, of substitution and of modus ponens for material implication are permissible in S.

THEOREM 11. *If X is a set of axioms and R is a set of derivation rules for a normal modal system S and S is a normal extension of B, then*

$$M^n\text{-}S = Cn(L^n R \cup \{L^n, L^n M^n\}, L^n X), \quad n \geq 1.$$

Since B and the systems T_n^+, $n \geq 1$, of Thomas are normal extensions of B, it follows from the Theorem 11 that all M^k-counterparts of these systems are axiomatizable. In particular, the M-counterparts or the *discussive logics* based on these systems are axiomatizable too.

THEOREM 12. $M^n\text{-}S4_n = Cn(L^n R_N \cup \{L^n, M^n\}, L^n A_{S4_n})$, $n \geq 1$.

The problem of the axiomatization of $M^k\text{-}S4_n$, $k < n$, is still open. It is very interesting that it is possible to give an axiomatics for any M^k-counterpart of T, namely (see Błaszczuk and Dziobiak 197+) the following theorem is true:

THEOREM 13. $M^n\text{-}T = Cn(L^n R_N \cup \{L^n, L^n M^n\}, L^n A_T)$, $n \geq 1$.

If we have $n = 1$, then from Theorem 13 we obtain an axiomatics for the discussive logic based on T. Furmanowski proved (his results are not yet published) that the discussive logic based on T has many interesting properties . For example, the mothodological theorems (which demonstrate that M-S5 is a very strong and useful system), proved in Jaśkowski 1948 , are also true in M-T.

5. A SEMANTICAL CHARACTERIZATION OF M-S5.

We proceed now to describe a Henkin semantics for M-S5 (or D_2). To begin with we need the following (modal) discussive connectives:

DEFINITION 7. $A \rightarrow_d B =_{def} MA \rightarrow B$ (*discussive implication*);

$A \wedge_d B =_{def} MA \wedge B$ (*left discussive conjunction*),

where $\wedge$ is the symbol of conjunction.

D_2 may be axiomatized as follows (cf. Costa and Dubikajtis 1977).

Primitive symbols: 1) Propositional variables; 2) connectives: $\rightarrow_d$, $\wedge_d$, $\vee$, and $\neg$; 3) Parentheses.

The notion of formula is defined as usual and the set of formulas will be noted F.

Axiom schemata:

A01) $A \rightarrow_d (B \rightarrow_d A)$.

A02) $(A \rightarrow_d B) \rightarrow_d ((A \rightarrow_d (B \rightarrow_d C)) \rightarrow_d (A \rightarrow_d C)))$.

A03) $((A \rightarrow_d B) \rightarrow_d A) \rightarrow_d A$.

A04) $(A \wedge_d B) \rightarrow_d A$.

A05) $(A \wedge_d B) \rightarrow_d B$.

A06) $A \rightarrow_d (B \rightarrow_d (A \wedge_d B))$.

A07) $A \rightarrow_d (A \vee B)$.

A08) $B \rightarrow_d (A \vee B)$.

A09) $(A \rightarrow_d C) \rightarrow_d ((B \rightarrow_d C) \rightarrow_d ((A \vee B) \rightarrow_d C))$.

A1) $A \rightarrow_d \neg\neg A$.

A2) $\neg\neg A \rightarrow_d A$.

A3) $\neg(A \vee \neg A) \rightarrow_d B$.

A4) $\neg(A \vee B) \rightarrow_d \neg(B \vee A)$.

A5) $\neg(A \vee B) \rightarrow_d (\neg A \wedge_d \neg B)$.

A6) $\neg(\neg\neg A \vee B) \rightarrow_d \neg(A \vee B)$.

A7) $(\neg(A \vee B) \rightarrow_d C) \rightarrow_d ((\neg A \rightarrow_d B) \vee C)$.

A8) $\neg((A \vee B) \vee C) \rightarrow_d \neg(A \vee (B \vee C))$.

A9) $\neg((A \rightarrow_d B) \vee C) \rightarrow_d (A \wedge_d \neg(B \vee C))$.

A10) $\neg((A \wedge_d B) \vee C) \rightarrow_d (A \rightarrow_d \neg(B \vee C))$.

A11) $\neg(\neg(A \vee B) \vee C) \rightarrow_d (\neg(\neg A \vee C) \vee \neg(\neg B \vee C))$.

A12) $\neg(\neg(A \rightarrow_d B) \vee C) \rightarrow_d (A \rightarrow_d \neg(\neg B \vee C))$.

AL3) $\neg(\neg(A \wedge_d B) \vee C) \rightarrow_d (A \wedge_d \neg(\neg B \vee C))$.

Derivation rule: MP_d : *If* A *and* $A \rightarrow_d B$, *then* B.

DEFINITIONS. $A \rightarrow B =_{def} \neg A \vee B$.

$A \wedge B =_{def} \neg(\neg A \vee \neg B)$.

$LA =_{def} \neg A \rightarrow_d \neg(A \vee \neg A)$.

$MA =_{def} \neg L \neg A$.

$IA =_{def} \neg MA$.

(Formal) proof, (formal) deduction and the symbol $\vdash$ are introduced as customary. And it is not hard to prove the theorems listed below.

THEOREM 14. *In* D_2 *we have:*

$\vdash (A \rightarrow_d B) \rightarrow_d ((A \rightarrow_d IB) \rightarrow_d IA)$,

$\vdash IA \rightarrow_d (A \rightarrow_d B)$,

$\vdash A \vee IA$.

Γ and Δ will denote subsets of F. The set $\{A \in F, \Gamma \vdash A\}$ is abbreviated to $\overline{\Gamma}$. Γ is said to be inconsistent if there is at least one formula A such that A and $\neg A$ belong to $\overline{\Gamma}$; otherwise Γ is called consistent. Γ is said to be trivial if $\overline{\Gamma} = F$; otherwise Γ is called nontrivial.

We say that a nontrivial set Γ is maximal if it is not properly contained in any other nontrivial set; i. e., if $\Gamma \subsetneq \Delta$, then $\overline{\Delta} = F$.

THEOREM 15. *If* Γ *is a nontrivial maximal set, then* ($\Rightarrow$ *and* $\Leftrightarrow$ *are the metalinguistic abbreviations of implication and equivalence*):

1) $\Gamma \vdash A \Rightarrow A \in \Gamma$;

2) $A \in \Gamma \Rightarrow IA \notin \Gamma$;

3) $IA \in \Gamma \Rightarrow A \notin \Gamma$;

4) $A \in \Gamma$ *or* $IA \in \Gamma$;

5) $\vdash A \Rightarrow A \in \Gamma$;

6) $A , A \rightarrow_d B \in \Gamma \Rightarrow B \in \Gamma$.

A *valuation* (or *interpretation*) of D_2 is a function $v: F \to \{0,1\}$ such that:

1) $v(A \to_d B) = 1 \iff v(A) = 0$ *or* $v(B) = 1$;

2) $v(A \wedge_d B) = 1 \iff v(A) = v(B) = 1$;

3) $v(A \vee B) = 1 \iff v(A) = 1$ *or* $v(B) = 1$;

4) *If* A *is one of the* Axioms A1-A13, *then* $v(A) = 1$;

5) $v(IA) = 1 \iff V(A) = 0$.

A valuation v is *singular* if there is a formula A such that $v(\neg A) = v(A) = 1$. If this condition is not satisfied by any formula whatsoever, v is *normal*.

A formula A of D_2 is valid if for every valuation v, $v(A) = 1$. A given valuation v is a *model* of a set of formulas Γ if $v(A) = 1$. For every formula A of Γ. If $v(A) = 1$ for every model v of Γ, we write $\Gamma \models A$. (In particular, $\models A$ means that A is valid.)

THEOREM 16. $\Gamma \vdash A \Rightarrow \Gamma \models A$.($\vdash A \Rightarrow \models A$.)

THEOREM 17. *Every nontrivial set of formulas of* D_2 *is contained in a maximal nontrivial set.*

THEOREM 18. *Any nontrivial set of formulas has a model.*

THEOREM 19. $\Gamma \models A \Rightarrow \Gamma \vdash A$. ($\models A \Rightarrow \vdash A$.)

THEOREM 20. *There are inconsistent sets of formulas of* D_2 *which are not trivial.*

THEOREM 21. *There are inconsistent (but nontrivial) sets of formulas which have models.* (Γ *has a model* $\iff$ Γ *is nontrivial.*)

With reference to the above semantical analysis of D_2 , three remarks are important: 1) Theorem 21 implies that there are singular valuations, and the existence of normal valuations is easy to be proved. 2) The value of a valuation v for a given formula A is not in general determined by the values of v for the propositional variables. 3) It is clear that we can modify the outlined semantical method in order to obtain semantics for almost all systems discussed in this paper (and for most modal calculi).

Finally it is worth noting that the semantical analysis of D_2 seems to be in agreement with the spirit of some views of the young Łukasiewicz. For instance, with the spirit of his hints for a definition of truth for logical systems in which the law of contradiction is not valid in general, as it is sketched in Łukasiewicz 1910, section 18, item (b').

6. Problems.

To finish, we formulate some open problems related to M^n-counterparts of modal systems and to the normal modal systems discussed above.

PROBLEM 1. Is M^n-$S4_n$, $k < n$, axiomatizable?

PROBLEM 2. If there exists a Kripke's semantics for a modal system S, then is there also a Kripke's semantics for M^n-S , $n \geq 1$?

PROBLEM 3. Are there nice algebrizations of C^*_n , T^*_n, and $S5_n$?

PROBLEM 4. Have C^*_n , T^*_n, and $S5_n$ the finite model property ?

PROBLEM 5. Is there a reasonable simple theory of the M^n-counterparts of non-normal modal systems, and in particular of M-S2, M-S3 and M-S3.5 ?

PROBLEM 6. Let S be a modal system having the properties of the systems described before Theorem 9. Is $S_{P'}$, $P' \subset P$, minimal among all normal modal logics containing $S^{P'}$ and having M^n-counterparts equal to M^n-S ? (This problem was formulated independently by J. Perzanowski.)

Suppose that S is a normal modal logic determined by frames, in which the relations are characterized by the independent conditions $\omega_1, \omega_2, \ldots, \omega_n$; sometimes there exists an axiomatics for S such that its axioms are independent and may be divided into two sets A and $\{A_1, A_2, \ldots, A_n\}$, where $A_1, A_2, \ldots, A_n$ correspond to $\omega_1, \omega_2, \ldots, \omega_n$, respectively, and the only derivation rules are substitution, detachement for material implication and Gödel's rule.

PROBLEM 7. Let S be a normal modal calculus axiomatizable as we have described. Are there axiomatics for $S^{P'}$, $P' \subset P$? If $S^{P'}$ is axiomatizable, then, is there an axiomatics for it which can be obtained in a simple way from the axiomatics for S ?

REFERENCES.

Błaszczuk, J. and W. Dziobiak

1975a, *Remarks on Perzanowski's modal system*, Bulletin of the Section of Logic, Polish Acad. of Sciences, 4, nº2, 57-64.

1975b, *Modal systems related to* $S4_n$ *of Sobociński*, Bulletin of the Section of Logic, Polish Acad. of Sciences, 4, nº 3, 103-108.

1976, *Modal systems placed in the "triangle"* $S5-T_1-T$, Bulletin of the Section of Logic, Polish Acad. of Sciences, 5, nº 1, 138-142.

197+, *Axiomatizations of* M^n*-counterparts of some modal systems*, to appear in Reports on Mathematical Logic.

Costa, N. C. A. da

1975, *Remarks on Jaśkowski's discussive logic*, Reports on Mathematical Logic, 4, 7-16.

Costa, N. C. A. da and L. Dubikajtis

1968, *Sur la logique discoursive de Jaśkowski*, Bulletin Acad. Polonaise des Sciences Math., Astr. et Phys., 16, 551-557.

1977, *On Jaśkowski's discussive logic*, these Proceedings.

Dziobiak, W.

197+, *Kripke's semantics for some modal systems*, to appear in Studia Logica.

Feys, R

1965, Modal Logic, Louvain, Nauwelaerts.

Furmanowski, T.

1975, *Remarks on discussive propositional calculus*, Studia Logica, 34, 39-43.

Hughes, G. E. and M. J. Cresswell

1968, An Introduction to Modal Logic, Methuen, London.

Jaśkowski, S.

1948, *Rachunek zdān dla systemōw dedukcyjnych sprzecznych*, Studia Societatis Scientiarun Torunensis, Sectio A, I , nº 5, 57-77.

Kotas, J.

1971, *On the algebra of classes of formulae of Jaśkowski's discussive system*, Studia Logica, 27, 81-90.

1974, *The axiomatization of S. Jaśkowski's discussive system*, Studia Logica, 33, 195-200.

Lemmon, E. I.

1966, *Algebraic semantics of modal logics* . I, The Journal of Symbolic Logic, 31, 46-65.

Łukasiewicz, J.

1910, *Über den Satz des Widerspruchs bei Aristoteles*, Bulletin International de l'Académie des Sciences de Cracovie, Classe de Philosophie, 15-38.

Perzanowski, J.

1975, *On M-fragments and L-fragments of normal propositional logics*, Reports on Mathematical Logic, 5, 63-72.

Instytut Matematyki
Uniwersytet Mikołaja Kopernika
Toruń, Poland;

and

Instituto de Matemática
Universidade de São Paulo
São Paulo, SP., Brazil.

Non-Classical Logics, Model Theory and Computability,
A.I. Arruda, N.C.A. da Costa and R. Chuaqui (eds.)

Infinite Rules in Finite Systems

by E. G. K. LÓPEZ-ESCOBAR

0. Introduction.

The idea of making use of rules of inference with infinitely many premisses in the discussion of finite formulae is certainly not new. In fact it can be argued that some of the paradoxes of antiquitity were paradoxes because of (implicit) uses of such infinite rules. Consider, for example, the paradox of Achilles and the Tortoise. Routledge formulated it in a modern style essentially as follows (see Routledge 1950):

Formal language of the theory A & T

(a) *Individual constants:* $\ulcorner r \urcorner$ for each rational number r ,

(b) *Relational constant:* P .

Axiomatization of A & T

(a) *Axiom:* $P(\ulcorner 0 \urcorner , \ulcorner 100 \urcorner)$.

(b) *Rule of inference:*

$$P(\ulcorner x \urcorner , \ulcorner y \urcorner) \Longrightarrow P(\ulcorner y \urcorner , \ulcorner y + \frac{1}{10}(y - x) \urcorner).$$

Interpretation: "$P(\ulcorner r \urcorner , \ulcorner s \urcorner)$" is interpreted (in the classical) model that Achilles is at a distance r from the starting point of the race when the tortoise is at a distance s from the starting point of the race.

An easy induction on the length of the derivations in A & T shows that for no number r is the sentence $P(\ulcorner r \urcorner , \ulcorner r \urcorner)$ provable in A & T. On the other hand it is well known (nowdays) that $P(\ulcorner 111\ 1/9 \urcorner , \ulcorner 111\ 1/9 \urcorner)$ is true in the classical model. The fact that $P(\ulcorner 111\ 1/9 \urcorner , \ulcorner 111\ 1/9 \urcorner)$ is true in some models but not provable in the formal system is,at least since 1930, not paradoxical and it is doubtful that it was the incompleteness of the formal system A & T that bothered the Greeks.

The problem had to do much more with the underivability in A & T of the following rule:

From: $P(\ulcorner x_0 \urcorner, \ulcorner y_0 \urcorner),\ P(\ulcorner x_1 \urcorner, \ulcorner y_1 \urcorner), \ldots,\ P(\ulcorner x_n \urcorner, \ulcorner y_n \urcorner),\ \ldots$

To conclude: $P(\ulcorner x \urcorner, \ulcorner y \urcorner)$

Where: $x = \lim_{n\to\infty} x_n$ *and* $y = \lim'_{n\to\infty} y_n$.

For it is precisely such a rule which focuses on both the problem of formally expressing the argument used in the paradox of Achilles and the Tortoise as well as on the problem of how a finitary statement requires infinitely many assumptions for its derivation.

At the present time such problems of how to carry out infinitely many acts in a finite amount of time are of no interest whatsoever. However, for quite different reasons, there is still some interest in rules of inference of the form:

(*) *From:* $F_0, F_1, F_2, \ldots, F_n, \ldots$

To conclude: G

The paradigm of such a rule being of course the ω - *rule* (also called Carnap's rule)

From: $A\underset{\sim}{0}, A\underset{\sim}{1}, A\underset{\sim}{2}, \ldots, A\underset{\sim}{n}, \ldots$

To conclude: $\forall x\, Ax$

and thus we shall call all rules of the form (*) ω-rules.

The renewed interest in such rules comes basically from two distinct, but not unrelated areas. The older one being "proof - theory" in which the ω - rule is sometimes added in order to have a more powerful consequence relation; see, for example, Grzegorczyk, Mostowski and Ryll - Nardzewski 1958. The newer area is the "theory of proofs" in which the ω- rule is used principally as a tool for making finite proofs infinite and thus distinguishing between concepts which are equivalent for proofs of finite length; see Kreisel, Mints, and Simpson 1974.

Irrespective of one's feeling about the validity of ω- rules, one certainly has to agree that the ω - rules are closely linked to formal proofs (or derivations). And since the usual connotation of "formal proof" is that of an *effective proof*, it didn't take long before attempts were made to make the ω - rule an "effective rule" of inference. Since the premisses of the ω - rule:

$$A\underset{\sim}{0}, A\underset{\sim}{1}, A\underset{\sim}{2}, \ldots, A\underset{\sim}{n}, \ldots$$

can be obtained effectively (primitively recursive) from the conclusion one has to look at something more than that the premisses be effectively given in order to make sure that the "ω-proofs" have still a modicum of effectiveness. A natural way to achieve the latter would be to require that the ω-proof be given by an effective function, and since to most people "effective" is synonymous with "general recursive", a natural way would be to require that the ω-proof be a general recursive function.

Now if Π is a general recursive function which is an ω-proof then it can easily be shown that to each application of the ω-rule in Π there corresponds a general recursive function ϕ such that ϕn is a Gödel number of the derivation of the n-th premise. In fact, it can be shown that the existence of a general recursive ω-proof is equivalent to restricting the applications of the ω-rule so that:

From: $\dfrac{A\underset{\sim}{0}\ , A\underset{\sim}{1}\ , A\underset{\sim}{2}\ , \ldots\ , A\underset{\sim}{n}\ , \ldots}{}$

To conclude: $\forall x\ A x$

Provided: *There is a general recursive function* ϕ *such that* ϕn *is a Gödel number of a proof of* An.

For obvious reasons the above rule is often called the *"recursively restricted ω-rule"*.

Shoenfield was the first to show that *PA* (first-order classical number theory) + ω-rule was a conservative extension of *PA* + (recursively restricted ω-rule) and also raised the question whether the analogous result would hold for full second-order number theory (see Shoenfield 1959). After a few false starts by various authors, Takahashi gave a positive answer to Shoenfield's question (see Takahashi 1970).

Takahashi's proof was of quite a different nature from the one given by Shoenfield and those given by others (Kent 1967, Nelson 1971) in that semantical concepts were completely avoided. In fact, looking over Takahashi's proofs, one gets the impression that the result would apply to any theory. However there are known (albeit not too many) counterexamples. The counterexamples are basically of two types: cut-free systems and intuitionistic systems. The reason why the usual axiomatizations of intuitionistic number produce a stronger theory with the unrestricted ω-rule than with the recursively restricted ω-rule is discussed in some detail in the section 7. The reason why in some (artificially constructed) cut-free systems the unrestricted ω-rule turns out to be more powerful than the recursively restricted ω-rule is of the same nature as the reason why the recursively restricted ω-rule is, occasionally, not a conservative extension of the primitive recursive ω-rule (for more details see Kreisel-Mints-Simpson 1973, page 82).

No one doubts that the ω-rules are concerned with formulae, and so eventually with their truth or falsity. However after a few moments of working with ω-rules one sees that the truth or falsity of the formulae has very little to do with the replacing unrestricted ω-rules by recursively restricted ω-rules (or recursive ones by primitive recursive ones). After a perfunctory remark to the effect that the rules of inference are sound (with respect to some chosen interpretation) one might as well forget that one is dealing with formulae. In fact it might be worthwhile to just consider (well-founded) trees of natural numbers so that the replacing of the unrestricted ω-rule by the recursively restricted ω-rule is reduced to the problem of replacing a well-founded tree of natural numbers by an "equivalent" recursive one.

Smullyan's book "Theory of Formal Systems" presents a very elegant way of avoiding semantical concepts in the discussion of formal theories and the question arose if the same could not be done for the ω-rules. Or more specifically: given a formal mathematical system (M) (as defined by Smullyan 1961), are there any combinatorial conditions on (M) which guarantee that $(M)^{\infty}$, that is (M) + unrestricted ω-rule, is a conservative extension of $(M)^{\infty}_{\mathrm{GR}}$, where $(M)^{\infty}_{\mathrm{GR}} = (M)$ + recursively restricted ω-rule?

In this paper we shall define such a combinatorial condition and give the proof that $(M)^{\infty}$ is then a conservative extension of $(M)^{\infty}_{\mathrm{GR}}$. The method of proof is but an abstraction of the proof given by Takahashi for classical full-second order number theory.

The only evidence that we have that the combinatorial condition is natural is that we can get as a corollary all the present known positive results concerning replacement of an unrestricted ω-rule by a recursively restricted ω-rule. It also sheds some light on the present negative results, for example, it shows that one of the stumbling blocks for intuitionistic number theory (as far as showing that $(HA)^{\infty}$ be a conservative extension of $(HA)^{\infty}_{\mathrm{GR}}$) is the underivability in $(HA)^{\infty}_{\mathrm{GR}}$ of

$$\forall x[Q \vee Px] \supset Q \vee \forall x Px$$

Perhaps of more interest would be to find necessary and sufficient conditions for $(M)^{\infty}$ to be a conservative extension of $(M)^{\infty}_{GR}$. However, in view of the generality of the definition of a mathematical system, it is doubtful that necessary and sufficient conditions could be found.

1. Preliminaries from recursive function theory.

For the most part we shall use the standard notations of recursive function theory, for example, we shall use $\{e\}$ for the e-th partial recursive function. However instead of using sequence numbers we shall use a (1-1) coding from the set of finite sequences of natural numbers onto the set of natural numbers; $\langle n_1,\ldots,n_k\rangle$ is the natural number which encodes the sequence $(n_1,\ldots,n_k)$. $\ell th(n)$ is a primitive recursive function giving the length of the sequence encoded by n, $(n)_i$ (for $i < \ell th(n)$) are the projection functions. $n*m$ is the associated concatenation function (also primitive recursive). The empty sequence is assigned the number 0, sometimes written "$\langle\rangle$". "$n \subseteq m$" expresses the condition that the sequence coded by n is an initial segment of the sequence coded by m.

A *spread* is a subset of the set N of natural numbers such that

(i) $\langle\rangle \in S$,

(ii) *if* $\langle x_0,\ldots,x_k\rangle \in S$ *then for all* $i < k$, $\langle x_0,\ldots x_i\rangle \in S$,

(iii) *if* $\langle x_0,\ldots,x_{k-1},x_k\rangle \in S$ *then for all* $z < x_k, \langle x_0,\ldots,x_{k-1},z\rangle \in S$,

(iv) *if* $\langle x_0,\ldots,x_k\rangle \in S$ *then* $\langle x_0,\ldots,x_k,0\rangle \in S$.

If S satisfies (i) - (iii) and instead of satisfying (iv) satisfies:

(iv') $\forall\alpha\exists k(\bar{\alpha}k \notin S)$

where α ranges over all unary functions and $\bar{\alpha}k = \langle\alpha 0,\ldots,\alpha(k-1)\rangle$, then S is a *tree*.

Finally we shall assume that to each $k > 0$ we have associated an effective function $\#q_1,q_2,q_3\#_k$ which maps $N \times \{0,1,\ldots,k+1\} \times N$ (1-1) onto N such that if $q = \#q_1,q_2,q_3\#$ then $q_1 \leq q$.

2. Formal mathematical theories.

In this paper we plan to restrict ourselves to languages with countably many formulae (or sequents). Thus we might as well assume that the well-formed expressions of our theories are natural numbers and thus avoid getting involved with tedious arithmetizations.

It is well known that natural numbers can also be used to code infinitely long formulae (see Takeuti - Kino 1963, López-Escobar 1967, Ohya 1970), but in such cases the set of codes of formulae do not form a recursive set. Thus we will consider a notion slightly more general than Smullyan's "Formal Mathematical System" (see definition 2.5).

2.1. DEFINITION. *By a theory T is meant a collection of at least the following items:*

(1) *a set W of natural numbers,*

(2) *a subset A of W,*

(3) *a finite sequence R of finitary relations on W.*

Given a theory $T = (W,A,R)$, then the elements of W are the *w.f.e.* (*well-formed expressions*) of T, the elements of A are the axioms of T and the elements of R are the *rules of inference* of T. If $R = (R_0,\ldots,R_r)$ then instead of writing "$(F_0,\ldots,F_u,E) \in R_i$" we shall often use one of the following more suggestive notations:

$$\frac{F_0,\ldots,F_u}{E} \; (R_i),$$

or:

$$F_0,\ldots,F_u \Longrightarrow_{R_i} E,$$

or just simply:

$$F_0,\ldots,F_u \Longrightarrow_i E.$$

We are interested in those theories to which an analogue of the ω-rule could be added. We shall call those theories "Carnap theories".

2.2. DEFINITION. *$C = (W,A,R,U,\sigma)$ is a Carnap theory if the following conditions are satisfied:*

(1) *(W,A,R) is a theory,*

(2) *$U \subseteq W$,*

(3) *σ is a function, $\sigma : U \times N \longmapsto W$, that is for each $E \in U$ and natural number k, $\sigma(E,k) \in W$.*

If we are considering a particular Carnap theory $C = (W,A,R,U,\sigma)$ then instead of writing "$\sigma(F,k)$" we shall write "$F[k]$". The Carnap theory C can be extended to the infinitary theory C^∞ by adding the following rule of inference applicable to all $F \in U$:

$$\text{(*) } \textit{From:} \quad \frac{F[0],\ F[1],\ldots,F[n],\ldots}{F} \quad \textit{To conclude:}$$

We shall call (*) the *ω - rule of* C^{∞}.[1]

Derivations in C and C^{∞} can then be defined in the usual tree form. Nevertheless in order to be explicit of the data (or analysis) placed at each node we shall give their definitions.

2.3. DEFINITION. *With each finite sequence* $E_0,\ldots,E_{k-1}$ *of w.f.e of C we associate the set* $\langle E_0,\ldots,E_{k-1}\rangle$ - Der *of derivations in C from* $E_0,\ldots,E_{k-1}$. $\langle E_0,\ldots,E_{k-1}\rangle$ - Der *is defined to be the least set of S of natural numbers such that:*

(i) $\langle E_i,0\rangle \in S$, *for all* $i < k$,

(ii) $\langle F,1\rangle \in S$, *for all* $F \in A$,

(iii) $\langle F,i+2\rangle \in S$ *whenever* $\langle E_0,d_0\rangle,\ldots,\langle E_u,d_u\rangle \in S$ *and* $E_0,\ldots,E_u \Longrightarrow_{R_i} F$.

2.4. DEFINITION . *A derivation in* C^{∞} *is a function α with the following properties:*

(i) $\{n : \alpha n \neq 0\}$ *is a tree,*

(ii) *if* $\alpha n \neq 0$, *then* αn *is of the form* $\langle F,d\rangle$ *with* $F \in W$ *and* $d \leq$ *length* (R) + 2.

(iii) *if* $\alpha n = \langle F,0\rangle$, *then* $F \in A$ *and* αn *is a terminal node, i.e.* $\alpha(n*\langle 0\rangle) = 0$,

(iv) *if* $\alpha(n) = \langle F,d\rangle$ *where* $0 < d \leq$ *length* (R) + 1 *then* $F \in W$ *and there are* $E_0,\ldots,E_u \in W$ *such that:*

(a) $E_0,\ldots,E_u \Longrightarrow_{d-1} F$

(b) *for all* $i \leq u$, $(\alpha(n*\langle i\rangle))_0 = E_i$

(c) *for all* $i > u$, $\alpha(n*\langle i\rangle) = 0$

(v) *if* $\alpha n = \langle F,d\rangle$ *and* $d =$ *length* (R) + 2 *then* $F \in U$ *and for all* i: $(\alpha(n*\langle i\rangle))_0 = \sigma(F,i) = F[i]$

A casual glance at definitions 2.3 and 2.4 should be enough to convince the reader that the data we are placing at each node is *only* the name of the rule used to obtain the w.f.e. at the node.

---

(1) The restriction to a single infinitary rule is clearly not essential.

The following abbreviations are self-explanatory:

$\mathrm{Der}_C(n; E_0, \ldots, E_{u-1} \vdash E) \equiv n \in \langle E_0, \ldots, E_{u-1}\rangle - \mathrm{Der}_C$ and $(n)_0 = E$.

$E_0, \ldots, E_{u-1} \vdash_C E \equiv \exists n \mathrm{Der}_C(n; E_0, \ldots, E_{u-1} \vdash E)$

$C \vdash E \equiv \exists n \mathrm{Der}_C(n; \vdash E)$.

$C^\infty \vdash E \equiv \exists \alpha(\alpha$ is a C^∞- derivation and $(\alpha 0)_0 = E)$.

$\mathrm{Der}_{C^\infty}(\alpha; E) \equiv \alpha$ is a C^∞ - derivation and $(\alpha 0)_0 = E)$.

$C^\infty_{GR} \vdash E \equiv \exists \alpha(\alpha$ is a C^∞-derivation, α is general recursive and $(\alpha 0)_0 = E)$.

We read "$C^\infty_{GR} \prec C^\infty$" as "*$C^\infty$ is a conservative extension of C^∞_{GR}*" and $C^\infty_{GR} \prec C^\infty$ *if and only if* $\forall F_{F \in W}(C^\infty \vdash F \equiv C^\infty_{GR} \vdash F)$.

The only requirements we have placed on the rules of inference of a (Carnap) theory is that they be finitary in the sense that each rule has finitely many premises. On the other hand, most of the formal systems considered in the literature of proof - theory usually have some effectiveness requirements. Smullyan distinguishes "mathematical systems" from "formal mathematical systems" by requiring that the set of theorems of the latter be recursively enumerable. Outright recursive enumerability of the theorems is too stringent requirement for some of the applications we have in mind. It suffices for our purposes that the set A of axioms, the set U, and the rules of inference be relatively solvable w.r.t. to the set of W of w.f.e. The following is a way to achieve it:

2.5. DEFINITION. *$F = (W, A, R, a, u, s, e)$ is a formal Carnap theory if the following conditions are satisfied:*

(1) *(W, A, R) is a theory,*

(2) *a, u, s, e are natural numbers*

(3) *if $U = \{n : \{u\}(n) = 0\} \cap W$, $\sigma = \{s\}$, then (W, A, R, U, σ) is a Carnap theory and domain $(\{u\}) \supseteq W$.*

(4) *e is of the form $\langle e_0, \ldots, e_r\rangle$ where $r =$ lenght (R) - 1, and for each $i \le r$, domain $(\{e_i\}) \supseteq W \times N$ and for all $E \in W$*

$$\{e_i\}(E,0), \ldots, \{e_i\}(E,n), \ldots$$

is an enumeration of the set $\{E\} \cup \{\langle F_0, \ldots, F_t\rangle : F_0, \ldots, F_t \Rightarrow_i E\}$, with each element repeated infinitely many times.

(5) *For all $E \in W$,* $\{a\}(E) = \begin{cases} 0 & \text{if } E \in A \\ 1 & \text{if } E \notin A \end{cases}$

The aim of this paper is to give some combinatorial conditions on a formal Carnap theory F which ensure that F^∞ is a conservative extension of F^∞_{GR}, or in simpler language: that the unrestricted ω - rule can be replaced by the recursively restricted ω - rule.

3. F^∞ AS A CONSERVATIVE EXTENSION OF F^∞_{GR}.

In this section we shall formulate a condition on a formal Carnap theory $F = (W, A, R, a, u, s, e)$ which ensures that F^∞ is a conservative extension of F^∞_{GR}. The condition is, loosely speaking, that there be an effective operation on the w.f.e. having some of the properties of disjunction (or union) and which commutes with the ω- rule.

3.1. DEFINITION. *A binary (partial) function* $\oplus$ *defined on* $W \times W$ *is called a disjunctor on* F *provided that for all* $E_0, \ldots, E_p, E, F, G \in W$

(i) $E \oplus E \vdash_F E$

(ii) $E \oplus F \vdash_F F \oplus E$

(iii) $(E \oplus F) \oplus G \vdash_F E \oplus (F \oplus G)$

(iv) *if* $E_0, \ldots, E_p \vdash_F E$ *then* $E_0 \oplus F, \ldots, E_p \oplus F \vdash_F E \oplus F$

(v) *if* $E \in U$, $F \in W$ *then there corresponds a* $G \in U$ *such that*
$G \vdash_F F \oplus E$,

$$F \oplus \sigma(E,n) \vdash_F \sigma(G,n),$$

where U *and* σ *are defined in terms of* u *and* s *as in* 2.5.

When F is a Hilbert-type system and U is a set of sentences of the form $\forall x Mx$ and $\sigma(\forall x Mx, k) = Mk$ then clause (v) corresponds to requiring that $\forall x(Q \vee Mx) \vdash Q \vee \forall x Mx$. Furthermore in a Hilbert-type formalism the conditions (i) - (v) are *effectively* satisfied by the disjunction operator. Thus we shall say that $\oplus$ is an *effective disjunctor of* F if in addition to $\oplus$ being a partial recursive function there are partial recursive functions $\tau_1, \ldots, \tau_7$ which satisfy the effective analogs of (i) - (v) i.e.:

(i') $Der_F(\tau_1(E); E \oplus E \vdash E)$,

(ii') $Der_F(\tau_2(E,F); E \oplus F \vdash F \oplus E)$,

(iii') $Der_F(\tau_3(E,F,G); (E \oplus F) \oplus G \vdash E \oplus (F \oplus G))$,

(iv') *if* $Der_F(n; E_0, \ldots, E_p \vdash E)$ *then* $Der_F(\tau_4(n,F);$
$E_0 \oplus F, \ldots, E_p \oplus F \vdash E \oplus F)$,

(v') *if* $E \in U$, *then* $\tau_5(F,E) \in U$ *and*

$Der_F(\tau_6(F,E);\ \tau_5(F,E) \vdash F \oplus E)$,

$Der_F(\tau_7(F,E,i);\ F \oplus \{s\}(E,i) \vdash \{s\}(\tau_s(F,E),i)$.

The proof of the following theorem will be completed in Section 6.

THEOREM 1. *If there is an effective disjunctor on a formal Carnap theory* F *then* F^∞ *is a conservative extension of* F^∞_{GR}.

4. APPLICATIONS OF THEOREM 1.

APPLICATION 1. Suppose *PA* is a formalized number theory (first-order, second-order or just about any recursive axiomatization with formulae of finite length) and that the formalization is a Hilbert-type axiomatization (i.e. an axiomatization involving the formulae themselves, with lots of axioms and with basically two types of rules of inference: modus ponens and universal generalization). Let U be the collection of formulae of the form $\forall xMx$ where x is a variable for natural numbers and let σ be the function such that $\sigma(\forall xMx, k) = M_x \underset{\sim}{k}$ (where $\underset{\sim}{k}$ is the numeral representing the natural number k). Then using U, σ the system *PA* can be considered as a formal Carnap theory *P*. P^∞ and P^∞_{GR} are then *PA* + (unrestricted ω-rule) and *PA* + (the recursively restricted ω-rule).

Consider the following mapping $\oplus$ on formulae of *PA*:

$$A \oplus B =_{Def} A' \vee B'$$

where A', B' are the universal closures of A and B respectively. Then it is a relatively simple matter to verify that $\oplus$ is an effective disjunctor on *P*; in order to satisfy condition 3 (v) associate with $\forall xMx$ and A the universal closure of the formula $\forall x(A' \vee Mx)$ where A' is the universal closure of A. Thus as a consequence of Theorem 1, we obtain that P^∞ is a conservative extension of P^∞_{GR}.

APPLICATION 2. Suppose that GA^∞ is a classical cut-free axiomatization for first-order number theory formalized using sequents and the unrestricted ω-rules. It is well known that it suffices to consider sequents of sentences, so we will assume that the sequents of GA^∞ are of sentences. Let GA^∞_{GR} be the corresponding theory with the recursively restricted ω-rule. Then Theorem 1 can also be applied to show that $GA^\infty \succ GA^\infty_{GR}$.

First of all, let U consist of all the sequents of the form:

$\Gamma \Longrightarrow \forall xMx, \Delta$ or $\Gamma, \exists xMx \Longrightarrow \Delta$, and let σ be the function such that $\sigma(\Gamma \Longrightarrow \forall xMx, \Delta, k) = \Gamma \Longrightarrow M_{\underset{\sim}{x}} k, \Delta$ and correspondingly for $\Gamma, \exists xMx \Longrightarrow \Delta$.

Then let G be the formal Carnap theory, obtained by deleting the ω-rules from (GA^{∞}, U, σ). Then $G^{\infty} = GA^{\infty}$ and $G^{\infty}_{GR} = GA^{\infty}_{GR}$.

Finally for $\oplus$ take the function acting on the sequents of GA such that:

$$(\Gamma_1 \Longrightarrow \Delta_1) \oplus (\Gamma_2 \Longrightarrow \Delta_2) = \Gamma_1, \Gamma_2 \Longrightarrow \Delta_2, \Delta_1 .$$

Then it can be verified that $\oplus$ is an effective disjunctor on G and so GA^{∞} is a conservative extension of GA^{∞}_{GR}.

It might be worthwhile observing that the cardinality requirement of the succedent for some of the rules of intuitionistic logic prevents the above $\oplus$ from being a disjunctor for intuitionistic number theory.

APPLICATION 3. Suppose GN^{∞} is a classical cut-free axiomatization for (full) 2nd-order number theory with the unrestricted ω-rule. Now, although we may assume that the sequents have no occurrences of free number variables, no such restriction may be assumed about the function variables (at least for the usual axiomatizations). In fact, one of the rules of inference of GN^{∞} is the rule of universal generalization $(\Longrightarrow \forall)$ which has the following form:

From $\dfrac{\Gamma \Longrightarrow A\beta, \Delta}{}$

To conclude: $\Gamma \Longrightarrow \forall \alpha A\alpha, \Delta$

Provided: The function variable β *does not occur free in the conclusion* $\Gamma \Longrightarrow \forall \alpha A\alpha, \Delta$.

It is precisely the requirement on the free function variable that prevents the function defined in Application 2 from being a disjunctor (and since the axiomatization is cut-free, we cannot first apply universal closures as was done in Application 1). However looking over the *proof* of Theorem 1 (see Sections 5 and 6) one sees that

(1) it suffices that the condition 3.1(iv) be satisfied only for those inferences that arise out of the canonical spread $\mathfrak{S}_E$ (see 5.6),

(2) there is enough latitude in the construction of the canonical spread $\mathfrak{S}_E$ so that not all the premisses of $\Gamma \Longrightarrow \forall \alpha A\alpha, \Delta$

need be included in $\mathfrak{S}_E$. It suffices to include only those which would satisfy the free-variable condition w.r.t. all the sequents occurring at that node.

Thus in effect the disjunctor of Application 2 could also have been used to conclude that $GN^\infty \succ GN^\infty_{GR}$.

We could have formulated a more general disjunctor to avoid having to appeal to the proof of Theorem 1 (rather than to the statement), but unfortunately our formulation was rather messy.

APPLICATION 4. Let $\mathcal{S}$ be the system of arithmetic, with constructively infinitely long expressions whose nesting number is finite, considered by Ohya 1970. The rules of $\mathcal{S}$ all have finitely many premisses except for the following rule:

From: $\dfrac{\Gamma \Longrightarrow \Delta, A_i \quad i < \omega}{\Gamma \Longrightarrow \Delta, \bigwedge_i A_i}$

To conclude:

By $\overline{\mathcal{S}}$ we understand the cut-free version of $\mathcal{S}$.

If we now pass to the arithmetized version and let $S = \langle W, A, R \rangle$ be the theory obtained by dropping the rules with infinitely many premisses, i.e. W is the set of Gödel numbers of sequents, $A \subseteq W$ the set of Gödel numbers of the axioms and R is the sequence of the arithmetized version of the finitary rules of inference.

Then let $\overline{S}$ be the theory obtained by dropping the rule in S corresponding to cut.

Glancing over the rules of $\overline{S}$ one sees that the premisses of a rule of inference can be effectively generated from the conclusion so that $\overline{S}$ can be considered as a formal Carnap theory. Applying the (analogue of the) disjunctor of Application 3 we obtain that $\overline{S}^\infty$ is a conservative extension of $\overline{S}^\infty_{GR}$ and thus that $\overline{\mathcal{S}}$ is a conservative extension of $\overline{\mathcal{S}}_{GR}$.

Using non-constructive methods it can be shown that $\mathcal{S}$ is a conservative extension of $\overline{\mathcal{S}}$ (see Ohya 1970) so we recapture Ohya's result that $\mathcal{S}$ is a conservative extension of $\mathcal{S}_{GR}$. Actually in order to be able to apply Theorem 1 it would suffice to have shown that $\mathcal{S}$ be conservative over the system obtained by requiring that in the cut-rule the cut-formula be restricted to a set of formulae which could be (uniformly) effectively generated from the conclusion.

5. CANONICAL SPREADS FOR EXPRESSIONS OF F^{∞}.

In the following two sections we shall be considering a fixed formal Carnap theory $F = (W, A, (R_0, \ldots, R_t), a, u, s, e)$ and all the definitions given below are supposed to be with respect to F. In order to avoid an overabundance of superscripts we shall omit "F". For the same reason we shall write "$\# x_1, x_3, x_3 \#$" instead of "$\# x_1, x_2, x_3 \#_t$".

To each w.f.e. E of W we now associate a spread $\mathfrak{S}_E$ which in a certain sense (which we shall not bother to make precise but which should be obvious from the construction), contains all the possible derivations of E in F^{∞} (provided there are any). Strictly speaking $\mathfrak{S}_E$ is a function such that $\{n : \mathfrak{S}_E(n) \neq 0\}$ is a spread and such that if $\mathfrak{S}_E(n) \neq 0$, then $\mathfrak{S}_E(n)$ is of the form $\langle E_1, \ldots, E_q \rangle$ where $q = \ell th(n) + 1$.

The definition of $\mathfrak{S}_E$ will be by recursion on the length of n.

BASIS STEP. $\mathfrak{S}_E(0) = \langle E \rangle$.

RECURSION STEP. Suppose that $\mathfrak{S}_E(n)$ has been defined and that it is equal to

$$\mathfrak{S}_E(n) = \langle E_1, \ldots, E_q \rangle,$$

where $q = \ell th(n) + 1$. We now give the instructions for computing $\mathfrak{S}_E(n * \langle i \rangle)$ for $i \in N$.

First determine the unique natural numbers such that

$$q = \# q_1, q_2, q_3 \# .$$

Proceed then by cases.

Case 1. $q_2 \leq t$. Compute $\{e_{q_2}\}(E_{q_1}, q_3)$. Suppose that the answer is $\langle F_0, \ldots, F_p \rangle$. Then define:

for $i \leq p$: $\mathfrak{S}_E(n * \langle i \rangle) = \mathfrak{S}_E(n) * \langle F_i \rangle$,

for $i > p$: $\mathfrak{S}_E(n * \langle i \rangle) = 0$.

Case 2. $p = t + 1$ and $\{u\}(E_{q_1}) = 0$. In that case $E_{q_1} \in U$. Then define for all $i \in N$:

$$\mathfrak{S}_E(n * \langle i \rangle) = \mathfrak{S}_E(n) * \langle \{s\}(E_{q_1}, i) \rangle .$$

Case 3. If neither Case 1 nor Case 2 applies, then define

$$\mathfrak{S}_E(n * \langle 0 \rangle) = \mathfrak{S}_E(n) * \langle E \rangle ,$$

and for all $i > 0$:

$$\mathfrak{S}_E(n * \langle i \rangle) = 0.$$

End of the definition of $\mathfrak{S}_E$.

The following definitions are included so that we can *talk* about $\mathfrak{S}_E$ (in contradistinction to *writing* about it).

(1) F *occurs at* (n, i) : $(\mathfrak{S}_E(n))_j = F$.

(2) f *is an occurrence*: for some F, n, j, $f = \langle F, n, j \rangle$ and F occurs at (n, j).

(3) *The occurrence* f *is active at* n: for some F, j, $f = \langle F, n, j \rangle$ and for some q_2, q_3 $lth(n) + 1 = \# j, q_2, q_3 \#$.

(4) *The rule* R_i *is acting on the occurrence* f *at* n: for some F, j, $f = \langle F, n, j \rangle$ for some q_3, $lth(n) + 1 = \# j, i, q_3 \#$ and $\{e_i\}(F, q_3)$ gives the premises of an application of the rule R_i with F as conclusion.

(5) *The* ω *-rule is acting on the occurrence* f *at* n: analogous to (4).

(6) *The occurrences* $(E_0, n*\langle 0 \rangle, k), \ldots, (E_p, n*\langle p \rangle, k)$ *yield the occurrence* $\langle F, n, h \rangle$: $\langle F, n, h \rangle$ is active at n, $k = lth(n) + 2$ and $E_0, \ldots, E_p \Longrightarrow_i F$.

Finally we shall often write "the spread $\mathfrak{S}_E$" when we should be writing "the spread $\{n : \mathfrak{S}_E n \neq 0\}$".

The following lemma summarizes the principal properties of the spread $\mathfrak{S}_E$:

5.1. LEMMA. *If* $E \in W$ *then* $\mathfrak{S}_E$ *is a total recursive function such that:*

(1) *if* F *occurs at* (n, j) *then for all successors* m *of* n *(i.e.* $n \subseteq m$ *and* $\mathfrak{S}_E(m) \neq 0$*),* F *occurs at* (m, j),

(2) *if* F *occurs at* (n, j) *then there is an* $l > lth(n)$ *such that for all successors* m *of* n *of* $lth(m) = l$ *the occurrence* $\langle F, m, j \rangle$ *is active at* m,

(3) *if* F *occurs at* (n, j) *and* $E_0, \ldots, E_p \Longrightarrow_i F$ *then there is an* $l > lth(n)$ *such that for all successors* m *of* n *of* $lth(m) = l$, *the occurrences* $\langle E_0, m*\langle 0 \rangle, l + 1 \rangle, \ldots, \langle E_p, m*\langle p \rangle, l + 1 \rangle$ *yield the occurrence* $\langle F, m, j \rangle$.

(4) *Correspondingly for the* ω *- rule.*

(5) *If* $F^{\infty} \vdash E$ *then in every branch of the spread* $\mathfrak{S}_E$ *there is an occurrence of an axiom.*

PROOF. (1) - (4) are more or less obvious from the definition of $\mathfrak{S}_E$. In fact, they were mentioned because they are helpful in visualizing the proof of (5). One way to prove (5) is to first assign ordinals to the derivations in F^{∞} and then use transfinite induction on the ordinal μ to show that $P(\mu)$ where

$P(\mu) \equiv \forall F \in \omega \forall n \forall j$ [if F occurs at (n,j) in $\mathfrak{S}_E$ and F has a derivation in F^{∞} of ordinal $\leq \mu$ then every branch of $\mathfrak{S}_E$ passing through n contains an axiom].

If $\mu = 0$ then $P(\mu)$ holds because (a) derivations of ordinal 0 are just the axioms of F and (b) because of (1). If $\mu > 0$ then using the induction hypothesis and either (3) or (4) one can conclude $P(\mu)$.

6. CANONICAL DERIVATIONS FOR THE THEOREMS OF F^{∞}.

We shall now proceed to show how given an effective disjunctor $\oplus$ on the formal Carnap theory F the spread $\mathfrak{S}_E$ can be transformed into a derivation of E in F^{∞}_{GR} (provided E had a derivation in F^{∞}). Loosely speaking the derivation is obtained by applying the disjunctor $\oplus$ to the w.f.e. occurring at the nodes of $\mathfrak{S}_E$. We need a function for that purpose: thus let Σ be the (partial) recursive function such that:

$$\Sigma(0) = 0,$$
$$\Sigma(<i>) = i,$$
$$\Sigma(n*<j>) = \Sigma(n) \oplus j \quad .$$

Thus $\Sigma(<E_0,\ldots,E_p>) = ((\ldots(E_0 \oplus E_1) \oplus \ldots) \oplus E_p)$.

We also need a function which tests whether we have reached an axiom or not. Thus let Δ be the partial recursive function such that:

$$\Delta(0) = 1,$$
$$\Delta(<i>) = \{a\}(i),$$
$$\Delta(n*<i>) = \Delta(n) \cdot \{a\}(i).$$

Hence for $<E_0,\ldots,E_p> = \mathfrak{S}_E(n)$, $\Delta(<E_0,\ldots,E_p>) = 0$ iff at least one of the $E_0,\ldots,E_p$ is an axiom of F. Consequently if $\Delta(\mathfrak{S}_E(n)) = 0$ then for all successors m of n, $\Delta(\mathfrak{S}_E(m)) = 0$.

6.1 DEFINITION. *For each* $E \in W$ *we define the function* D_E *as follows:*

BASIS STEP. $D_E(0) = E$.

RECURSION STEP. *Suppose that* $D_E(n)$ *has been defined.*

Case 1. $D_E(n) = 0$. *Then for all* $i \in N$, *let*

$$D_E(n*<i>) = 0$$

Case 2. $D_E(n) \neq 0$. *Then compute* $\mathfrak{G}_E(n)$

Subcase 2.1. $\Delta(\mathfrak{G}_E(n)) = 0$. *Then define for all* $i \in N$

$$D_E(n*<i>) = 0.$$

Subcase 2.2. $\Delta(\mathfrak{G}_E(n)) = 1$. *Then define for all* $i \in N$

$$D_E(n*<i>) = \Sigma(\mathfrak{G}_E(n*<i>)).$$

It is an immediate consequence of the definition that D_E will be well-founded (i.e. $\{n : D_E(n) \neq 0\}$ is a tree) iff every branch of $\mathfrak{G}_E$ contains an axiom. Furthermore, if $E \in W$ then D_E is a total recursive function irrespective of whether D_E is well founded or not. Thus if D_E is well founded, then applying the recursion lemma (fixed point theorem) of recursive function theory (see Rogers 1965) it can be shown that every w.f.e. occurring in a node of D_E is derivable in F^{∞}_{GR}. Rather than give the actual proof we find it much more worthwhile to explain the role of the conditions (i')-(v') of Section 3. Thus suppose that we are at a node n of D_E, that $F \in U$ and that the occurrence $<F,n,j>$ is active at n because of the ω-rule. Then in D_E we have the following kind of situation (where we are writing neither the parentheses nor $\oplus$):

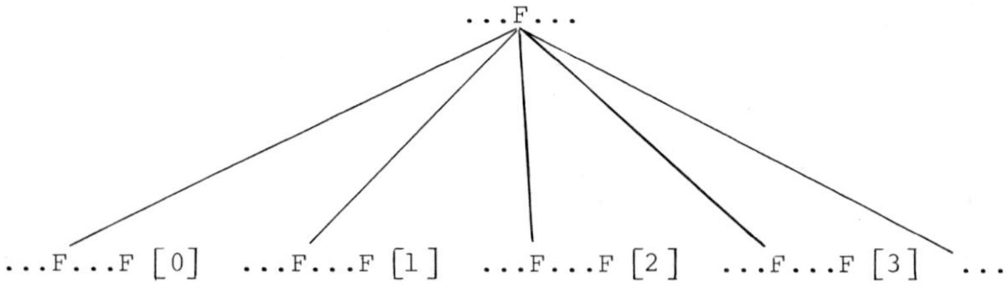

Suppose that we have a recursive listing of derivations in F^{∞}_{GR} of ...F...F [0], ...F...F [1], ..., that is, we have a z such that for all k, $Der_{F^{\infty}}(\{\{z\}(k)\};$ ...F...F $[k])$. In order to be able to apply the recursion lemma we must be able to effectively determine a natural number $\hat{z}$ such that $\mathrm{Der}_{F^\infty}(\{\hat{z}\};\ \ldots F\ldots)$. $\hat{z}$ can be obtained as follows : First obtain $G \in U$ such that

(1) $G \vdash_F \ldots F\ldots F$,

(2) $\ldots F\ldots F[k] \vdash_F G[k]$, for all k.

Then using z and the effectiveness of (2) (as formulated in (v')) we can determine a z_1 such that

(3) $\mathrm{Der}_{F^\infty}(\{z_1\};\ G)$.

Combining it with (1) we can determine a z_2 such that

(4) $\mathrm{Der}_{F^\infty}(\{z_2\};\ \ldots F\ldots F)$.

Then by repeated applications of (i') - (iii') (the exact number of applications can be read off from n) we can then obtain a $\hat{z}$ such that

$$\mathrm{Der}_{F^\infty}(\{\hat{z}\};\ \ldots F\ldots).$$

The gist of the above remarks is that if in every branch of $\mathfrak{S}_E$ there occurs an axiom of F then there is a recursive derivation of E in F^∞; that is $F^\infty_{GR} \vdash E$. Thus, in order to prove Theorem 1 (of Section 3)it suffices to show that if $F^\infty \vdash E$ then in every branch of $\mathfrak{S}_E$ there occurs an axiom of F. But the latter is true for any (formal) Carnap theory (see Lemma 5.1). Thus the proof of Theorem 1 is now complete.

7. An unfaithful axiomatization of intuitionism.

The only (non-contrived) case that we know of an unrestricted ω-rule which cannot be replaced by the recursively restricted one is in intuitionistic number theory. Take, for example, first-order intuitionistic number theory HA formalized as in Kleene 1952 and let HA^∞, HA^∞_{GR} be the extensions of HA obtained by adding the ω-rule and the recursively restricted ω-rule. Let $\mathfrak{N}$ be the structure $(\omega, +, \cdot, 0, ')$. Then an easy induction shows that for all sentences A of HA:

(.1) *if* $\mathfrak{N} \models A$ (*i.e.* A *is true in* $\mathfrak{N}$) *then* $HA^\infty \vdash A$,

(.2) *if* $\mathfrak{N} \models \neg A$, *then* $HA^\infty \vdash \neg A$.

Furthermore, the methods used in the last chapter of Kleene 1952 show that for all sentences A of HA:

(.3) *if* $HA^\infty_{GR} \vdash A$ *then* A *is realizable.*

Since there are sentences which are true but not realizable (e.g.

$\forall x\,[\exists y T(x,x,y) \vee \forall y \rightarrow T(x,x,y)])$ it follows that HA^{∞} is not a conservative extension of HA^{∞}_{GR}.

The derivations of (.1) - (.3) are carried out in classical mathematics and thus not too relevant to intuitionism and since HA is supposed to be a formalization of (a part of) intuitionism, the counterexample is also somewhat contrived. On the whole it is much more natural to use an intuitionistic metatheory when discussing intuitionistics systems. Intuitionistic systems are very often closed under Church's Rule (CR) and thus the following argument of Kreisel gives a very simple solution to the problem of replacing the ω-rule by the recursively restricted ω-rule (when the metatheory is closed under CR).

For simplicity let us consider HA, HA^{∞} and HA^{∞}_{GR} and let the metatheory M be closed under Church's rule. Then Kreisel's formulation of the problem is to show that:

(*) $\{A : M \vdash (\exists\alpha)\mathrm{Prf}_{HA^{\infty}}(\alpha,\ulcorner A\urcorner)\} \subseteq \{A : M \vdash (\exists\alpha)\mathrm{Prf}_{HA^{\infty}_{GR}}(\alpha,\ulcorner A\urcorner)\}$

where $\mathrm{Prf}_{HA^{\infty}}$ and $\mathrm{Prf}_{HA^{\infty}_{GR}}$ are (canonical) representations in M of the proof predicates of HA^{∞} and HA^{∞}_{GR} respectively. The proof of (*) is as follows:

Suppose that $M \vdash (\exists\alpha)\ \mathrm{Prf}_{HA^{\infty}}(\alpha,\ulcorner A\urcorner)$. Then since M is closed under CR, $M \vdash (\exists\alpha)_{GR}\mathrm{Prf}_{HA^{\infty}}(\alpha,\ulcorner A\urcorner)$. But then, because of the form of the predicates $\mathrm{Prf}_{HA^{\infty}}$ and $\mathrm{Prf}_{HA^{\infty}_{GR}}$, $M \vdash [(\exists\alpha)_{GR}\mathrm{Prf}_{HA^{\infty}}(\alpha,\ulcorner A\urcorner) \longrightarrow (\exists\alpha)\mathrm{Prf}_{HA^{\infty}_{GR}}(\alpha,\ulcorner A\urcorner)]$. Thus $M \vdash (\exists\alpha)\mathrm{Prf}_{HA^{\infty}_{GR}}(\alpha,\ulcorner A\urcorner)$.

In spite of the above solution there is still some room for improvement. It seems to us that if one is to be completely faithful to the principles embodied in HA then one should try to prove, *in the intuitionistic metatheory* M, the following sentence:

(**) $\forall x(\mathrm{Fmla}_{HA}(x) \wedge (\exists\alpha)\mathrm{Prf}_{HA^{\infty}}(\alpha,x) \longrightarrow (\exists\alpha)\mathrm{Prf}_{HA^{\infty}_{GR}}(\alpha,x))$

or at least the schema:

(***) $(\exists\alpha)\mathrm{Prf}_{HA^{\infty}}(\alpha,\ulcorner A\urcorner) \longrightarrow (\exists\alpha)\mathrm{Prf}_{HA^{\infty}_{GR}}(\alpha,\ulcorner A\urcorner)$.

Although we do not doubt that it is rewarding to be faithful, in this particular instance we find that if we follow the intuitionistic metatheory (and thus look for solutions of the kind (*) - (***)) we do not really determine which characteristics of the usual formalization of HA produce the result that HA^{∞} is not a conservative extension of HA^{∞}_{GR}.

The point of view we prefer to adopt (for the purpose of this paper) is that HA is just a formal system amongst many and then to try to find out whay HA, as a formal system, has the property that HA^{∞} is not a conservative extension of HA^{∞}_{GR}.

The definitive answer still eludes us. However we have reason to believe that it has something to do with the fact that a formal proof in HA is an attempt to represent the intuitive proof or construction. That is a formal proof in HA of a sentence A is closely related to the (effective) construction that validates A. HA^{∞}_{GR} still maintains a resemblance of effectiveness, HA^{∞} does not (unless the metatheory is itself constructive). In order to support this belief we now introduce another formalization for first-order intuitionistic number theory which has the same class of theorems as HA, but whose proofs try to conceal as best as possible the constructive content of the theorem. It is for the latter reason that we shall call it UHA (Unfaithful Heyting's arithmetic). It has the property that UHA^{∞} is a conservative extension of UHA^{∞}_{GR}.

The formulae of UHA are just the formulae of HA, however proofs in UHA will not involve the formulae themselves but rather finite, non-empty, sequences of formulae which we shall call *c-sequents* (read: confused sequents).

AXIOMS OF UHA are those c-sequents in which there occurs at least one axiom of HA.

STRUCTURAL RULES OF UHA are the rules of the form:

From: / *To conclude:*
$$\frac{A_0, \ldots, A_n}{A_{\Pi_0}, \ldots, A_{\Pi_n}}$$

Provided: Π *is a permutation of* $\{0, \ldots, n\}$

DECIDING RULES OF UHA are given by the schema:

From: / *To conclude:*
$$\frac{A_0, \ldots, A_{n-1}, B, B}{A_0, \ldots, A_{n-1}, B}$$

ANCIENT RULES OF UHA are given by the following schemata:

From: / *To conclude:*
$$\frac{A_0, \ldots, A_{n-1}, B_0 \quad \text{and} \quad A_0, \ldots, A_{n-1}, B_1}{A_0, \ldots, A_{n-1}, C}$$

Provided: $\dfrac{B_0, B_1}{C}$ is a rule of HA.

and correspondingly for the one premiss rules of HA.

Derivations of c-sequents in UHA are then defined in the usual tree form. We say that a formula A is provable in UHA and express it in symbols: $UHA \vdash A$, just in case the one-termed c-sequent (A) is derivable in UHA.

7.1 LEMMA. *For any formula* A *of* HA *the following conditions are equivalent:*

(1) $HA \vdash A$.

(2) $UHA \vdash A$.

PROOF. That (1) $\Longrightarrow$ (2) is trivial. To prove that (2) $\Longrightarrow$ (1) let $P(n)$ stand for the following:

$\forall\Pi$ [*if* Π *is a derivation in* UHA *of length* $\leq n$ *then at least one of the formulae in the end c-sequent of* Π *is provable in* HA].

A straight forward induction on n shows that $P(n)$ holds for all n. Suppose next that Π is a derivation in UHA of the one termed c-sequent (A). Then $HA \vdash A$.

Thus as far as the class of theorems is concerned HA and UHA are equivalent. The only difference between the proof in HA and UHA is that in UHA the actual provable formula can be hidden with extraneous formulae. In the finite case this characteristic of UHA causes no problem because it is relatively simple to search through finite sets.

However it does make an enormous difference when we add the ω-rule.

From: $A_0, \ldots, A_{n-1}, B_x\underset{\sim}{k}$ *for all* $k < \omega$

To conclude: $A_0, \ldots, A_{n-1}, \forall xB$

For let UHA^∞, UHA^∞_{GR} be UHA + ω-rule, UHA + (recursively restricted ω-rule) respectively. By appropiate changes in the proof of Lemma 7.1 we obtain the following lemma:

7.2 LEMMA. *For any sentence* A *of* HA *the following conditions are equivalent:*

(1) $HA^\infty \vdash A$,

(2) $UHA^\infty \vdash A$.

On the other hand the spread constructed in section 5 (see Lemma 5.1 (5)) gives us the following:

7.3 LEMMA. *For all sentences* A *of* HA

$$UHA^{\infty} \vdash A \text{ if and only if } UHA^{\infty}_{GR} \vdash A.$$

Thus we obtain that HA^{∞}, UHA^{∞} and UHA^{∞}_{GR} are all equivalent as far as the classes of theorems are concerned.

We doubt that UHA^{∞} (or its equivalent UHA^{∞}_{GR}) is of much use as far as proof-theory is concerned, although there might be applications in the case of the theory of species where it is sometimes customary to have a classical ω-complete first-order subtheory (cf. Smorynski 1973). Systems like *UHA* and UHA^{∞} are made for the theory of proofs and probably should be considered in conjunction with the α-well founded proofs discussed in Kreisel-Mints-Simpson 1974.

References.

Grzegorczyk, A., A. Mostowski, and C. Ryll-Nardzewski

1958. *The classical and the ω-complete arithmetic*, The Journal of Symbolic Logic, vol. 23, pp. 188-206.

Kent, C. F.

1967. *Restricted ω-rule for arithmetic*, Notices of the American Mathematical Society, vol. 14, pp. 665.

Kleene, S. C.

1952. Introduction to Metamathematics, D. van Nostrand Publ. Co. New York, x + 550 pp.

Kreisel, G., G. E. Mints, and S. G. Simpson

1974. *The Use of Abstract Language in Elementary Metamathematics: Some Pedagogic Examples*, Article in Logic Colloquium, Lecture Notes in Mathematics, vol. 453, Springer Verlag Publ. Co.

López-Escobar, E. G. K.

1967. *Remarks on an infinitary language with constructive formulas*, The Journal of Symbolic Logic, vol. 32, pp. 305-318.

Nelson, G. C.

1971. *A further restricted ω-rule*, Colloquium Mathematicum, vol. 23, pp. 1-3.

Ohya, T.

1970. *On recursive restriction of proofs in a system with constructive infinitely long expressions*, Comment. Math. Univ. St. Paul, vol. 18, pp. 43-52.

Rogers, H.

1967. Theory of Recursive Functions and Effective computability, McGraw-Hill Publ. Company, xix + 482 pp.

Routledge, N. A.

1950. *Greek Mathematics*. Article in Eureka, The Archimedeans' Journal, October, pp. 3-4.

Shoenfield, J. R.

1959. *On a restricted ω-rule*, Bulletin de l'Academie Polonaise des Sciences, Serie des Sci. Math., Astr., et Phys., vol. 7, pp. 405-407.

Smorynski, C.

1973. Chapter V of Metamathematical Investigation of Intuitionistic Arithmetic and Analysis, Lecture Notes in Mathematics, vol. 344, Springer-Verlag Publ. Co.

Smullyan, R. M.

1961. Theory of Formal Systems, Annals of Mathematics studies, vol. 47, 142 pp., Princeton University.

Takahashi, M.

1970. *A theorem on the second order arithmetic with the ω-rule*, Journal of the Mathematical Society of Japan, vol. 22, pp. 15-24.

Takeuti, G. and A. Kino

1963. *On predicates with constructively infinitely long expressions*, Journal of the Mathematical Society of Japan, vol. 15, pp. 176-190.

Department of Mathematics
University of Maryland
College Park, Maryland, U.S.A.

ADDED IN PRINT (Jan. 1977).

(1) The use of "Formal" in 2. is misleading. A better choice would have been "Recursively definable".

(2) I would like to thank Professor A. Troelstra for pointing out that D.

Prawitz in the article "*Comments on Gentzen-type procedures and the classical notion of truth*" (Proof Theory Symposium, Lecture Notes in Mathematics, vol. 500) also considers the effects of adding a non-iterative classical disjunction to intuitionistic logic.

(3) G. C. Mints has shown that there is no partial recursive function τ mapping formulae A of intuitionistic arithmetic to locally correct recursive proof figures of $HA+\omega$-rule such that:

$\exists p \in Rec\ Prov_{HA+\omega}(p,A) \rightarrow \tau(A)$ is defined and well-founded.

Non-Classical Logics, Model Theory and Computability,
A.I. Arruda, N.C.A. da Costa and R. Chuaqui (eds.)

Some Remarks on Discussive Logic

by *L. H. LOPES DOS SANTOS*

S. Jaškowski (Jaškowski 1948) introduced his discussive logic D_2 as the set of all formulas A such that MA belongs to the modal propositional logic S5, where M is the usual possibility operator. Generalizing, the discussive logic J(K) associated with any modal logic K may be defined as the set {A : MA ∈ K}. An axiomatics for J(S5) is presented in Costa 1975, and in T. Furmanowski 1975 is proved that J(S4) coincides with J(S5). In the first part of this paper, we present axiomatics for B (the Brouwerian system), S4, S5, J(B) and J(S5) specially appropriate to reflect their mutual inclusion and non-inclusion relations.

The propositional logic J(S5) is not closed under all rules of inference which are universally valid in the classical propositional logic, not even under all those which are valid in the positive propositional logic; particularly, it is not closed under material detachement (cf. Jaškowski 1948). But if we put $A \rightarrow \mathcal{D}$ (A *discussively implies* $\mathcal{D}$) for $MA \supset \mathcal{D}$ and $A \wedge \mathcal{D}$ (A and B are *discussively conjoined*) for $MA \,\&\, \mathcal{D}$, where ⊃ means classical implication and & means classical conjunction, then all valid laws and rules of the positive propositional logic come to be valid in J(S5), when ⊃ and & are replaced respectively by → and ∧ (and usual disjunction is maintained). In fact, N.C.A. da Costa and L. Dubikajtis (see Costa and Dubikajtis, these Proceedings) showed how to translate J(S5) in a language whose primitive connectives, besides classical negation and classical disjunction, are just discussive implication and discussive conjunction; they also showed how to axiomatize it by means of the discussive counterparts of the well-known axioms and rules of the positive propositional calculus (including Peirce law), plus a finite set of axiom schemas relating negation to the other connectives. In the second part of this paper, we extend this axiomatics in order to obtain a natural one for the discussive logic associated with the predicate logic S5 (with the Barcan formulas). Finally we note that this discussive logic is identical to the discussive logic associated with the predicate logic S4.

PART I.

1. LANGUAGE.

The Logics considered in Part I are to be thought as formulated in a language whose primitive symbols are denumerably many propositional letters $p_1, p_2, \ldots,$ the connectives $\neg$ (negation) and $\vee$ (disjunction), the necessity operator L, and parentheses. The following definitions are adopted:

Df.1. $MA =_{df.} \neg L \neg A$.

Df.2. $(A \supset C) =_{df.} (\neg A \vee C)$.

Throughout this paper, the letters A, C and $\mathcal{D}$, with or without numerical subscripts, will be employed as syntactical variables for formulas; for the sake of brevity, the most external parentheses of a formula will be often omitted.

2. THE LOGICS B, J(B) AND J(S5).

The logic B is included in J(B), which is included in J(S5). The following theorems reflect axiomatically the situation.

THEOREM 1. B *is axiomatizable by means of the following axiom schemas and rules:*

(1) LA, *if* A *is a tautology,*

(2) $L(LA \supset A)$,

(3) $L(L(A \supset C) \supset (LA \supset LC))$,

(4) $L(A \supset LMA)$,

(5) *infer* LC *from* $L(A \supset C)$ *and* LA,

(6) *infer* LLA *from* LA,

(7) *infer* A *from* LA.

PROOF: It is well known that B is axiomatizable by taking as axioms all tautologies, all instances of the schemas $LA \supset A$, $L(A \supset C) \supset (LA \supset LC)$, $A \supset LMA$, and as rules material detachment and Gödel's rule. Now let $A \in B$; then A is a thesis of this well-known axiomatics; we show easily, by induction, that LA is a thesis of the axiom system (1) - (7); hence so is A itself, by (7). Conversely, schemas (1) - (4) and rules (5) - (7) can be clearly derived in the well-known axiomatics for B; therefore, all theses of the axiom system (1) - (7) belong to B.

THEOREM 2. J(B) *is axiomatizable by means of* (1) - (7) *and:*

(8) *infer* A *from* LMA.

PROOF: Let $A \in J(B)$; then $MA \in B$, by definition of J(B), and $LMA \in B$, for B is closed under Gödel's rule; therefore, LMA is a thesis of the axiom system (1) - (8), by Theorem 1, and so is A, by (8). On the other hand, let A be a thesis of the axiom system (1) - (8); it can be easily shown, by induction, that $MA \in B$; therefore $A \in J(B)$, by definition of J(B).

It is worth noting that rule (7) is superfluous in the axiom system (1) - (8):

THEOREM 3. J(B) *is axiomatizable by means of* (1) - (6) *and* (8).

PROOF: By means of (1) - (6) and (8), any formula A can be derived from LA as follows:

(i) LA ;

(ii) $L(A \supset LMA)$, by (4);

(iii) LLMA, by (i), (ii) and (5);

(iv) $L(LMA \supset MA)$, by (2);

(v) LMA, by (iii), (iv) and (5);

(vi) A, by (v) and (8).

THEOREM 4. J(S5) *is axiomatizable by means of* (1) - (6), (8) *and*

(9) $L(LA \supset LLA)$.

PROOF: It is well known that S5 can be axiomatized by taking as axioms all tautologies, all instances of the schemas $LA \supset A$, $L(A \supset C) \supset (LA \supset LC)$,

$A \supset LMA$, $LA \supset LLA$, and as rules material detachment and Gödel's rule; by induction on the length of a derivation of A in this axiomatics, we show easily that LA is a thesis of the axiomatics proposed for J(S5) if $A \in$ S5. Now let $C \in$ J(S5); then $MC \in$ S5, by definition of J(S5), and LMC is a thesis of the axiom system proposed for J(S5); hence so is C, by (8). Conversely, if C is a thesis of the axiom system proposed for J(S5), then it is easy to show, by induction, that $MC \in$ S5; therefore $C \in$ J(S5), by definition of J(S5).

The logic B is indeed a *proper* part of J(B), which is also a *proper* part of J(S5). The non-inclusion relations can be thus established, taking into account Theorems 1, 2 and 4:

THEOREM 5. B *is not closed under the Rule* (8).

PROOF: With the help of simple Kripke diagrams, we may verify that $LM(Mp_1 \supset p_1) \in$ B and $Mp_1 \supset p_1 \notin$ B.

THEOREM 6. *Not all instances of* (9) *belong to* J(B).

PROOF: By means of a simple Kripke diagram, we may verify that $ML(Lp_1 \supset LLp_1) \notin$ B; hence $L(Lp_1 \supset LLp_1) \notin$ J(B), by definition of J(B).

3. THE LOGICS S4, S5 AND J(S5).

The logic S4 is a subset of S5, which is a subset of J(S5). This chain of inclusions may be axiomatically founded on Theorems 7 - 9 below.

THEOREM 7. S4 *is axiomatizable by means of* (2) *and*

(10) $LLLA$, *if* A *is a tautology;*

(11) $L(L(A \supset C) \supset L(LA \supset LC))$;

(12) *infer* C *from* $L(A \supset C)$ *and* A.

PROOF: All theses of this axiomatics belong to S4, for (2) and (10)-(12) are clearly valid in S4. Proving the converse requires some lemmas.

LEMMA 7.1. *The following rule is valid in the axiom system proposed for* S4 *in Theorem* 7 : *infer* A *from* LA.

PROOF: Trivial, by (2) and (12).

LEMMA 7.2. *The following rule is valid in the axiom system proposed for* S4 *in Theorem 7: infer* LC *from* $L(A_1 \supset (\ldots \supset (A_n \supset C))\ldots)$, $LA_j (1 \leq j \leq n)$.

PROOF: By induction on n.
CASE 1. $n=1$; from $L(A_1 \supset C)$, we derive $L(LA_1 \supset LC)$, by (11)-(12); from this and LA_1, we derive LC, by (12).

CASE 2. $n > 1$; from LA_1 and $L(A_1 \supset (\ldots \supset (A_n \supset C))\ldots)$, we derive $L(A_2 \supset (\ldots \supset (A_n \supset C))\ldots)$, by Case 1 above; from this, induction hypothesis and LA_j $(1 \geq j \geq n)$, we derive LC.

COROLLARY. *If* C *is a tautological consequence of* $A_1, \ldots, A_n$, *then* LC *is derivable from* LA_j $(1 \leq j \leq n)$ *in the axiom system proposed for* S4 *in Theorem* 7.

LEMMA 7.3. $L(LA \supset LLA)$ *is a valid schema of the axiom system proposed for* S4 *in Theorem* 7.

PROOF: Let A be any formula, let C be the formula $\neg A \vee A$. By (10) and Lemma 7.1, $L(A \supset (C \supset A))$ is a thesis of the axiom system referred above. From this, by (11)-(12), we derive

(i) $L(LA \supset L(C \supset A))$.

The formulas

(ii) $L(L(C \supset A) \supset L(LC \supset LA))$,

(iii) $L(L(LC \supset LA) \supset L(LLC \supset LLA))$

are instances of (11), and

(iv) $L(L(LLC \supset LLA) \supset (LLC \supset LLA))$

is an instance of (2). From (i)-(iv), by the Corollary of Lemma 7.2, we derive $L(LLC \supset (LA \supset LLA))$; from this, by (11)-(12), we get $L(LLLC \supset L(LA \supset LLA))$. But LLLC is an instance of (10); therefore, $L(LA \supset LLA)$ is a thesis of the axiom system referred above, by (12).

COROLLARY. *The following rule is valid in the axiomatics proposed for* S4 *in Theorem* 7: *infer* LLA *from* LA.

Now we are able to complete the proof of Theorem 7. It is well known that S4 is axiomatizable by means of all tautologies, all instances of the schemas $LA \supset A$ and $L(A \supset C) \supset L(LA \supset LC)$, material detachement and Gödel's rule. Now let $A \in S4$; it is not difficult to show, by induction, that LA is a thesis of the axiom system proposed for S4 in the body of Theorem 7, with the help of Lemma 7.3; hence so is A, by Lemma 7.1.

THEOREM 8. S5 *is axiomatizable by means of* (2), (10)-(12) *and*

(13) *infer* LA *from* MLA.

PROOF: Similar to the proof of Theorem 7, considering that S5 can be axiomatized by means of all tautologies, all instances of the schemas $LA \supset A$, $L(A \supset C) \supset L(LA \supset LC)$ and $A \supset LMA$, material detachement and Gödel's rule. Lemmas 7.1-7.3 are to be called for, as well as:

LEMMA 8.1. $L(A \supset LMA)$ *is a valid schema of the axiom system proposed for* S5 *in Theorem* 8.

PROOF: It is easily verifiable that all instances of the schema $ML(A \supset LMA)$ belong to S4; consequently, by Theorem 7, they are all derivable by means of (2) and (10)-(12); therefore, all instances of the schema $L(A \supset LMA)$ are theses of the axiom system proposed for S5 in Theorem 8, by (13).

THEOREM 9. J(S5) *is axiomatizable by means of* (2), (10)-(12) *and*

(14) *infer* A *from* MA.

PROOF: It is known that J(S5) is axiomatizable by means of (1)-(2), (4), (9), (11)-(12) and (14) (cf. Theorem 4 of Costa 1975). Given Lemmas 7.1-7.3 and 8.1, it is easy to show the equivalence of both axiom systems.

The axiom system presented for J(S5) in Theorem 9 is really an extension of the one presented for S5 in Theorem 8, for any application of rule (13) is an application of rule (14). The logic S5 is indeed a proper part of J(S5), since we have:

THEOREM 10. S5 *is not closed under rule* (14).

PROOF: Simple Kripke diagrams show us that $M(p_1 \supset Mp_1) \in S5$ and $p_1 \supset Mp_1 \notin S5$.

Although S4 is a proper part of S5, J(S4) is not a proper part of J(S5). As a matter of fact, J(S4) is identical to J(S5). This identity, algebraically established in T. Furmanowski 1975, is an immediate consequence of Theorem 9 and

THEOREM 11. J(S4) *is axiomatizable by means of* (2), (10)-(12) *and* (14).

PROOF: Let A ∈ J(S4); then MA ∈ S4, by definition of J(S4), and MA ∈ S5, for S5 includes S4; so A ∈ J(S5), by definition of J(S5); therefore, A is a thesis of the axiomatics proposed for J(S4), by Theorem 9. Conversely, let A be a thesis of this axiomatics; by induction, we show that MA ∈ S4; therefore A ∈ J(S4), by definition of J(S4).

If m is a modality and K is a modal logic, the set {A: mA ∈ K} is called the m-fragment of K. By definition, J(K) is the M-fragment of K, for any K. It is not difficult to verify that B and S5 are the L-fragments respectively of J(B) and J(S5). Nevertheless, the analogous assumption concerning S4 is not true: together with Theorem 11, it would lead us to the false conclusion that S4 coincides with S5.

4. THE LOGICS S4, S5 AND J(B).

The logics S4 and S5 neither are subsets of J(B) nor have it as a subset. On the one hand, the formula $Mp_1 \supset p_1$ belongs to J(B) and does not belong to S5; so J(B) is a subset neither of S5 nor of S4. On the other hand, $L(Lp_1 \supset LLp_1)$ belong to S4 and it does not belong to J(B); so J(B) includes neither S4 nor S5.

PART II

1. THE AXIOM SYSTEM J*.

We shall conceive the axiom system J* as expressed in a language whose primitive symbols are: a denumerable infinite set of individual variables; a nonvoid set of n-ary predicate letters, for each natural number n,

$n \geq 1$; the classical connectives $\neg$ and V; the necessity operator L; the existential quantifier $\exists$; and parentheses. The set of formulas is recursively delimited as usual. The letters x and y will be employed as syntactical variables for individual variables of the object-language. Subject to the standard restriction, the notation $A(x/y)$ will refer to the formula obtained from A by replacing each free occurrence of x in A by an occurrence of y. Finally we introduce the following definitions:

Df. 1. $(A \supset C) =_{df.} (\neg A \vee C)$,

Df. 2. $(A \,\&\, C) =_{df.} \neg(\neg A \vee \neg C)$,

Df. 3. $(A \equiv C) =_{df.} ((A \supset C) \,\&\, (C \supset A))$,

Df. 4. $MA =_{df.} \neg L \neg A$,

Df. 5. $(A \rightarrow C) =_{df.} (MA \supset C)$,

Df. 6. $(A \wedge C) =_{df.} (MA \,\&\, C)$.

J* has the following axiom schemas and rules:

J*1. LA, *if* A *is a tautology.*

J*2. $L(LA \supset A)$.

J*3. $L(L(A \supset C) \supset L(LA \supset LC))$.

J*4. $L(A \supset LMA)$.

J*5. *Infer* C *from* $L(A \supset C)$ *and* A.

J*6. *Infer* A *from* MA.

J*7. $L(A(x/y) \supset \exists x A)$.

J*8. *Infer* $L(\exists x A \supset C)$ *from* $L(A \supset C)$, *if* x *is not free in* C.

The notation $J^* \vdash A$ means that A is a thesis of J*.

THEOREM 12. J* *is an adequate axiomatics for the discussive logic associated with the modal predicate logic* S5 (*with the Barcan formulas*).

PROOF: In Theorem 6 of Costa 1975 an axiomatics for the discussive logic associated with the modal predicate logic S5 is presented; the equiva-

lence between this axiomatics and J^*, with respect to the usual difinitions of the quantifiers, can be easily shown.

2. The axiom system $\tilde{J}^*$

The language of the axiom system $\tilde{J}^*$ must comprise as primitive symbols those of J^* with the exception of the necessity operator, which is to be replaced by the binary discussive connectives $\rightarrow$ and $\wedge$. In $\tilde{J}^*$ we retain Definitions 1-3 and add:

Df. 7. $LA =_{df.} (\neg A \rightarrow \neg(A \vee \neg A))$,

Df. 8. $MA =_{df.} \neg L \neg A$,

Df. 9. $OA =_{df.} \neg(A \vee \neg A)$.

The set of theses of $\tilde{J}^*$ is characterized by the following axiom schemas and rules:

Ax.01. $A \rightarrow (C \rightarrow A)$.

Ax.02. $(A \rightarrow (C \rightarrow D)) \rightarrow ((A \rightarrow C) \rightarrow (A \rightarrow D))$.

Ax.03. $((A \rightarrow C) \rightarrow A) \rightarrow A$.

Ax.04. $(A \wedge C) \rightarrow A$.

Ax.05. $(A \wedge C) \rightarrow C$.

Ax.06. $A \rightarrow (C \rightarrow (A \wedge C))$.

Ax.07. $A \rightarrow (A \vee C)$.

Ax.08. $C \rightarrow (A \vee C)$.

Ax.09. $(A \rightarrow D) \rightarrow ((C \rightarrow D) \rightarrow ((A \vee C) \rightarrow D))$.

Ax.1. $A \rightarrow \neg\neg A$

Ax.2. $\neg\neg A \rightarrow A$.

Ax.3. $\neg(A \vee \neg A) \rightarrow C$.

Ax.4. $\neg(A \vee C) \rightarrow \neg(C \vee A)$.

Ax.5. $\neg(A \vee C) \rightarrow (\neg A \wedge \neg C)$.

Ax.6. $\neg(\neg\neg A \vee C) \rightarrow \neg(A \vee C)$.

Ax.7. $(\neg(A \vee C) \rightarrow D) \rightarrow ((\neg A \rightarrow C) \vee D)$.

Ax.8. $\neg((A \vee C) \vee D) \rightarrow \neg(A \vee (C \vee D))$.

Ax.9. $\neg((A \rightarrow C) \vee D) \rightarrow (A \wedge \neg(C \vee D))$.

Ax.10. $\neg((A \wedge C) \vee D) \rightarrow (A \rightarrow \neg(C \vee D))$.

Ax.11. $\neg(\neg(A \vee C) \vee D) \rightarrow (\neg(\neg A \vee D) \vee \neg(\neg C \vee D))$.

Ax.12. $\neg(\neg(A \rightarrow C) \vee D) \rightarrow (A \rightarrow \neg(\neg C \vee D))$.

Ax.13. $\neg(\neg(A \wedge C) \vee D) \rightarrow (A \wedge \neg(\neg C \vee D))$.

Ax.14. $\neg(A(x/y) \supset \exists x A) \rightarrow \neg(A \supset A)$.

Ax.15. $\neg(\exists x A \supset C) \rightarrow \exists x \neg(A \supset C)$ *if x is not free in C.*

R1. *Infer C from $A \rightarrow C$ and A.*

R2. *Infer $\exists x A \rightarrow C$ from $A \rightarrow C$, if x is not free in C.*

The notation $\tilde{J}^* \vdash A$ means that A is a thesis if $\tilde{J}^*$

3. Equivalence between J^* and $\tilde{J}^*$.

Two axiom systems A and A' are said to be equivalent with respect to a given set of definitions when: (i) this set is enough to assure general and univocal translability of the language of one system to the language of the other; (ii) any formula of A is a thesis of A if and only if it abbreviates in A', according to the definitions, a thesis of A'; (iii) any formula of A' is a thesis of A' if and only if it abbreviates in A, according to the definitions, a thesis of A. We will prove that J^* and $\tilde{J}^*$ are equivalent with respect to Definitions 5-7.

Condition (i) above is obviously fulfilled by Definitions 5-7. If A is a formula of J^*, let $\mathrm{T\tilde{r}}A$ be the formula of $\tilde{J}^*$ abbreviated by A according to Definition 7; if C is a formula of $\tilde{J}^*$, let $\mathrm{Tr}C$ be the formula of J^* abbreviated by C according to Definitions 5-6. Conditions (ii) and (iii) may now be so rewritten: (ii') for any formula A of J^*, $J^* \vdash A$ if and only if $\tilde{J}^* \vdash \mathrm{T\tilde{r}}A$; (iii) for any formula C of $\tilde{J}^*$, $\tilde{J}^* \vdash C$ if and only if $J^* \vdash \mathrm{Tr}C$. Proving their fulfillment requires some lemmas.

LEMMA 1. *The following rule is valid in $\tilde{J}^*$: infer $A \rightarrow D$ from $A \rightarrow C$*

and $C \rightarrow D$.

PROOF: It is well known that the rule is derivable by means of Axioms 01-02 and R1.

LEMMA 2. $\tilde{J}^* \vdash L(A(x/y) \supset \exists x A)$.

PROOF: By Axiom 4 and Definition 1,

(i) $\tilde{J}^* \vdash \neg(A \supset A) \rightarrow \neg(A \vee \neg A)$;

by (i), Axiom 3 and Lemma 1,

(ii) $\tilde{J}^* \vdash \neg(A \supset A) \rightarrow O(A(x/y) \supset \exists x A)$;

by Axiom 14, (ii) and Lemma 1,

(iii) $\tilde{J}^* \vdash \neg(A(x/y) \supset \exists x A) \rightarrow O(A(x/y) \supset \exists x A)$.

Applications of Definitions 7 and 9 on (iii) complete the proof.

LEMMA 3. *The following rule is derivable in* $\tilde{J}^*$: *infer* $L(\exists x A \supset C)$ *from* $L(A \supset C)$ *if* x *is not free in* C.

PROOF: By Definitions 7 and 9, $L(A \supset C)$ is

(i) $\neg(A \supset C) \rightarrow O(A \supset C)$;

by Axiom 3 and Definition 9,

(ii) $\tilde{J}^* \vdash O(A \supset C) \rightarrow O(\exists x A \supset C)$;

so we derive from (i), with the help of (ii) and Lemma 1,

(iii) $\neg(A \supset C) \rightarrow O(\exists x A \supset C)$.

Let us suppose that x is not free in C; then it is not free in $O(\exists x A \supset C)$; therefore, we can derive

(iv) $\exists x \neg(A \supset C) \rightarrow O(\exists x A \supset C)$

from (iii), by R2; from (iv) we get, by Axiom 15 and Lemma 1, $\neg(\exists x A \supset C) \rightarrow O(\exists x A \supset C)$, which is $L(\exists x A \supset C)$, by Definitions 7 and 9.

LEMMA 4. *Understanding* L *as a defined symbol of* $\tilde{J}^*$, *the schemas* J^*1-J^*6 *are valid in* $\tilde{J}^*$.

PROOF: Cf. Costa and Dubikajtis, these Proceedings, Part II, Theorems 7, 9-11, 13, 20, 23, 26-29.

LEMMA 5. *If* $J^* \vdash A$, *then* $\tilde{J}^* \vdash Tr A$.

PROOF: Straightforward by Lemmas 2-4.

LEMMA 6. *If* $\tilde{J}^* \vdash C$, *then* $J^* \vdash \mathrm{Tr}C$.

PROOF: Understanding $\rightarrow$ and $\wedge$ as defined symbols of J^*, Axioms 01 – A15 and R1-R2 are easily shown to be valid in J^*.

LEMMA 7. *If* $\mathcal{D}$ *is a tautological consequence of* A *and* C, *then* $L\mathcal{D}$ *is derivable from* LA *and* LC *in* $\tilde{J}^*$

PROOF: Let us suppose that $\mathcal{D}$ is a tautological consequence of A and C ; by Lemma 4 ,

(i) $\tilde{J}^* \vdash L(A \supset (C \supset \mathcal{D}))$,

(ii) $\tilde{J}^* \vdash L(L(A \supset (C \supset \mathcal{D})) \supset L(LA \supset L(C \supset \mathcal{D})))$,

(iii) $\tilde{J}^* \vdash L(L(C \supset \mathcal{D}) \supset L(LC \supset L\mathcal{D}))$:

By (i), (ii) and Lemma 4,

(iv) $\tilde{J}^* \vdash L(LA \supset L(C \supset \mathcal{D}))$.

From LA, (iv) and Lemma 4, we derive

(v) $L(C \supset \mathcal{D})$;

from (iii), (v) and Lemma 4 we get

(vi) $L(LC \supset L\mathcal{D})$;

finally we derive $L\mathcal{D}$ from LC and (vi), by Lemma 4.

LEMMA 8. $\tilde{J}^* \vdash L((A \rightarrow C) \equiv (MA \supset C))$ *and*

$\tilde{J}^* \vdash L((A \wedge C) \equiv (MA \,\&\, C))$.

PROOF: Trivial, by Lemma 7 and Theorems 30-33 in Part II of Costa and Dubikajtis, these Proceedings.

LEMMA 9. $\tilde{J}^* \vdash L((\tilde{\mathrm{Tr}}\,\mathrm{Tr}A \rightarrow \tilde{\mathrm{Tr}}\,\mathrm{Tr}C) \equiv \tilde{\mathrm{Tr}}\,\mathrm{Tr}(A \rightarrow C))$ *and*

$\tilde{J}^* \vdash L((\tilde{\mathrm{Tr}}\,\mathrm{Tr}A \wedge \tilde{\mathrm{Tr}}\,\mathrm{Tr}C) \equiv \tilde{\mathrm{Tr}}\,\mathrm{Tr}(A \wedge C))$.

PROOF: Considering that $\tilde{\mathrm{Tr}}\,\mathrm{Tr}(A \rightarrow C)$ and $\tilde{\mathrm{Tr}}\,\mathrm{Tr}(A \wedge C)$ are respectively $(M\tilde{\mathrm{Tr}}\,\mathrm{Tr}A \supset \tilde{\mathrm{Tr}}\,\mathrm{Tr}C)$ and $(M\tilde{\mathrm{Tr}}\,\mathrm{Tr}A \,\&\, \tilde{\mathrm{Tr}}\,\mathrm{Tr}C)$, the lemma is an immediate consequence of Lemma 8.

LEMMA 10. *The following rule is valid in* $\tilde{J}^*$: *infer* $L(\exists x A \equiv \exists x C)$ *from* $L(A \equiv C)$.

PROOF: From $L(A \equiv C)$ we derive, by Lemma 7,

(i) $L(A \supset C)$ and $L(C \supset A)$.

But Lemma 2 assures that

(ii) $\tilde{J}^* \vdash L(C \supset \exists x C)$ and $\tilde{J}^* \vdash L(A \supset \exists x A)$;

so we derive

(iii) $L(A \supset \exists x C)$ and $L(C \supset \exists x A)$

from (i) and (ii), by Lemma 7. From (iii), by Lemma 3, we get

(iv) $L(\exists x A \supset \exists x C)$ and $L(\exists x C \supset \exists x A)$;

from this, by Lemma 7, we derive $L(\exists x A \equiv \exists x C)$.

LEMMA 11. *The following rule is derivable in* $\tilde{J}^*$: *infer* $L(MA \equiv MC)$ *from* $L(A \equiv C)$

PROOF: From $L(A \equiv C)$ we derive, by Lemma 7,

(i) $L(\neg A \supset \neg C)$ and $L(\neg C \supset \neg A)$;

by Lemma 4,

(ii) $\tilde{J}^* \vdash L(L(\neg A \supset \neg C) \supset L(L\neg A \supset L\neg C))$,

(iii) $\tilde{J}^* \vdash L(L(\neg C \supset \neg A) \supset L(L\neg C \supset L\neg A))$.

By the same lemma, we derive

(iv) $L(L\neg A \supset L\neg C)$ and $L(L\neg C \supset L\neg A)$

from (i)-(iii); from (iv), by another application of the lemma, we get $L(\neg L\neg A \equiv \neg L\neg C)$, which is $L(MA \equiv MC)$ by Definition 8.

LEMMA 12. $\tilde{J}^* \vdash A$ *if and only if* $\tilde{J}^* \vdash \tilde{Tr}\, Tr\, A$.

PROOF: With the help of Lemmas 7-11, we show easily that $\tilde{J}^* \vdash L(A \equiv \tilde{Tr}\, Tr A)$, by induction on the length of A. (Remember that $\tilde{Tr}\, Tr \neg C$, $\tilde{Tr}\, Tr\, (C \vee D)$ and $\tilde{Tr}\, Tr\, \exists x C$ are respectively $\neg \tilde{Tr}\, Tr\, C$, $(\tilde{Tr}\, Tr\, C \vee \tilde{Tr}\, Tr D)$ and $\exists x \tilde{Tr}\, Tr C$). The lemma follows from this result by Lemma 7.

LEMMA 13. *If* D *is a tautological consequence of* A *and* C, *then* LD *is derivable from* LA *and* LC *in* J^*.

PROOF: It is enough to make obvious modifications in the proof of Lemma 7.

LEMMA 14. $J^* \vdash L(LA \equiv (\neg A \Rightarrow \neg(A \vee \neg A)))$.

PROOF: It is easy to verify that the considered schema is valid in the predicate logic S5, which is included in the set of theses of J^*.

LEMMA 15. *The following rule is valid in* J^* : *infer* $L(\exists x A \equiv \exists x C)$ *from* $L(A \equiv C)$.

PROOF: It is enough to make obvious modifications in the proof of Lemma 10.

LEMMA 16. $J^* \vdash L(L\,\mathrm{Tr}\,\tilde{\mathrm{Tr}}A \equiv \mathrm{Tr}\,\tilde{\mathrm{Tr}}A)$.

PROOF: Considering that $\mathrm{Tr}\,\tilde{\mathrm{Tr}}LA$ is $(\neg\mathrm{Tr}\,\tilde{\mathrm{Tr}}A \Rightarrow \neg\mathrm{Tr}\,\tilde{\mathrm{Tr}}A \vee \neg\mathrm{Tr}\,\tilde{\mathrm{Tr}}A))$, the lemma is an immediate consequence of Lemma 14.

LEMMA 17. *The following rule is derivable in* J^*: *infer* $L(LA \equiv LC)$ *from* $L(A \equiv C)$.

PROOF: From $L(A \equiv C)$, by Lemma 13, we get

(i) $L(A \supset C)$ and $L(C \supset A)$;

from (i) and conveniently chosen instances of J^*3, we derive

(ii) $L(LA \supset LC)$ and $L(LC \supset LA)$,

by J^*5. From (ii) and Lemma 13, we derive $L(LA \equiv LC)$.

LEMMA 18. $J^* \vdash A$ *if and only if* $J^* \vdash \mathrm{Tr}\,\tilde{\mathrm{Tr}}A$.

PROOF: By the same method employed in the proof of Lemma 12, with the help of Lemmas 13-17.

THEOREM 12. $J^* \vdash A$ *if and only if* $\tilde{J}^* \vdash \tilde{\mathrm{Tr}}A$;

$\tilde{J}^* \vdash C$ *if and only if* $J^* \vdash \mathrm{Tr}\,C$.

PROOF: If $J^* \vdash A$, then $\tilde{J}^* \vdash \tilde{\mathrm{Tr}}A$, by Lemma 5; conversely, if $\tilde{J}^* \vdash \tilde{\mathrm{Tr}}A$, then $J^* \vdash \mathrm{Tr}\,\tilde{\mathrm{Tr}}A$, by Lemma 6, and $J^* \vdash A$, by Lemma 18; hence $J^* \vdash A$ if and only if $\tilde{J}^* \vdash \tilde{\mathrm{Tr}}A$. We show similarly that $\tilde{J}^* \vdash C$ if and only if $J^* \vdash \mathrm{Tr}\,C$, with the help of Lemmas 5, 6 and 12.

We have thus seen that $\tilde{J}^*$ is equivalent to the discussive logic J^*

associated with the predicate logic S5 . It is worth noting that $\tilde{J}^*$ is also equivalent to the discussive logic associated with the predicate logic S4. This discussive logic is indeed identical to J^*, as we may verify by the same methods employed in Part I in order to show that J(S4) coincides with J(S5).

REFERENCES.

Costa, N. C. A. da

1975, *Remarks on Jaśkowski discussive logic*, Reports on Mathematical Logic, 4, 7-16.

Costa, N. C. A. da and L. Dubikajtis

197+, *On Jaśkowski discussive logic*, these Proceedings.

Furmanowski, T.

1975, *Remarks on discussive propositional calculus*, Studia Logica, 34, 39-43.

Jaśkowski, S.

1948, *Rachunek zdań dla systemów dedukcyjnych sprzecznych*, Studia Societatis Scientiarum Torunensis, Sectio A, I, nº 5, 57-77. (An English translation of this paper appeared in Studia Logica, 24 (1969), 143-157.)

Centro de Lógica, Epistemologia e História da Ciência
Universidade Estadual de Campinas
Campinas, São Paulo, Brazil.

PART II

MODEL THEORY

Non-Classical Logics, Model Theory and Computability,
A.I. Arruda, N.C.A. da Costa and R. Chuaqui (eds.)

SOME DIRECTIONS IN MODEL THEORY (*)

by *MIROSLAV BENDA*

This paper attempts to expose and alert a student of model theory to new areas of research. We, of course, have to make the usual disclaimer, that because of the tremendous growth of model theory nothing close to completeness can be done in an hour's lecture. To give an example, we do not discuss the interesting concept of recursive model theory proposed in Barwise and Schlipf 1976. (We mention it because of a still unsolved problem which is simple to state and warrants greater dissemination: if every countable model of a theory is recursively saturated, is the theory ω-categorical ?)

The choice has been made firstly by the freshness of the material and secondly by the desire to bring out as many connections to other fields (set theory, probability, computer science) as possible. We avoided therefore well-kown directions like the outcomes of Morley-Baldwin-Lachlan categoricity work, forcing, etc. The vague problem in the last section is aimed at the philosophically minded members of the audience; however, we wish that a possible outcome were precise and technical.

To give an idea of the work in model theory done in the past we discuss briefly the major problems (and again we just choose) model theory has been facing for some time. More complete lists may be found in Friedman 1975, and for model theory in particular in Robinson 1973.

(*) Preparation of this paper was partially supported by Fundação de Amparo à Pesquisa do Estado de São Paulo (FAPESP) and Financiadora de Estudos e Projetos (FINEP), Brazil.

THREE OUTSTANDING PROBLEMS.

Vaught's conjecture: *The number of countable models of a theory is* $\leq \omega$ *or* 2^{ω}. *For a more general question replace "theory" by a sentence of* $L_{\omega_1 \omega}$.

This conjecture has stimulated and is stimulating a lot of research. Our work on modeloids, on which we reported before the congress, was partially influenced by the desire to prove the conjecture. The first general result is due to Morley (see Morley 1970). The number of countable models of a sentence of $L_{\omega_1 \omega}$ is $\leq \omega_1$ or 2^{ω} . Morley's result and its extensions are now well understood thanks mainly to the work of Vaught (Vaught 1974), Burgess-Miller (Burgess and Miller 1975), and others. Closest to proving the conjecture came G. Sacks whose work is not yet published but an account of it may be found in Harnik and Makkai 1976. Harnik and Makkai, in fact, proved independently a result close to Sacks' which is easy to state: a $PC_{\omega_1 \omega}$ class whose countable members have only countably many automorphisms has either $\leq \omega$ or 2^{ω} countable models. Because of indicernibles, no first-order theory satisfies the assumptions of this theorem. Sacks' result has much milder condition on the automorphism satisfiable in some EC_{Δ} classes. In this respect a result of M. Rubin and S. Shelah (see Rubin and Shelah 1975) (very) weakly complements the above: If a theory T has Skolem functions and a linear ordering of the universe then it has 2^{ω} countable models. (This extends our earlier result that such a theory has $\geq \omega$ models.) Most models of these theories have 2^{ω} automorphisms.

The spectrum problem: *A spectrum is the set of cardinalities of finite models of a sentence* ϕ. *Are spectra closed under complementation ?*

This problem was posed by H. Scholz in 1952. Most of the research went into establishing that the spectra are very effective sets of natural numbers. This direction was culminated by Jones and Selman who showed that the spectra are exactly the sets of natural numbers which are recognizable by non-deterministic Turing machines time bounded by 2^{cx}, c a constant, x length of input (see Jones and Selman 1972). This links the now famous P = NP problem to the spectrum problem: If P = NP then spectr are closed under complementation. On the model-theoretic side the problem was recently investigated by Fagin (Fagin 1976 contains basic references) who among other things proved that for any sentence ϕ the spectrum of ϕ or $\neg\phi$ is cofinite. In fact the set of sentences which are eventually true in almost all models of cardinality n is a complete and decidable theory (studied earlier by Gaifman).

Morley's conjecture: *Let T be a theory. The larger the uncountable cardinal, the more models of T of that cardinality.*

All the results we know of in this direction are due to Shelah and it is difficult to give a simple account of them. Rather we refer the reader to his Tarski's Symposium lecture and to Shelah 197+ (specifically the section on p. 283).

FINITE MODEL THEORY.

Classical model theory has the ultimate result about finite models. Two finite models are elementarily equivalent iff they are isomorphic. Of course, this is an overstatement, as we have seen in the discussion of the Scholz' problem. But the fact remains that most methods and results of model theory apply almost exclusively to infinite models. The first concentrated effort on finite models was initiated by Keisler; we shall report on some of his results which open unexpected connections. The basic idea is to study infinite models which appear finite in some model of mathematics. The idea of using non-standard models to get standard results was used by A. Robinson and even earlier by C. Ryll-Nardzewski. Briefly, let M be mathematics, by which we mean the collection of objects and relations in the public domain of mathematicians. This can be considered as an ill-defined structure; model theory showed how we can get en extension of it, M* , which could be characterized by saying that the following holds about M*:

whatever could be, is.

Speaking precisely, M* is saturated or at least ω_1-saturated and the phrase "*object could be*" means that the existence of the object is not forbidden by a first order formula. (These explanations are the main reason for the slow acceptance of non-standard methods). So M* has infinite natural numbers because there is no first-order formula which says that every natural number is either 0 or 1 or ... . M* is richer than M in some aspects but in others it is poorer. M* does not recognize the set of standard natural numbers, for example; the objects it does recognize are called *internal*. A theorem about an internal object can be transformed into an information about properties of standard objects (think of M* as an ultrapower). Keisler puts it by saying "M*-*finite models exhibit the limiting behavior of finite models*".

But the internal model theory of M* is the same as in M so we have not gained much. And it is here where we should ask ourselves the question: "Why do we want to study finite models". The answer to this is clear, finite models are everywhere. Let us take an example:

EXAMPLE. We have subjects and want to teach them to distinguish black from white. We can get only statistical results so we should study larger groups and we in fact make the leap to study an infinite group A of subjects but one which is finite in M*. Let $F(a)$ be the probability that a subject a learns this on the first trial, and assume that F is internal so that we can transfer the results back to M. Questions which might be asked about this situation are: (1) what is the probability that a subject learns the task after two sessions (assume he does not forget it). (2) what is the average probability of learning the task after two sessions.

The answer to (1) is

$$G(a) = F(a) + (1 - F(a)).F(a),$$

and to the second question,

$$\sum_{a\in A} G(a).\frac{1}{|A|}$$

or if we denote $\mu(a) = \frac{1}{|A|}$, the uniform measure on $P(A)$ (taken in M*) we have

$$\int G\, d\mu\ .$$

Of course, the task of model theory is not in finding $F(a)$ nor even in making the computations. The task is to abstract from such examples and seek results about the abstractions.

Keisler proposes to study models (see Keisler 1976) $(A, \mu, F_i)_{i\in I}$ where A is internal to M*, M*-finite but infinite; μ is a probability measure on $P(A)$ (taken in M*) and F_i, $i \in I$, is an internal M*-finite sequence of functions of various numbers of arguments from A into $[0,1]^*$

(the reals of M*).

We take a hint from the example and build out of the functions F_i terms using continuous functions from R^n into R, sup, inf and integration with respect to the measure (see Keisler for details). This is a powerful language stronger then $L_{\omega\omega}$ but weaker than $L_{\omega_1\omega}$: if F_i admits only 0 and 1 as values and we interpret 1 as "*true*" then we have an ordinary relational structure and any first-order statement about it can be expressed by saying that a certain term has value 1; use multiplication instead of conjunctions, $1-x$ instead of negation, and sup instead of the existencial quantifier. Thus, the continuous functions act as connectives, sup and inf as quantifiers. The integral is a kind of qualitative quantifier not expressible in $L_{\omega\omega}$. It allows us to make statements like "about half of the elements satisfy ϕ" by writing

$$\frac{1}{2} + \int \phi \, d\mu$$

($\phi = \phi(x)$ is a term here $=1$ if ϕ is true and 0 otherwise): if this expression is ≈ 1 we, as before, think of it as true and it does mean that about half, according to μ, elements satisfy ϕ.

The simplest most useful result in model theory is the Skolem-Löwenheim theorem. To formulate the result in the present set-up Keisler defines as substructure of (A, μ, F_i) a structure $\mathfrak{A}|B = (A, \mu_B, F_i)$ where $B \subseteq A$ is internal and μ_B is the measure:

$$\mu_B(a) = \begin{cases} \frac{1}{|B|} & \text{if } a \in B \\ 0 & \text{if not.} \end{cases}$$

So only the expressons using integration (the qualitative quantifier) may have changed meaning. When they do not we call the substructure elementary: $\mathfrak{A}|B$ is an elementary substructure of $\mathfrak{A}$ if for any term $t(v_1 \ldots v_n)$ and any $a_1 \ldots a_n \in A$ (repeat A see the reasons above) t evaluated in $\mathfrak{A}$ is infinitesimally close to t evaluated in $\mathfrak{A}|B$ (at $a_1 \ldots a_n$).

SKOLEM-LÖWENHEIM THEOREM (Keisler):

Let $H = |A| + |I|$ *where* $\mathfrak{A} = (A, \mu, F_i)_{i \in I}$ *is as before. Let* n *be such that* $\frac{\log H}{n} \approx 0$ *and* $\frac{n^2}{|A|} \approx 0$. *Then there is a* $B \subseteq A$ *such that* $|B| = n$ *and* $\mathfrak{A}|B$ *is an elementary substructure of* $\mathfrak{A}$. *In fact almost every set* $B \supseteq A$ *(in the uniform measure on* $\{B \subseteq A \mid |B| = n\}$*) is such.*

Note that $\mu(B) \approx 0$. The import of the theorem to standard models in a special situation is this: assume we have a constant term t and models $\mathfrak{A}_n = (n, \mu_n, f_1 \ldots f_k)$ such that $t^{\mathfrak{A}_n} = t^{\mathfrak{A}_m}$. Then given $\varepsilon > 0$, there is n_o such that is $n \geq n_o$ then $t^{\mathfrak{A}_n} = t^{\mathfrak{A}_n|B}$ for some $B \subseteq n$ and $|B| < n.\varepsilon$.

The proof of the theorem uses the weak law of large numbers and the Stone-Weierstrass theorem . The theorem may be proved without the assumption $n^2 / |A| \approx 0$ but the proof is easier with it. Under this assumption almost all $f : n \to A$ are one-to-one. The computation of this is simple and presents a tiny bit of the use of non-standard probability used throughout Keisler's work so we show it in full: we think of the function $f : n \to A$ as choosing n-times an element from A replacing it after each choice. The probability of choosing a one-to-one sequence is

$$1.(1 - \frac{1}{|A|}) . (1 - \frac{2}{|A|}) \ldots (1 - \frac{n}{|A|}) \geq (1 - \frac{n}{|A|})^n .$$

But

$$(1 - \frac{n}{|A|})^n = (1 - \frac{1}{\frac{|A|}{n}})^{\frac{|A|}{n} \cdot \frac{n^2}{|A|}} \approx (e^{-1})^{\frac{n^2}{|A|}} \approx 1$$

as $(1 - \frac{1}{n})^m \approx e^{-1}$ for infinite m.

But the law of large numbers is used in trying to show that

$(\int t(x)\, d\mu)^{\mathfrak{A}}$ is close to $(\int t(x) d\mu)^B$

for almost all $B \subseteq A$, $|B| = n$. B is obtained by n choices from A. With k-th choice we associate a random variable $\chi_k = t(k\text{-th choice})$. The expectation of each χ_k is

$$E = (\int t(x) d\mu)^{\mathfrak{A}}$$

and since we are replacing, χ_k's are independent. The law of large numbers says that

$$\frac{1}{n} \cdot \sum_{k \leq n} \chi_k \approx E.$$

The sum on the left-hand side is almost the integral of t in $\mathfrak{A}|B$. In fact to make the induction go through we need Bernstein's inequality and eventually use saturation but this gives a sample of the ideas used.

Many other classical theorems (indicernibles, elimination of quantifiers ...) have counterparts in this set-up. An important direction is conversion of the non-standard model into a standard probability structure using the work of P. Loeb (see Loeb 197+).

COST ALGEBRAS.

This concept was inspired by Silver's work (Silver 197+) which in turn arose from a desire to simplify Jensen's 1972. Silver presented his work in terms of machines but some of his remarks (see especially Footnote 5 on p. 22) indicate that he was thinking in terms of cheap Skolem closures as we would put it in our terminology. The concept is very natural, one could say it attempts to bring something of the real world into mathematics.

DEFINITION. *A cost algebra is a pair* (A, C) *where:*

(a) $A = (A, F_n)_n$ *is an algebra, i. e. each* F_n *is a function on* A^k *into* A *(may be partial);*

(b) $C = ((B, \leq), C)$ *where* $(B, \leq)$ *is a linear order and* C *assigns for each operation* f *of the algebra* A *and for each argument* $a_1 \ldots a_k$ *of* f *an element of* B. *Thus* $C(f, a_1 \ldots a_k) \in B$.

We think of $C(f, a_1 \ldots a_k)$ as the cost of computing f at $a_1 \ldots a_k$. The linear order on B gives us a way to compare the cost of computing various operations at various arguments.

EXAMPLE 1. Let A be the algebra of Turing machines, i.e. we view a Turing machine as a (partial) operation on the natural numbers. Let the cost of computing on an input be the number of steps required to complete the computation; let it be ω if the machine does not halt. The linear order in this case is $\omega + 1$.

EXAMPLE 2. Let G be a group generated by a set of generators. For every element of G there are many words formed from the generators which are equal to the element; let the height of an element a be the least number n such that if in a reduced word only exponents k, $-k$ with $0 < k < n$ appear then the word is not equal to a. The generators g get assigned 1, and g^2 has 2 unless it is equal to a generator. The cost of computing $x.y$ is defined as the maximun of the heights of the elements; the cost of computing x^{-1} is the height of x.

Each class of algebras has its own concepts and techniques but the notion of a subalgebra is present in all algebraic considerations. This notion generalizes to cost algebras in the following form.

DEFINITION. *Let* (A,C) *be a cost algebra* $b \in B$ *(the linear order). A set* $S \subseteq A$ *is called a b-subalgebra of* A *if for any operation* f *of* A *and any argument* $a_1, \ldots, a_k \in S$ *of* f *we have:*

if $C(f, a_1, \ldots a_k) < b$ *then* $f(a_1 \ldots a_k) \in S$.

In Example 2, an n-subgroup of G is a group iff the index (order) of G is $< n$. This shows that a set $S \subseteq A$ which is not a subalgebra may be a b-subalgebra for some $b \in B$. Intuitively, for a pragmatist who cannot spend more than b a subalgebra and a b-subalgebra are the same thing.

A b-subalgebra can be made, in a natural way, into a cost algebra with a shorter cost scale. The operations when reduced to S will be in general partial; that is a reason for starting with partial operations already on the original algebra. Also, in order to be able to speak of real subalgebras, B should have a largest element which is not a cost of any computation. Variants of the definition involving subsets of the cost scale etc. could also be explored.

The intent of Jensen's and Silver's work is preserved. In fact the whole of Silver's argument is devoted to a detailed study of a particular cost algebra which is defined on the ordinals which also form the cost scale. The major tool in the work are δ-subalgebras which enable us to have a close watch on what sets of ordinals get constructed. In Silver's words " the constructible universe slowly reveals itself".

More interesting variant arises if we look at Example 1 and compare it with the current interest in computer sciences. The n-subalgebras limit the number of steps the Turing machine may perform by n and in this case the operations simply can not apply to large numbers $(> 2^n)$ because the machine cannot even read the input. We, therefore, consider the number of steps required to finish the computation but in dependence on the input. The most natural abstract setting would be algebras on A^* the set of finite sequences of members of A.

In order to give an example of the use of cost subalgebras we show that they were implicitly used in an old proof of Ryll-Nardzewski 1952.

THEOREM (Ryll-Nardzewski). *Peano arithmetic is not finitely axiomatizable.*

PROOF: Assume the contrary. Then we have finitely many axioms and because the theory has definable Skolem functions (via the well-ordering) what we have, essentially, is finitely many functions and axioms without quantifiers which tell us the properties of the functions. The consequence of this is that any set in any model of arithmetic which is closed under the functions is a model of those finitely many axioms. The natural numbers together with the functions give rise to a cost algebra somewhat like the algebra in Example 2. If we take finitely many numbers, among them 0, every number can be expressed as a term in the functions and the finitely many "generators". We define the cost of computing f at n as the complexity of n (as a term) plus 1. A simple, but crucial, observation is that there are finite n-subalgebras for every $n \in N$. Now everything said up to this point is expressible in arithmetic, in particular the formula $\phi(n)$ saying that there is a bounded n-algebra. This formula has the property that

$$\phi(0) \wedge (\forall n)(\phi(n) \rightarrow \phi(n+1))$$

holds in all models of the finitely many axioms. Consider the model of these axioms which is generated in a non-standard model of true number theory by an infinite element ω, using those finitely many functions. It is evident that in this model any ω-subalgebra is in fact the whole model, therefore $\neg\phi(\omega)$ holds there.

This theorem could be formulated for abstract cost algebras. It would claim that a cost algebra satisfying certain conditions (the cost defined by complexity of terms and expressible in first order etc.) has an elementarily equivalent cost algebra where subalgebras and n-subalgebras are the same for some n in the cost scale. More useful direction though would be to keep the cost scale fixed and explore the cost functions which are tied to the complexity of terms. In this connection an axiom to consider is

$$C(f,\ g(a)) \geq C(f\circ g, a)$$

assuming the operations are closed under substitutions.

A somewhat more concrete problem is to characterize the cost algebra constructed by Silver in more general terms; perhaps the fact that algebras associated with Skolem functions on L_α are Jōnsson is relevant here (see the proofs of 7.4.9 and 7.4.10 in Chang and Keisler 1973).

SAMPLINGS.

There are two kinds of tools: those of general use and those for specific operations. The method of what we call samplings has been used so far in the latter capacity but we think that its right place is in the first category among ultraproducts, indicernibles etc. The method consists of extracting information about a structure from small samples of the structure in the sense described below. The first result of this kind was obtained by Kueker; it is related below. The general problem in this area is best expressed in Shelah 197+ : "Kueker in ... defined some filters ... . We can easily suggest more". We would like to add that rather than filters more samplings should be constructed; they more or less canonically define the filters.

DEFINITION. *A sampling of* A *is a set* $S \subseteq P(A)$ *such that for every finite* $a \subseteq A$ *and any algebra on* A *of countably many operations we can find* $s \in S$ *which includes* a *and is closed under the operations.*

The definition may be stated equivalently for structures and elementary substructures instead of algebras and subalgebras.

EXAMPLE 1. The Löwenheim-Skolem theorem (for countable languages) is equivalent to saying that $P_{\omega_1}(A)$, the countable subsets of A, forms a sampling of A. More generally, if $\omega \leq \kappa < |A|$ then $P_\kappa(A)$ is a sampling of A.

EXAMPLE 2. Chang's conjecture for (κ^{++},κ^{+}) (κ^{+},κ) is equivalent to saying that the sets of order type κ^{+} forms a sampling of κ^{++}. (see 7.3.4 in Chang and Keisler 1973).

It is clear that there are no interesting samplings of countable sets because a countable set with a "successor" function has no proper elementary substructure. Intuitively, sampling is a collection of small subsets of A and there are so many of them that one gets an idea of the set from the properties of the samples. To make this clearer we associate with every sampling a filter of large sets of samples.

DEFINITION. *Let* S *be a (non-trivial) sampling of* A. *By* F_S *we denote the filter on* S *generated by*

$$\{s \mid s \text{ closed under } f_1, \ldots, f_n, \ldots\}$$

where $(A, f_1, \ldots, f_n, \ldots)$ *is an algebra.*

In the case when $S = P_{\omega_1}(A)$, F_S is just the filter generated by closed unbounded subsets of $P_{\omega_1}(A)$. In general, it is countably complete and normal. To express normality and other things it is convenient to introduce a quantifier: we write $(\forall_S s)(\ldots s \ldots)$ if

$$\{s \in S \mid \ldots s \ldots\} \in F_S \,;$$

for $\neg(\forall_S s)\neg$ write $(\exists_S s)$ $\forall_S$ and $\exists_S$ mean in the case of $S = P_{\omega_1}(A)$ that the set in question is closed-unbounded and stationary respectively. Normality then means that it is possible to switch (partially) $\forall$ and $\forall_S$:

$$(\forall x \in A)(\forall_S s)\, R(xs) \rightarrow (\forall_S s)(\forall x \in s)\, R(xs)$$

where $R \subseteq A \times S$. This is the property on which most theorems (but not all, see below) hinge. It enables us to prove a result on omitting of types which loosely speaking says that if a type can be sampled by subtypes all of which can be omitted then the type itself can be omitted (see Benda 197+b for details; the type is countable so it has no sampling but we can relativize the notion).

Another result proved using normality is essentially the first result in this field if we do not count the powerful samplings provided by supercompact cardinals (see Kueker 1972):

THEOREM (Kueker). *Let* ϕ *be a sentence of* $L_{\omega_1\omega}$, S *a sampling of* A *and* $\mathfrak{A}$ *a structure on* A. *Then:*

$$\mathfrak{A} \models \phi \quad \text{iff} \quad (\forall_S s)\ \mathfrak{A}|s \models \phi \,.$$

In Barwise 1974 these results are extended. A use of sampling is implicit also in the work of Shelah on Whitehead conjecture (Shelah 197+).

He defines there what amounts to a sampling of a strong limit cardinal λ (satisfying some additional properties) but he does not use the filter defined above; instead of countably many functions he uses κ ($<\lambda$) functions with the effect that the filter becomes κ-complete. The quantifier associated with it satisfies:

if $(\exists_S s)(A|s$ is free$)$ then A is free

for an algebra A (say a group) of cardinality λ.

It would be useful to know how the preservation of different properties varies from sampling to sampling. Specifically, is the theorem above false for some of the canonical samplings we discussed before ? These questions may later appear as naive as the question whether ultrafilters on ω have different properties.

Samplings play a role in set theory as well. Assume that we have a model of ZF with a set A in it and we want to extend it elementarily so that A stays in the extension but the image of A is properly larger than A (situations like this occur frequently in large cardinal questions). What are the properties A will have in the extension ? A partial answer is this: if $P(.)$ is a property such that for every sampling S (non-trivial) of A we have $(\forall_S s)\ P(s)$ then $P(A)$ is true in the extension. Let us denote by ϕ_A the properties defined in the last sentence.

QUESTION: Is ϕ_A a type ? When is it complete ?

Note that if the axiom of determinateness holds ϕ_{ω_1} is a complete type. (There might be some problems with the axiom of choice here but the alternative definition of a sampling is effective). A simple fact we noticed is that $\phi_{\omega_1} \neq \phi_{\omega_2}$, because ϕ_{ω_1} contains the property being a countable ordinal while ϕ_{ω_2} does not contain the property. Is $\phi_{\omega_2} = \phi_{\omega_3}$?

CAN MODEL THEORY LIVE UP TO ITS NAME ?

J. Silver once related to me a story about a man from an industry who called him wishing to consult him on model theory with prospects of using it in marketing. Blunders like this are fairly common, people think (rightly ?) that model theory refers to the important and fundamental activity of getting at facts and predictions by means of models. The model theory we study is, of course, not suited, by its own definition, for producing quantitative results sought in applied areas. But it is not always the numbers that we seek, sometimes we need a qualitative judgment. This thought is most effectively expressed in the following quotation from R. Thom 1969 (p. 333):

> "But as soon as we run into difficulties, contradictions, (like in Elementary Particle theory now) or when we feel overwhelmed by the mass of empirical data then the need arises for some conceptual guidance in order to classify the data and to find out the most significant phenomena. If scientific progress is to be achieved by other means than pure chance and lucky guess it relies necessarily on a qualitative understanding of the process studied".

Many problems which need a qualitative understanding could be readily suggested; we shall describe one which seems intimately connected with model theory and what we discussed above. The problem is to imitate on general structures the process by which natural scientists obtain information about the world we live in. The value of better understanding of this process is clear.

Let us be more specific. In the first approximation we see that the information is obtained by taking finite samples of the structure (i.e. the Universe). This is a process we understand to some degree if we take countable samples and we also understand that, in general, a finite sample has almost no information about the structure from which it is taken. If we take, for an example, an infinite linear order , a finite sample of it gives absolutely no information about the order because finite linear orders are the same everywhere. This example is unfair to linear orders because in fact all structures exhibit to a large degree such behavoir. (Ramsey's theorem being partly responsible for this.) But if we pull a real out of a model of set theory we know that it may contain a lot of information about the model. It may happen that the real is $0^{\#}$ which codes up information about a fairly large part of the whole structure. The key to finding out about the model from the real is to interpret it the right way. And this is exactly the methodology used in science. The finite sample is interpreted, generalized, idealized into a pattern which is subsequently checked on the universe by an experiment. Can this process be usefully modelled on abstract structures ? This would involve defining the possible interpretations and specifying the permissible experiments for verifying the conclusions of these interpretations. This essentially asks for some measure of the amount of information about a structure which can be obtained from small parts of it.

We are quite familiar with the reverse procedure, injecting information into a structure. A tipical example is the proof of Morley's Upward Categoricity result using Keisler's two cardinal theorem. We have a structure which is not saturated and we inject this information into it so that, to put it antropomorphically, the structrure "*knows*" that it is not saturated and "*remembers*" it after passages to equivalent structures.

Can we turn this technique around ? Perhaps we should not be too frugal in the begining and allow countable samples because it is not finiteness which is important; it is the interpretations of the sample. And here we come to the beginings of model theory which started with the study of interpretations; however, these are interpretations of syntactical objects by semantical ones. The interpretations under discussion are just the opposite, they seek to illuminate a semantical information by syntactical means. This is exactly the practice used in science. The processes of the real world like the behavior of the electron or the development of an embrio are interpreted by formal means like Hilbert spaces and stability theory respectively (the popularity of the catastrophe theory is mainly due to this kind of interpretation). What model theory has studied and studies is the relation of the theories encountered above, or theories simply invented, to abstract structures which pass as a substitute for the real world. Perhaps by turning some of its attention to the first step (real world - theory and checking) will model theory fully live up to its name.

References.

Barwise, K. J.

1974, *Mostowski's collapsing function*, Fund. Math. 82, 95-103.

Barwise, K. J. and J. S. Schlipf

1976, *An introduction to recursively saturated and resplendent models*, The Journal of Symbolic Logic, 41, nº 2, 531-536.

Benda, M.

197+a, *Modeloids*, in preparation.

197+b, *Compactness for omitting of types*, to appear.

Burgess, J. and D. Miller

1975, *Remarks on invariant descriptive set theory*, Fund. Math. 90, nº 1, 53-75.

Chang, C. C. and H. J. Keisler

1973, Model Theory , North-Holland, Amsterdam.

Friedman, H.

1975, *One hundred and two problems in mathematical logic*, The Journal of Symbolic Logic, 40, nº 1, 113-129.

Fagin, R.

1976, *Probabilities on finite models*, The Jounal of Symbolic Logic, 41, nº 1, 50-58.

Hanick, V. and M. Makkai

1976, *Vaught sentences and the covering theorem*, The Journal of Symbolic Logic, 41, nº 1, 171-187.

Jensen, B. R.

1972, *The fine structure of constructible hierarchy*, Annals of Math. Logic, 4, nº 3, 229-308.

Jones, N. D. and A. L. Selman

1972, *Turing machines and the spectra of first-order formulas*, Proceedings of the 4th Symposium on Theory of Computing, Denver, 185-196.

Keisler, H. J.

1976, Model Theory of Finite Structures, (lecture notes), Spring of 1976.

Kueker, D. W.

1972, *Löwenheim-Skolem and interpolation theorems*, Bull. Amer. Math. Soc., 78, 211-215.

Loeb, P. A.

197+, *Conversion from nonstandard to standard measure spaces*, to appear.

Morley, M. D.

1970, *The number of countable models*, The Journal of Symbolic Logic, 35, nº 1, 14-18.

Robinson, A.

1973, *Metamathematical problems*, The Journal of Symbolic Logic, 38, nº 3, 500-516.

Rubin, M. and S. Shelah

1975, *On linearly ordered models*, Notices AMS, oct. , A646.

Ryll-Nardzewski, C.

1952, *The role of the axiom of induction in elementary arithmetic*, Fund. Math., 39, 239-263.

Shelah, S.

197+, *A compactness theorem for singular cardinals, free algebras, Whitehead problem, and transversals*, (seminar notes) Louvain.

Silver, J.

197+, *How to eliminate the fine structure from the work of Jensen*, to appear,

Thom, R.

1969, *Topological models in biology*, Topology, 8, nº 3, 313-335.

Vaught, R. L.

1974, *Invariant sets in topology and logic*, Fund. Math. 82, 269-293.

Department of Mathematics
University of Washington
Seattle, Washington, U.S.A.

and

Departamento de Matemática
Universidade de São Paulo
São Paulo, SP., Brazil.

Non-Classical Logics, Model Theory and Computability,
A.I. Arruda, N.C.A. da Costa and R. Chuaqui (eds.)

A SEMANTICAL DEFINITION OF PROBABILITY

by *ROLANDO B. CHUAQUI*

When one speaks about the Foundations of Probability, there are two subjects that come to mind. On the one hand, the axiomatic foundations of the Calculus of Probability, which is a well - developped, independent mathematical discipline; on the other hand, the study of possible interpretations for probability statements. As is well - known, there are several conflicting interpretations of these statements held by the different schools in the subject. In this paper, based on the interpretation of Probability relating it to truth, I present a new definition of a probability measure in semantical terms. Thus, most of the content of the paper deals with the second of the foundational subjects mentioned above.

Expressions of the form 'It is probable that...' guide belief in the same way as, although more guardedly than, 'It is true that...'. Thus, both expressions have gerundive force. (For an illuminating discussion see Lucas 1970, Ch. I, II). This kinship naturally leads to attempt a definition of Probability similar to Tarski's semantical definition of truth (see Tarski 1935).

I believe that for many common language uses of the word 'probable' there is no adequate measure of probability. However, with these I shall not be concerned. My aim is to attain a definition of a probability measure, satisfying the axioms of the Calculus, that can account for all, or at least most, of its applications in current scientific and statistical practice.

My conception is connected with Carnap's definition (see Carnap 1950 or Carnap and Jeffreys 1971) - actually, the picture on page 297 of his 1950 book inspired some of my ideas - because I use model - theoretical methods.

However, since I see Probability in the same category as truth and not logical truth, there are important differences in the two conceptions. In particular, Carnap intends his measure to be defined on the space of all models, whereas I limit the space to those models that picture a possible outcome. Which outcomes are possible is determined by the laws of the phenomenon involved. Thus, 'possible' is meant here in the sense of physical possibility and not logical possibility.

In order to obtain a probability measure in the space of possible outcomes, I use an equal likelihood or equiprobability relation. Being based on equiprobability, my definition is also related to the classical conception. However, my equal likelihood relation will be well-defined and will not suffer from the pitfalls of the classical one.

The paper has an introductory first section that gives a general description of the procedures involved, and a second section with some mathematical preliminaries. It continues in a third section with the formal definition of the simple probability structures and the corresponding equiprobability relation, notions which arise when there are no sequences of outcomes. The next section gives necessary and sufficient conditions for the existence of a probability measure compatible with an equivalence relation having the same properties as the equal likelihood relation defined previously. This section of the paper contains the most interesting new mathematical results of the paper: necessary and sufficient conditions for the existence of a measure on a field of sets, invariant under a group of transformations. The fifth section discusses compound probability structures, which arise when there are sequences of outcomes, and ways of defining a probability measure for them. The paper concludes, in section six, with examples and methods for applying the probability models given. A reader not interested in the technical material could read sections one and six to understand the main ideas.

A brief outline of an earlier version of these ideas appeared in Chuaqui 1975, although some of them were already in Chuaqui 1965.

1. GENERAL DESCRIPTION.

We assume a language L with some set of nonlogical constants. For the moment we leave the exact description of L undetermined. It might be a finitary first-order language, an infinitary language, or a language of some other type. For the definition of truth, we usually interpret L in *relational systems* $\mathfrak{A} = \langle A, R_1, \ldots, R_{n-1} \rangle$, where A is a nonempty set and $R_1, \ldots, R_{n-1}$ is a sequence of relations over A of the appropiate type so as to match the nonlogical symbols in L. For any sentence ϕ we can define, by Tarski's method, 'ϕ *is true in* $\mathfrak{A}$'. In a system of this type everything relating to the symbols in L is determined. Thus, for every sentence of L, ϕ is true or false in $\mathfrak{A}$. Systems represent, it may be said, possible states of the world. When we interpret L in $\mathfrak{A}$, we determine completely which sentences are true and which are false.

When probability statements apply, the precise state of the world which obtains is not completely determined; therefore we must change the notion of an interpretation of the language L. Our interpretations will no longer be single systems but certain classes of systems with some additional properties.

The Calculus of Probability is usually applied to happenings or occurrences, which may be experiments, observations, or natural phenomena. What interests us in all these cases is the *result* or *outcome* of the phenomenon. The "theory" (in a wide sense of the term) is what determines which are the possible outcomes. In some cases there is only one possible outcome; in others, many. It is in these latter cases that the Calculus of Probability is important.

Let us take as our first example the choosing of a sample S of size m from a finite population P. When we say 'S has n red things' we mean that one of the properties of the outcome was that the sample had n red objects. The same outcome has many different properties, which can be described in many different ways. We can think of an ideal approximation of an outcome, namely a relational system that represents a possible model of the situation involved. In the case we are looking at we can schematize the possible outcomes as systems $\mathfrak{A}_S = \langle P, R_1, \ldots, R_{n-1}, S \rangle$, where P is a fixed finite set, $R_1, \ldots, R_{n-1}$ are fixed subsets of P that represent the properties we are interested in (for instance, 'red'), and S is any subset of P of m members (the sample). For each subset S of m members there is a corresponding system $\mathfrak{A}_S$; hence the set of possible outcomes consists of all models $\mathfrak{A}_S$ of the form described above.

To speak about these outcomes we need a language L with nonlogical symbols $\overline{R}_1, \ldots, \overline{R}_{n-1}, \overline{S}$. We now use as interpretations for L not systems, but classes, $\mathbb{K}$ of relational systems appropiate for L. $\mathbb{K}$, then, will be the set of possible outcomes.

In the case that we shall first consider, i.e. the *simple probability structures*, we can take $\mathbb{K}$ to be a set of relational systems with a common universe. These simple cases may be characterized as those in which there are no sequences of outcomes. For instance, in the example just mentioned, $\mathbb{K}$ is the set of systems $\mathfrak{A}_S$. We shall later analyze cases in which sequences of outcomes appear; then our probability structures will be more complicated. However, in order to build these *compound probability structures* we shall need the simple ones.

Properties of outcomes are usually called *events*. These events can be represented by sentences. If two sentences are logically equivalent, they represent the same event. Thus we may take events as equivalence classes of sentences determined by the relation of logical equivalence. Thus, if we take ϕ to be the sentence 'there are n red things in S', the event that ϕ obtains is the class of sentences logically equivalent to ϕ. We shall define a probability measure on these equivalence classes. However, when there is no danger of confusion we shall speak of the probability of ϕ meaning the probability of the corresponding equivalence class.

We know that for relational system $\mathfrak{A}$ and sentences ϕ we can define ϕ is true in $\mathfrak{A}$. For simple probability structures $\mathbb{K}$, we say that ϕ *is true* (or *false*) *in* $\mathbb{K}$, if ϕ is true (or false) in every relational system in $\mathbb{K}$. Similarly, we shall define the *probability of* ϕ *in* $\mathbb{K}$. We understand this probability as sort of a measure of the degree of "partial truth" of a sentence. Thus, if ϕ is true in $\mathbb{K}$, it should be the case that the probability of ϕ in $\mathbb{K}$ is one; if ϕ is false in $\mathbb{K}$, the probability of ϕ in $\mathbb{K}$ should be zero; and in all other cases this probability should be between zero and one. This degree of "partial truth" is vaguely proportional to the size of the set of models of ϕ that are in $\mathbb{K}$. If ϕ is true in $\mathbb{K}$, this set is $\mathbb{K}$ itself; and if ϕ is false in $\mathbb{K}$, this set is empty. The natural thing to do, then, is to obtain a probability measure defined on a collection of subsets of $\mathbb{K}$ that includes the sets of models in $\mathbb{K}$ of sentences of the appropiate language. Having done this, we may define the probability of ϕ in $\mathbb{K}$ as the measure of the set of models of ϕ that are in $\mathbb{K}$.

In order to define this measure I shall introduce an *equal likelihood* or *equiprobability* relation between events, i.e. between equivalence classes of logically equivalent sentences. This equiprobability relation is derived from a *symmetry* relation between sets of models, i.e. subsets of $\mathbb{K}$. Intuitively, two sets of models are *symmetrical*, if they are symmetrical with respect to the laws of the correspending phenomenon. For determining these symmetries, we consider the *group of transformations* (functions) *that are "invariant under these laws"*. Two subsets of $\mathbb{K}$ will be symmetrical if one can be transformed into the other by one of these functions. Then, two sentences (or, more accurately, two equivalence classes of logically equivalent sentences) ϕ and ψ are equally likely in $\mathbb{K}$, if the corresponding sets of models in $\mathbb{K}$ of ϕ and ψ are symmetrical.

The laws of the phenomenon determine the simple probability structure $\mathbb{K}$; $\mathbb{K}$, in its turn, gives the group of transformations. Let us see how this group, call it G_K, is obtained in our example. G_K is a subgroup of the group of permutations of the universe, P. For a permutation f to be in G_K it must satisfy some additional requirements. Each relational system $\mathfrak{A}_S$ can be decomposed in two parts. One that is the same for all systems in $\mathbb{K}$, namely the constant part $\mathfrak{C}_K = \langle P, R_0, \ldots, R_{n-1} \rangle$; and another, the variable part $\mathfrak{A}_{S,V} = \langle P, S \rangle$. Now, the first requirement on f is that it should not take a system in $\mathbb{K}$ outside of $\mathbb{K}$. That is, if we apply f to the variable part of a system, $\mathfrak{A}_{S,V}$, transforming it into another system $\mathfrak{A}_{S',V}$ (S' is the set of elements $f(x)$ for x in S, i.e. the image of S by f), then $\mathfrak{A}_{S',V}$ should also be the variable part of a system in $\mathbb{K}$ (i.e. $\langle P, R_0, \ldots, R_{n-1}, S' \rangle$ should be in $\mathbb{K}$). In the particular case of the example, since the only condition on S is its cardinality, all permutations of P satisfy this requirement. In general, however, this is not so.

A second condition imposed on the functions f in G_K appears when there are distinguished elements (denoted by individual constants in the language) in the systems in $\mathbb{K}$. If we transform the set of distinguished elements by f, the new elements should satisfy the same sentences as the old ones. To state precisely this requirement we need some technical machinery that will be introduced in the next section. The last condition imposed is that the image by f of a set of models of a sentence should be the set of models of (possibly another) sentence.

Having this group G_K, we can define the symmetry relation between subsets of $\mathbb{K}$. Let A, B be subsets of $\mathbb{K}$; we say that A and B are symmetrical if there is a function f in G_K such that the variable parts $\mathfrak{B}_V$ of systems $\mathfrak{B}$ in B are exactly those obtained by the action of f on the variable parts $\mathfrak{A}_V$ of systems $\mathfrak{A}$ in A.

We are now left with the mathematical problem of obtaining a probability measure invariant under this equivalence relation. Namely, a measure that assigns the same number to symmetrical sets of models. I shall discuss in section four, possible solutions to this problem, in particular, some necessary and sufficient conditions for the existence of such a measure.

The simple probability structures are, according to my views, the basic structures. All probability measures are derived from probabilities defined on these simple structures. I shall not analyze all cases of derived probability. However a very important case appears when there are sequences of outcomes. For this situation I introduce the *compound probability structures*. They are built from the simple ones by adding an ordered system that represents the ordering in the sequence of outcomes. At each point in the order, the preceding outcomes determine a simple probability structure and the next outcome is from this structure. The probability measure for compound structures is computed from the measures on the simple ones as is usually done with conditional probability.

These compound structures will be discussed in some detail in section five. However the need for them may be seen from the following example:

Suppose we have two urns, urn one and urn two. Urn one has one black ball, and urn two has two white balls, ball one and ball two. Suppose, furthermore, that we choose an urn at random, and then a ball from that urn also at random. It is clear that the event of choosing a white ball is equiprobable to the event of choosing a black ball. However, the first event can be divided into two events while the second cannot. It is easy to see that the definition of equal likelihood I have given for simple structures implies the following:

If A is equally likely to B and A can be decomposed into two disjoint events A_1, A_2, then B can also be decomposed into disjoint events B_1, B_2 that are equally likely respectively to A_1, A_2.

In the case we have just discussed this is not true, unless we artificially divide the event of choosing a black ball into two events. The explanation for this phenomenon is that we are confronted with compound outcomes. We first choose an urn and then, depending on this choice, we choose a ball. The probability structures we have to consider are, thus, more complicated. They should be of the following type:

Since there are only two succesive outcomes, the basic order of sequence of outcomes can be given by the numbers 0 and 1 in their natural order. We have two sets of relational systems $\mathbb{K}_0$ and $\mathbb{K}_1$. $\mathbb{K}_0$ consists of models $\mathfrak{U}_C = \langle U, C \rangle$ where U is the set of urns and C contains the one chosen. Thus, $\mathbb{K}_0$ consists of $\mathfrak{U}_1$, where urn one is chosen, and $\mathfrak{U}_2$, where urn two is chosen. $\mathbb{K}_1$ contains systems of the form $\langle A, B, W, D \rangle$, where A is the set of balls in an urn, B the black balls, W the white balls, and D the chosen ball. $\mathbb{K}_1$ is divided in two parts, $\mathbb{K}_{\mathcal{U}_1}$ and $\mathbb{K}_{\mathcal{U}_2}$. $\mathbb{K}_{\mathcal{U}_1}$ contains just one system $\mathcal{O}\!\mathcal{L} = \langle A_1, B_1, W_1, D\rangle$, where A_1 contains just the black ball in urn one, $B_1 = A_1 = D$, and W_1 is empty.

$\mathbb{K}_{\mathcal{U}_2}$ contains two models $\mathcal{O}\!\mathcal{L}_2 = \langle A_2, B_2, W_2, D_1\rangle$ and $\mathcal{O}\!\mathcal{L}_3 = \langle A_2, B_2, W_2, D_2\rangle$. A_2 contains the white balls in urn two, B_2 is empty, $W_2 = A_2$, D_1 contains ball one, and D_2 contains ball two.

A compound outcome here is a function f whose domain is $\{0, 1\}$ and such that

$$f(0) \in \mathbb{K}_0 \quad \text{and} \quad f(1) \in \mathbb{K}_{f(0)}.$$

To speak about these outcomes we need a language that allows us to talk about sequences of elements of the order. This is not difficult to do as will be seen in section five. Once we have this language we have our events. For instance, the event of choosing a white ball has as models the two functions f, g such that:

$$f(0) = g(0) = \mathcal{U}_2, \quad f(1) = \mathcal{O}\!\mathcal{L}_2 \quad \text{and} \quad g(1) = \mathcal{O}\!\mathcal{L}_3.$$

In order to define a probability measure we proceed as for the simple probability structures and define if for $\mathbb{K}_0$, $\mathbb{K}_{\mathcal{U}_1}$, and $\mathbb{K}_{\mathcal{U}_2}$ independently. Then by usual conditional probability procedures, we define it for the compound events. For instance, let F be the set of models consisting of the function f given above. Then, as the probability in $\mathbb{K}_0$ of $\mathcal{U}_2$ is $\frac{1}{2}$ and the probability of $\mathcal{O}\!\mathcal{L}_2$ in $\mathbb{K}_{\mathcal{U}_2}$ is also $\frac{1}{2}$, we have that the probability of F is $\frac{1}{2} \cdot \frac{1}{2} = \frac{1}{4}$.

2. SET-THEORETICAL AND METALOGICAL PRELIMINARIES.

We use throughout various familiar set-theoretical notions and symbols. $\cup$, $\cap$, $-$, $\subseteq$ will denote *union, intersection, set-theoretical difference*, and the *inclusion relation*. By ω we denote the *set of all natural numbers* (nonnegative integers), which are identified with the finite ordinals. ω_1 denotes the first uncountable ordinal. Each ordinal is identified with the set of all smaller numbers:

$$\alpha = \{\beta : \beta < \alpha\}$$

the formula $\alpha < \beta$ is thus equivalent to the condition $\alpha \in \beta$ and α, β are ordinals. In particular, the number zero, 0, is the empty set $\emptyset$, $1 = \{0\}$, $2 = \{0,1\}$, etc...

${}^{B}A$ denotes the *set of all functions on* B *into* A (i.e. with domain B and range included in A). If f is a function, $\mathcal{D}of$ denotes its *domain*, f^{-1} its *converse*, $\mathcal{D}of^{-1}$ its *range*, and for any $x \subseteq \mathcal{D}of$, $f^{*}x$ is the *image of* x *under* f, and $f \mid x$ is f *restricted in its domain to* x. The *value of a function* f *at the element* i of its domain is denoted by $f(i)$ or f_i. If $F \in {}^{I}A$, we write in some cases $F = \langle F_i : i \in I\rangle$ and call it a *system*.

The members of ${}^{\omega}A$ are what are called *simple infinite sequences with all terms in* A. In case $n \in \omega$, the members of ${}^{n}A$ are referred to as *finite* n*-termed sequences* again *with terms in* A. If $f \in {}^{n}A$, we write $f = \langle f_0,\ldots,f_{n-1}\rangle$; in particular $\langle a, b\rangle$ is the two-termed sequence whose terms are a, b. The symbol ${}^{\underset{\smile}{\omega}}A$ denotes the *set of all finite sequences with terms in* A (i.e. ${}^{\underset{\smile}{\omega}}A = \bigcup\{{}^{n}A : n \in \omega\}$).

An n*-ary relation* R *on* A is any subset of ${}^{n}A$. For n-ary relations R we write $Rx_0,\ldots,x_{n-1}$ instead of $\langle x_0,\ldots,x_{n-1}\rangle \in R$. If R is binary, aRb stands for $\langle a, b\rangle \in R$. A *relational system* is a system $\mathfrak{A}$ of the form $\mathfrak{A} = \langle A,R,a\rangle$, where: A is a nonempty set called the *domain* or *universe* of the system $\mathfrak{A}$, $R = \langle R_i : i \in I\rangle$ is a system of relations on A, and $a = \langle a_j : j \in J\rangle$ is a system of elements of A called the *distinguished elements of* $\mathfrak{A}$. I and J should be disjoint; $I \cup J$ is called the *index set of* $\mathfrak{A}$. We also write $\mathfrak{A} = \langle A,R_i, a_j\rangle_{i \in I,\ j \in J}$. We use german letters for relational systems and the corresponding capital italic letter for their universes. (For details see Tarski 1954).

It is useful to define the *similarity type* of a relational system. If $s = \langle\langle m_i : i \in I\rangle, J\rangle$, where the m_i's are natural numbers for every $i \in I$, then a relational system $\mathfrak{A} = \langle A,R_i, a_j\rangle_{i \in I,\ j \in J}$ is of similarity type s if, for each $i \in I$, R_i is an m_i-ary relation. We say that two relational systems are *similar* if they are of the same type.

If $\mathfrak{A} = \langle A,R_i, a_j\rangle_{i \in I,\ j \in J}$ and $K \subseteq I \cup J$, we call the system obtained by omitting all relations and distinguished elements whose index is not in K the restriction of $\mathfrak{A}$ to K, in symbols:

$$\mathfrak{A} \mid K = \langle A,R_i, a_j\rangle_{i \in I \cap K,\ j \in J \cap K}.$$

We now consider the concept of *isomorphism* of two similar systems. Let $\mathfrak{A} = \langle A,R_i, a_j\rangle_{i \in I,\ j \in J}$ and $\mathfrak{B} = \langle B,S_i, b_j\rangle_{i \in I,\ j \in J}$ be two similar relational systems. Then $\mathfrak{A}$ and $\mathfrak{B}$ are isomorphic if there is a one - one function f from A onto B such that:

(i) for every $i \in I$ and $x_0, \ldots, x_{m_i - 1} \in A$ we have, $R_i x_0, \ldots, x_{m_i - 1}$ if and only if $S_i f(x_0), \ldots, f(x_{m_i - 1})$,

(ii) for every $j \in J$, $f(a_j) = b_j$.

Under these conditions f is called an *isomorphism* and $\mathfrak{B}$ the *isomorphic image of* $\mathfrak{A}$ *under* f, in symbols $\mathfrak{B} = f * \mathfrak{A}$. It is clear that if $\mathfrak{A}$ is a system and f a one - one function with domain A, then there is one and only one relational system $\mathfrak{B}$ such that $\mathfrak{B} = f * \mathfrak{A}$. If $\mathbb{B}$ is a set of relational systems with a common universe A, and f is a one - one function with domain A, then $\mathbb{B}^f$ denotes the *set of systems obtained by the action of* f in $\mathbb{B}$, in symbols:

$$\mathbb{B}^f = \{f * \mathfrak{A} : \mathfrak{A} \in \mathbb{B}\} .$$

We also use some terminology from General Algebra. For more details about this and relational systems see Henkin, Monk, and Tarski 1971, Preliminaries and Ch. 0.

For any $\alpha \in \omega$ or $\alpha = \omega$, an α - *ary partial operation* Q on a set A is a function with domain included in ${}^{\alpha}A$ and range included in A. If $DoQ = {}^{\alpha}A$, then we say that Q is a *total* α - *ary operation*, or simply an α - *ary operation, on* A. A total 0 - ary operation on A represents a distinguished element of A. A *partial algebra* is a system $\mathfrak{A}$ of the form $\mathfrak{A} = \langle A, Q\rangle$ where A is a nonempty set and $Q = \langle Q_i : i \in I\rangle$ is a system of partial operations on A. $\mathfrak{A}$ is an *algebra* if every partial operation in the system is total. An equivalence relation R on A is a *congruence relation* on $\mathfrak{A}$ if for every $a, b \in {}^{\alpha}A$ such that Q_j is α - ary, $a_i R b_i$ for every $i \in \alpha$, and $Q_j a$ and $Q_j b$ exist, we have $Q_j a \, R \, Q_j b$, for every $j \in I$. It is well-known that it is possible to obtain, for each congruence relation R, the *quotient (partial) algebra* $\mathfrak{A} / R$.

We also consider certain kinds of algebras and partial algebras. A *Boolean algebra* (BA) is an algebra $\mathbb{B} = \langle B, \vee, \wedge, -, 0, 1\rangle$, where $\vee, \wedge$ are the binary operations of *join* and *meet*, - is the unary operation of *complementation*, and 0 and 1 are distinguished elements of B. BA's are characterized by well - known axiom systems. If $A \subseteq B$, we designate by $\bigvee\{x : x \in A\}$, $\bigwedge\{x : x \in A\}$ the *supremum* or *infimum* of A if they exist in B. When we have an infinite sequence $x \in {}^{\omega}B$, we write $\bigvee\{x_i : i \in \omega\}$, $\bigwedge\{x_i : i \in \omega\}$. If all suprema and infima exist for countable subsets A, we say that $\mathbb{B}$ is *countably complete*. The notions of *atom*, *ideal*, and *countably complete ideal* in a BA will also be needed. If I is an ideal in a BA $\mathbb{B}$, we write $\mathbb{B}/I$ for the quotient algebra determined by I. For more details about BA's see Sikorski 1969.

A particular kind of BA is a *field of subsets of a set* X, $\mathbb{F} = \langle F, \cup, \cap, -, \emptyset, X \rangle$, where F contains subsets of the set X, and the operation are *union, intersection, complement relative to* X, the *empty set*, and X itself. A *countably complete field of sets* is a field of sets that contains all countable unions and intersections. A *group of transformations of* $\mathbb{F}$ is a subset *of the group of permutations* (*one-one, onto functions*) *of* X, that transform elements of F into elements of F.

The set of nonnegative real numbers extended by adding ∞ is denoted by $\overline{R}$. $[0,1]$ is the closed interval between 0 and 1. We use addition $+$, and countable addition Σ among real numbers.

A BA $\mathbb{B}$ is isomorphic to the *closed-open sets of a totally disconnected compact topological space, the Stone space of* $\mathbb{B}$; in symbols $S(\mathbb{B})$. If X is a Stone space, $C(X,\overline{R})$ denotes the *set of continuous functions with compact support on* X *and values in* $\overline{R}$. Binary addition $+$ of functions is pointwise addition. Countable addition Σ is the continuous limit of the finite sums, which differs from the pointwise limit on a set of first category.

A *generalized cardinal algebra* (GCA) is a partial algebra $\mathfrak{A} = \langle A, +, \Sigma \rangle$, with a partial binary operation $+$ and a partial countable operation Σ (see Tarski 1949). A *cardinal algebra* (CA) is a GCA in which the partial operations are total. A particular kind of GCA is the disjunctive BA $\dot{\mathbb{B}}$ associated with the BA $\mathbb{B}$. $\dot{\mathbb{B}} = \langle B, +, \Sigma \rangle$ is obtained as follows: $a + b = c$ if and only if $a \vee b = c$ and $a \wedge b = 0$, $\Sigma_{i \in \omega} x_i = c$ if and only if $\bigvee\{x_i : i \in \omega\} = c$ and $x_i \wedge x_j = 0$ for all $i < j \in \omega$. (See Tarski 1949, def. 15.14).

By a *finitely additive* (*probability*) *measure on a BA* $\mathbb{B}$ we understand, as usual, a function μ satisfying the following conditions:

1. $Do\ \mu = B$ and $Do\ \mu^{-1} \subseteq [0,1]$;
2. if $a,b \in B$ and $a \wedge b = 0$, then $\mu(a+b) = \mu(a) + \mu(b)$;
3. $\mu(1) = 1$.

We say that μ is a *countable additive* (*probability*) *measure* or, simply, a *measure*, if in addition μ satisfies:

4. If $x \in {}^{\omega}B$ and $\Sigma_{i \in \omega} x_i$ exists, then $\mu(\Sigma_{i \in \omega} x_i) = \Sigma_{i \in \omega} \mu(x_i)$.

A measure μ is strictly positive if:

5. $\mu(a) = 0$ implies $a = 0$, for every $a \in B$.

Throughout this paper we shall consider two different first-order languages of similarity type $s = \langle\langle m_i : i \in I\rangle, J\rangle$, a *finitary language* L^s and an *infinitary language* $L^s_{\omega_1\omega}$. The language L^s has a denumerable set of *variables*, and $L^s_{\omega_1\omega}$ has a set of variables of cardinality ω_1. The *constants* are divided into *logical* and *nonlogical* ones. Both languages have logical constants $\wedge, \vee, \neg, \forall, \exists$ and $=$, standing for (finite) *conjunction*, *disjunction*, *negation*, *universal* and *existential quantification*, and *identity*. In addition the infinitary language $L^s_{\omega_1\omega}$ has logical constants $\bigwedge$ and $\bigvee$ standing for denumerable conjunctions and disjunctions, respectively. The nonlogical constants for both languages are the *predicates* (or *relation symbols*) and the *individual constants*. With every predicate a natural number is correlated, which is called the *rank* of the symbol. For every $i \in I$ there is a predicate of rank m_i. For each $j \in J$, an individual constant. The correspondence is such that different indices correspond to different symbols. The identity symbol, though regarded as a logical constant, is included in the set of binary predicates (predicates of rank two). The variables and the individual constants are the *individual symbols*.

The *expressions* of L^s are finite sequences of symbols; the expressions for $L^s_{\omega_1\omega}$ are transfinite sequences of symbols of length less than ω_1. Among expressions we distinguish the *formulas*. The simplest, so-called *atomic formulas*, are obtained by combining n arbitrary individual symbols by means of a predicate of rank n; compound formulas are built from simpler ones by means of sentential connectives and *quantifier expressions* (i.e. quantifiers followed by variables such as $\forall v$ or $\exists v$). An occurrence of a variable in a formula may be either *free* or *bound*; a formula in which no variable occurs free is called a *sentence*. The set of sentences of L^s is designated by S^s, and the set of sentences of $L^s_{\omega_1\omega}$ by $S^s_{\omega_1\omega}$.

We assume known Tarski's definition of *satisfaction* and *truth* (for L^s see Tarski 1954, and for $L^s_{\omega_1\omega}$ see Keisler 1971). When there is no danger of confusion we shall use the same metalogical symbols for both languages.

An *assignment in* $\mathfrak{A}$ is a function from the variables into A. For x an assignment in a relational system $\mathfrak{A}$ of similarity type s and ϕ a formula of L^s or $L^s_{\omega_1\omega}$ we assume defined:

$$\mathfrak{A} \models \phi\,[x] \quad \text{(i.e. } x \textit{ satisfies } \phi \textit{ in } \mathfrak{A}\text{)};$$

and, also, if ϕ is a sentence we have:

$$\mathfrak{A} \models \phi \text{ (i.e. } \phi \textit{ is true in } \mathfrak{A}).$$

Two relational systems $\mathfrak{A}, \mathfrak{B}$ of similarity type s are *elementarily* (*or* $L_{\omega_1\omega}$) *equivalent*, if they satisfy exactly the same sentences of L^s (or $L^s_{\omega_1\omega}$); i.e. $\mathfrak{A} \equiv \mathfrak{B}$ if and only if for every sentence ϕ of L^s (or $L^s_{\omega_1\omega}$)we have, $\mathfrak{A} \models \phi$ if and only if $\mathfrak{B} \models \phi$.

For any class $\mathbb{K}$ of relational systems of similarity type s and with common universe A, any formula ϕ of L^s (or $L^s_{\omega_1\omega}$), and any assignment x in A, we define:

$$Mod_{K,x}(\phi) = \{\mathfrak{A} : \mathfrak{A} \in \mathbb{K},\ \mathfrak{A} \models \phi[x]\}.$$

If ϕ is a sentence, we write $Mod_K(\phi)$ for $Mod_{K,x}(\phi)$; $Mod_K(\phi)$ *is the set of models of* ϕ *that are in* $\mathbb{K}$.

If Σ is a set of sentences and ϕ is a sentence, then ϕ *is a consequence of* Σ if ϕ holds is all systems in which all sentences of Σ hold, and we write $\Sigma \models \phi$. ϕ *is valid* if it is a consequence of the empty set, and we write $\models \phi$. For both languages L^s and $L^s_{\omega_1\omega}$ we choose a standard system of deduction, and we write $\Sigma \vdash \phi$ if ϕ *is derivable from* Σ. ϕ *is a theorem* if it is derivable from the empty set, and we write $\vdash \phi$ (for $L^s_{\omega_1\omega}$ consult Keisler 1971). By the well-known Completeness Theorem of finitary first-order logic we have for every $\Sigma \subseteq S^s$ and every $\phi \in S^s$, $\Sigma \models \phi$ if and only if $\Sigma \vdash \phi$. This is not true for $L^s_{\omega_1\omega}$; however we still have "weak" completeness in the sense that for every $\phi \in S^s_{\omega_1\omega}$, $\models \phi$ if and only if $\vdash \phi$.

We call two sentences *equivalent* if $\vdash \phi \longleftrightarrow \psi$ (where, as usual, $\phi \leftrightarrow \psi$ stands for $(\phi \wedge \psi) \vee (\neg\phi \wedge \neg\psi)$). It is well-known that this is an equivalence relation, and the equivalence classes form a BA, the so-called *Lindenbaum-Tarski algebra of sentences*. In the infinitary case, this BA is countably complete. We denote, for any sentence ϕ, by $\phi/\vdash$ the equivalence class of ϕ, and $S^s/\vdash$ and $S^s_{\omega_1\omega}/\vdash$ are the corresponding BA's. By the weak Completeness Theorem these algebras are isomorphic to a field of sets of models.

We shall define probability measures on $S^s/\vdash$ and $S^s_{\omega_1\omega}/\vdash$. By the isomorphism mentioned above, this is equivalent to defining them on a field of sets of models as Carnap does in Carnap and Jeffreys 1971.

For $S^s / \vdash$ we have the following theorem (Scott and Krauss 1966, Lemma 7.1):

THEOREM 2.1. *Every finitely additive measure on* $S^s / \vdash$ *is countably additive.*

3. Simple probability structures.

As explained in section one, to introduce probability we have to change the concept of an interpretation. Instead of interpreting our languages in relational systems, we do it in more complicated structures. I introduce, now, the simple probability structures:

A *simple probability structure* is a set $\mathbb{K}$ of relational systems with a common similarity type, say $s = \langle\langle m_i : i \in I\rangle, J\rangle$, that satisfies the following conditions:

(i) For each $\mathfrak{A}, \mathfrak{B} \in \mathbb{K}$, $\mathfrak{A} \mid J = \mathfrak{B} \mid J$.

(ii) Let K be the largest subset of $I \cup J$ such that for any $\mathfrak{A}, \mathfrak{B} \in \mathbb{K}$, $\mathfrak{A} \mid K = \mathfrak{B} \mid K$; then if there is a formula with one free variable ϕ such that b is the only element of A with $\mathfrak{A} \mid K \models \phi[b]$, then there is a $j \in J$ such that $b = a_j$.

It is clear that condition (i) implies that all systems in $\mathbb{K}$ have the same universe and the same distinguished elements. It is also easy to see that there is a largest set K with the properties required for (ii). $\mathfrak{A} \mid K$ will be called the *common part of* $\mathbb{K}$; for further reference we shall designate by $\mathfrak{A}_v$ (the *variable part of* $\mathfrak{A}$) $\mathfrak{A} \mid (I \cup J) - K$; and, for any $B \subseteq \mathbb{K}$, $B_v = \{\mathfrak{A}_v : \mathfrak{A} \in B\}$. Condition (ii) expresses the fact that any definable element in the common part of $\mathbb{K}$ is designated by an individual constant.

It is to be remarked that the only condition essential for the rest of our work is that all systems in $\mathbb{K}$ have a common universe. The other requirements seem to be natural, but they are not indispensable. In the first place, it would not be natural to have as one of our possible choices the designation of an element of the universe, Hence, the individual constants should have an invariant interpretation in all the systems. The naturalness of the second condition will be clear from the definition of the group of transformations that preserve the "laws of the phenomenon".

This *group of functions* G_K, contains all permutations f of the common universe A that satisfy simultaneously the following conditions:

(1) For any $\mathfrak{A} = \langle A, R_i, a_j \rangle_{i \in I,\ j \in J} \in \mathbb{K}$ we have:

(a) $\langle A, R_i, a_j \rangle_{i \in I,\ j \in J} \equiv \langle A, R_i, f(a_j) \rangle_{i \in I,\ j \in J} \equiv \langle A, R_i, f^{-1}(a_j) \rangle_{i \in I,\ j \in J}$.

(b) If $\langle A, R_i, a_j \rangle_{i \in I,\ j \in J} \equiv \langle A, R_i, b_j \rangle_{i \in I,\ j \in J}$, where $b_j = a_{g(j)}$ for some permutation g of J , then $\langle A, R_i, f(a_j) \rangle_{i \in I,\ j \in J} \equiv \langle A, R_i, f(b_j) \rangle_{i \in I,\ j \in J}$.

(c) $f^{*}(\mathfrak{A}_V)$ and $f^{-1*}(\mathfrak{A}_V) \in \mathbb{K}_V$.

(2) If $B = Mod_K(\phi)$ for some sentence ϕ , $C_V = B_V^f$ and $D_V = B_V^{f^{-1}}$, then there are sentences ψ , θ such that $C = Mod_K(\psi)$ and $D = Mod_K(\theta)$.

I think that because of requirements (a), (b) above, condition (ii) in the definition of a simple probability structure is natural.

It is easy to prove that G_K is always a group of transformations.

We need to define a relation of *symmetry* between subsets of $\mathbb{K}$ of the form $Mod_{K,x}(\phi)$. We can do it for any subsets B , C of $\mathbb{K}$ by:

$B \underset{f}{\simeq} C$ if and only if $C_V = B_V^f$;

$B \simeq C$ if and only if there is an $f \in G_K$ such that $B \underset{f}{\simeq} C$.

Events are considered to be elements of $S^s / \vdash$ (or $S^s_{\omega_1\omega} / \vdash$). We now define when two events $\phi / \vdash$ and $\psi / \vdash$ are *equally likely in* $\mathbb{K}$:

$\phi / \vdash \ \underset{K}{\simeq} \ \psi / \vdash$ if and only if $Mod_K(\phi) \simeq Mod_K(\psi)$.

This is well-defined as if ϕ is equivalent to ψ , then $Mod_K(\phi) = Mod_K(\psi)$.

Let us now consider the fields of subsets of $\mathbb{K}$: $\mathbb{M}$, $\mathbb{M}_{\omega_1\omega}$, $\mathbb{M}'$, and $\mathbb{M}'_{\omega_1\omega}$; where $\mathbb{M}$ and $\mathbb{M}_{\omega_1\omega}$ have as universes the family of sets of the form $Mod_K(\phi)$ for ϕ in S^s or $S^s_{\omega_1\omega}$ respectively; and $\mathbb{M}'$, $\mathbb{M}'_{\omega_1\omega}$ have the family of sets of the form $Mod_{K,x}(\phi)$ for ϕ a formula in L^s or $L^s_{\omega_1\omega}$, and x an assignment. $\mathbb{M}_{\omega_1\omega}$ and $\mathbb{M}'_{\omega_1\omega}$ are countably complete.

Suppose we have a measure μ defined on $\mathbb{M}_{\omega_1\omega}$. Then we can define a probability measure P_K on $S^s_{\omega_1\omega} / \vdash$ by:

$$P_K(\phi/\vdash) = \mu(Mod_K(\phi)).$$

As there is no danger of confusion we may write:

$\phi \mathrel{\dashv_K} \psi$ for $\phi/\vdash \mathrel{\dashv_K} \psi/\vdash$, and $P_K(\phi)$ for $P_K(\phi/\vdash)$.

If we want a measure on $S^{s}/\vdash$, then we only need, because of Theorem 2.1, a finitely additive measure on $\mathbb{M}$.

If we have a measure defined on $\mathbb{M}'_{\omega_1\omega}$, we can extend the probability measure to a language that has a new individual constant t_a for every element a of the common universe A. We call this language $L^{s}_{\omega_1\omega}(T_K)$ and its set of sentences $S^{s}_{\omega_1\omega}(T_K)$. We extend P_K to $S^{s}_{\omega_1\omega}(T_K)/\vdash$. Let $\phi \in S^{s}_{\omega_1\omega}(T_K)$ and let $t_{a_0},\ldots,t_{a_{n-1}}$ be all the constants from T_K that appear in ϕ; let ψ be the formula obtained from ϕ replacing $t_{a_0},\ldots,t_{a_{n-1}}$ by new variables $v_0,\ldots,v_{n-1}$; and let x be an assignment such that $x(v_i) = a_i$ for all $i < n$. We define:

$$P_K(\phi/\vdash) = \mu(Mod_{K,x}(\psi)).$$

Similar considerations can be applied to $S^{s}/\vdash$ and $\mathbb{M}'$.

In order to have an adequate probability measure, it must be invariant under $\dashv_K$. That is:

If $\phi \mathrel{\dashv_K} \psi$, then $P_K(\phi) = P_K(\psi)$.

Thus, μ also has to be invariant under $\sim$, i.e. for B, C in one of our fields of sets, we must have:

If $B \sim C$, then $\mu(B) = \mu(C)$.

Our problem, then, is to find a measure μ in one of the fields $\mathbb{M}, \mathbb{M}', \mathbb{M}_{\omega_1\omega}$, or $\mathbb{M}'_{\omega_1\omega}$ that has this property. I shall use $\mathbb{F} = \langle F, \cup, \cap, -, \emptyset, \mathbb{K}\rangle$ for any of these fields.

I now reformulate these ideas in order to have a more perspicuous mathematical problem. We consider the field of subsets of $\mathbb{K}_V$, $\mathbb{F}_V = \langle F_V, \cup, \cap, -, \emptyset, \mathbb{K}_V\rangle$ given by,

$$F_V = \{B_V : B \in F\}.$$

$\mathbb{F}_V$ is obviously isomorphic to $\mathbb{F}$. Thus if we define a measure $\bar{\mu}$ on $\mathbb{F}_V$ we can obtain μ on $\mathbb{F}$ by,

$$\mu(B) = \bar{\mu}(B_V).$$

Let $G_K^* = \{f^*: f \in G_K\}$, where f^* applied on an element $\mathfrak{a}_V$ of $\mathbb{K}_V$ gives the isomorphic image of $\mathfrak{a}_V$ by f, $f^*(\mathfrak{a}_V)$. G_K^* is, then, a group of transformations of $\mathbb{F}_V$.

The group G_K^* determines an equivalence relation $\sim_{G_K^*}$ on F_V by:

$B \sim_{G_K^*} C$ if and only if $f^*(B) = C$ for some $f \in G_K^*$ (notice that if $f \in G_K^*$, then $f = g^*$ for some $g \in G_K$ and $f^*(B) = B^g$).

Also we have for B , $C \in F$, $B \sim C$ if and only if $B_V \sim_{G_K^*} C_V$.

Thus, what we need is a measure $\bar{\mu}$ on $\mathbb{F}_V$ such that:

if $B \sim_{G_K^*} C$, then $\bar{\mu}(B) = \bar{\mu}(C)$, for all B , $C \in F_V$.

$\bar{\mu}$ is what is called a measure on $\mathbb{F}_V$ *invariant under the group of transformations* G_K^* (see Tarski 1949, p. 229). We shall discuss in the next section necessary and sufficient conditions for the existence of such measures.

4. Existence of measures.

Our task now is to find a measure on the field of sets $\mathbb{F}_V = \langle F_V , \cup , \cap , - , \emptyset , \mathbb{K}_V \rangle$ invariant under the group of transformations G_K^* . Necessary and sufficient conditions for the existence of finitely additive measures invariant under a group of functions are given in Tarski 1949. p. 231. Hence I shall concentrate on (countably additive) measures.

To discuss this kind of measures it is convenient to have a countably complete field of sets. If our language is $L^s_{\omega_1\omega}$ (i.e. it contains countable conjunctions and disjunctions) $\mathbb{F}_V$ is countably complete. But if our language is L^s (finitary) we work with the smallest countably complete field of subsets of $\mathbb{K}_V$ generated by $\mathbb{F}_V$.

Thus, the mathematical problem, in all its generality, is to obtain a measure μ on a countably complete field of sets $\mathbb{F} = \langle F, \cup , \cap , - ,\emptyset , X \rangle$, invariant under a group of permutations G of the unit set X ; namely μ should satisfy:

(*) if $A , B \in F$ and $A = f^*B$ for some $f \in G$, then $\mu(A) = \mu(B)$.

A measure μ that satisfies (*) is called G-*invariant*. All results about these measures presented in this section are also valid for quasi-groups instead of groups (see Tarski 1949, p. 142).

In order to solve this problem of existence of measures, I shall prove some theorems that apply to arbitrary BA's. Thus, if $\mathbb{B}$ is a BA and R an equivalence relation on B we need a measure μ on $\mathbb{B}$ that is R-*invariant*, i.e.,

$$a, b \in B \quad \text{and} \quad aRb \quad \text{imply} \quad \mu(a) = \mu(b).$$

To fit our problem in this framework, we define the equivalence relation $\rightleftharpoons_G$ on $\mathbb{F}$ by:

$$A \rightleftharpoons_G B \text{ if and only if there is an } f \in G \text{ such that } f^*A = B.$$

By (*) our measure μ must be $\rightleftharpoons_G$-invariant. If μ is $\rightleftharpoons_G$-invariant, then it also has to be invariant under the following equivalence relation:

$A \simeq_G B$ if and only if there are $Y, Z \in {}^{\omega}F$ such that $A = \Sigma_{i<\omega} Y_i$, $B = \Sigma_{i<\omega} Z_i$, and $Y_i \rightleftharpoons_G Z_i$ for every $i < \omega$.

A measure μ is $\rightleftharpoons_G$-invariant if and only if it is $\simeq_G$-invariant.

This equivalence relation $\simeq_G$ has the following properties:

(i) $\simeq_G$ is *a congruence relation of* $\dot{\mathbb{F}} = \langle F, +, \Sigma\rangle$, namely,

(a) if $A, B, C, D \in F$ with $A \cap B = \emptyset = C \cap D$ and $A \simeq_G C$, $B \simeq_G D$, then $A + B \simeq_G C + D$,

(b) if $Y, Z \in {}^{\omega}F$ with $Y_i \cap Y_j = \emptyset = Z_i \cap Z_j$ and $Y_i \simeq_G Z_i$ for all $i \neq j \in \omega$, then $\Sigma_{i<\omega} Y_i \simeq_G \Sigma_{i<\omega} Z_i$.

(ii) $\simeq_G$ is (*finitely*) *refining*, namely, if $A, Y_0, Y_1, B \in F$, $A \simeq_G B$ and $A = Y_0 + Y_1$, then there are $Z_0, Z_1 \in F$ such that $B = Z_0 + Z_1$, $Y_0 \simeq_G Z_0$, and $Y_1 \simeq_G Z_1$.

(For a discussion and proof of these properties see Tarski 1949, pp. 144, 145, 222).

There are some elements of F that must have measure zero. Let us say that an element $A \in F$ is G-*negligible* if there is a sequence of disjoint elements of F, $Y \in {}^{\omega}F$, such that $A \simeq_G Y_i$ for every $i < \omega$. Now, μ has to be invariant also for the equivalence relation $\equiv_G$ defined by:

$A \equiv_G B$ if and only if there are G-negligible elements of F, C, D, C', D' and elements $A', B' \in F$ such that $A' \simeq_G B'$, $A \cup C = A' \cup C'$, and $B \cup D = B' \cup D'$. $\equiv_G$ is, again, a refining congruence relation of $\dot{\mathbb{F}}$.

So, finally, we get:

μ is G-invariant if and only if μ is $\equiv_G$-invariant.

As we require the measure μ to have property (3) i.e. $\mu(X) = 1$, then for an invariant measure to exist, X should not be G-negligible, that is there should not exist disjoint sets $A, B \in F$ such that $X = A \cup B$ and $X \cong_G A \cong_G B$ (it is easy to see that this condition is equivalent for X to the definition of negligible given previously).

The set of negligible elements I is a countably complete ideal in $\mathbb{F}$; thus, $\mathbb{F}/I$ is again a countably complete BA. From the relation $\cong_G$ it is possible to obtain a corresponding equivalence relation on $\mathbb{F}/I$ defined by:

$$A/I \simeq_G B/I \text{ if and only if } A \cong_G B.$$

This new relation $\simeq_G$ is now a refining, congruence relation on $(\mathbb{F}/I)$, which has the additional property (iii) if and only if X is not G-negligible:

(iii) $\simeq_G$ is *strictly positive*, i.e. if a is a nonzero element of F/I, then there is no sequence of disjoint elements $x \in {}^{\omega}(F/I)$ such that $a \simeq_G x_i$ for every $i < \omega$.

For a proof of these facts see Chuaqui 1977.

Thus, we are left with the general problem of the existence of a measure on BA $\mathbb{B}$ that is R-invariant under a strictly positive, refining, congruence relation R on $\dot{\mathbb{B}}$. In what follows we use the terms congruence relation, refining, and strictly positive for arbitrary equivalence relations R on BA's.

We shall first discuss this problem when $\mathbb{B}$ is a finite BA. Although the general solution given later includes this case, finite BA's will enable us to obtain some examples of measures and see their characteristics. For the finite case, conditions (i) and (iii) for the equivalence relation may be simplified:

(i') R is a congruence relation of $\langle B, +\rangle$, i.e. if $a, b, c, d \in B$, $a \wedge b = 0 = c \wedge d$, aRc, and bRd, then $a + bRc + d$.

(iii') R is strictly positive, i.e. $aR0$ if and only if $a = 0$.

When $\mathbb{B}$ is finite there always exists a strictly positive measure invariant under a strictly positive, refining, congruence relation on $\dot{\mathbb{B}}$. Moreover, the measure μ might have the additional property:

$$\text{if } \mu(a) = \mu(b) \text{ then } aRb.$$

Before proving this, we obtain the following:

LEMMA 4.1. *If $\mathbb{B}$ is a finite BA and R a strictly positive, refining congruence relation on $\dot{\mathbb{B}}$, then:*

aRb implies that there are $n < \omega$ and sequences of atoms of $\mathbb{B}$,

$x, y \in {}^nB$ such that $a = \Sigma_{i<n} x_i$, $b = \Sigma_{i<n} y_i$, and $x_i R y_i$ for every $i < n$ (i.e. a and b contain the same number of respectively equivalent atoms).

PROOF: Suppose $a = \Sigma_{i<n} x_i$ with the x_i's atoms, and aRb. By refinement, $b = \Sigma_{i<n} y_i$ with $x_i R y_i$ for every $i < n$. Using refinement and strict positivity it is easy to see that the y_i's are also atoms.

THEOREM 4.2. *Suppose $\mathbb{B}$ is a finite BA and R a refining congruence relation on $\dot{\mathbb{B}}$; suppose further that $a_0 / R, \ldots, a_{n-1} / R$ are the different equivalence classes of atoms that are not equal to $0 / R$, and m_i the number of atoms in a_i / R. Let $p_0, \ldots, p_{n-1}$ be any sequence of nonnegative real numbers such that $\Sigma_{i<n} m_i \cdot p_i = 1$. Then there is an R-invariant measure such that*

$$(*) \qquad \mu(a_i) = p_i \qquad \text{for all} \quad i < n$$

if and only if not OR1.

Moreover, if not OR1, *the measure satisfying* (*) *is unique.*

PROOF: Theorem 4.2 is obtained immediately from Lemma 4.1 and properties of BA's.

THEOREM 4.3. *Suppose $\mathbb{B}$ is a finite BA and R a refining congruence relation on $\dot{\mathbb{B}}$. Then a necessary and sufficient condition for the existence of a strictly positive measure on $\mathbb{B}$ such that*

$$(+) \qquad \mu(a) = \mu(b) \qquad \text{if and only if} \qquad aRb$$

is that R be strictly positive.

PROOF: Suppose R is strictly positive and let $a_0 / R, \ldots, a_{n-1} / R$ be the different equivalence classes of atoms. Suppose, further, that a_i/R contains m_i atoms for every $i < n$. Define μ' recursively on the atoms of $\mathbb{B}$ by:

$$\text{if} \quad b \in a_0 / R, \quad \text{then} \quad \mu'(b) = 1,$$

$$\text{if} \quad b \in a_{p+1} / R, \quad \text{then} \quad \mu'(b) = \Sigma_{i<p} m_i \cdot \mu'(a_i).$$

Obtain μ on the atoms by,

$$\mu(b) = \mu'(b) / (\Sigma_{i<n} m_i \cdot \mu'(a_i)).$$

Then extend μ to B by Theorem 4.2. It is not difficult to prove, using Lemma 4.1 and simple properties of natural numbers, that μ has the desired properties.

The converse implication is obvious.

In a sense, the measure μ defined above is quite unnatural as a probability measure. If a and b are atoms that are not equivalent, then $\mu(a)$ may be very different from $\mu(b)$ and we may choose $\mu(a) > \mu(b)$ or $\mu(b) > \mu(a)$. To avoid this problem we need to impose a stronger requirement than invariance on the measure μ:

We say that a measure μ is *strongly R - invariant* if for any $a, b \in B$ we have,

$\mu(a) \leq \mu(b)$ if and only if there is a $c \in B$ such that aRc and $c \leq b$.

Here, the second $\leq$ is the usual partial ordering in $\mathbb{B}$. We write '$a \leq_R b$' for 'there is a $c \in B$ such that aRc and $c \leq b$'. We have the following:

THEOREM 4.4. *Let $\mathbb{B}$ be a finite BA and R a strictly positive, refining, congruence relation on $\dot{\mathbb{B}}$. Then there is a strongly R - invariant measure on $\mathbb{B}$ if and only if for any atoms $a, b \in B$, we have aRb. Moreover, if this last condition is fulfilled the strongly R - invariant measure is unique.*

PROOF: By Lemma 4.1 it is clear that the measure defined in the proof of 4.2 is strongly invariant if all atoms are equivalent (it is enough, in this case, to define $\mu(a) = 1/m$, where m is the number of atoms, for any atom a).

Suppose, now, that μ is strongly R - invariant, and a, b are atoms with $\mu(a) \leq \mu(b)$. Then, $aRc \leq b$ for some c. But c has to be an atom and, so, necessity follows.

It is easy to see that this measure is R - invariant even if not all atoms are equivalent, though in this case it is neither strongly R - invariant nor satisfies (+) of 4.3. Notice, also, that the condition of equivalence for all atoms could be replaced by '$\leq_R$ is a simple ordering', i.e. '$a \leq_R b$ or $b \leq_R a$ for all $a, b \in B$'.

Let us now consider the general case. $\mathbb{F}$ is a countably complete field of subsets of a set X, G a group of transformations of $\mathbb{F}$, and I a subset of F. We want a measure μ that is G - invariant and that vanishes exactly on the elements of I. We call measures that satisfy this last requirement *I - positive* (i.e. μ is I - positive if, $\mu(A) = 0$ if and only if $A \in I$). In order to formulate necessary and sufficient conditions for the existence of such measures we shall use the requirements obtained in Kelley 1959, for the existence of a strictly positive measure on a Boolean Algebra. We need some definitions, which apply to any countably complete BA $\mathbb{B}$:

Let $x = \langle x_i : i < n \rangle$ be a finite sequence in B. We define $i(x) = m / n$ where m is the largest integer $k \leq n$ such that $x_{i_0} \wedge \ldots \wedge x_{i_{k-1}} \neq 0$ for $i_0 < i_1 < \ldots < i_{k-1} < n$. Then, if $A \subseteq B$ we define the *intersection number of* A, in symbols $i(A)$ by,

$$i(A) = \inf \{ i(a) \ : a \in {}^nA \ \text{ for some } \ n < \omega \} .$$

We say that $\mathbb{B}$ has the *Kelley property* if $B - \{0\}$ is a countable union of sets with positive intersection number. $\mathbb{B}$ is *weakly countably distributive* if for every double sequence $x \in {}^{\omega\times\omega}B$ such that $x_{i,j+1} \leq x_{ij}$ for every $i, j < \omega$, we have

$$\vee\{ \wedge \{x_{ij} : j \in \omega\} : i \in \omega\} = \wedge\{\vee\{x_{i,\phi(i)} : i \in \omega\} : \phi \in {}^{\omega}\omega\}.$$

We have the following:

THEOREM 4.5. *Let* $\mathbb{F}$ *be a countably complete field of subsets of a set* X, G *a group of transformations of* $\mathbb{F}$, *and* I *a subset of* F. *Then the following conditions are necessary and jointly sufficient for the existence of a countably additive,* G*-invariant, and* I*-positive measure on* $\mathbb{F}$:

(i) I *is a countably complete proper ideal in* $\mathbb{F}$;

(ii) *all* G*-negligible sets belong to* I;

(iii) *if* $A \in I$ *and* $A \precsim_G B$, *then* $B \in I$;

(iv) $\mathbb{F} / I$ *has the Kelley property and is weakly countably distributive.*

A complete proof of this theorem is too long to be included here and will appear elsewhere (Chuaqui 1977). I shall give only a brief sketch. The necessity of the conditions is clear, given Kelley's necessary conditions for the existence of a strictly positive measure on a BA (see Kelley 1959, and Horn and Tarski 1948).

In order to construct the measure μ with the required properties we proceed as follows:

We start with the disjunctive BA $\dot{\mathbb{B}}$, where $\mathbb{B} = \mathbb{F} / I$. $\dot{\mathbb{B}}$ is a GCA. Using the refining congruence relation R on $\dot{\mathbb{B}}$ defined by:

aRc if and only if there are $A, C \in F$ such that $a = A / I$, $c = C / I$, and $A \eqsim_G C$;

we obtain a new GCA $\mathfrak{a} = \dot{\mathbb{B}} / R$ (see Tarski 1949, Th. 6.10). We close $\mathfrak{a}$

to $\overline{\mathfrak{A}}$ (Tarski 1949, Ch. 7). Using the properties of $\mathbb{B}$ assumed by the theorem, it is possible to prove the following for $\overline{\mathfrak{A}} = \langle A, +, \Sigma \rangle$:

(1) if $a, b \in A$, then $a \wedge b$ exists,

(2) if $a \in A$, then $a \leq \infty h$, where $h = (X/I)/R$ is finite.

By a theorem of Fillmore 1965, it is possible to prove that $\overline{\mathfrak{A}}$ is isomorphic to a subalgebra of $\langle C(S(\mathbb{D}), \overline{R}), +, \Sigma \rangle$ where $\mathbb{D}$ is the BA of idemmultiple elements of $\overline{\mathfrak{A}}$. From condition (iv) on $\mathbb{B}$, we deduce that $\mathbb{D}$ also has the properties included in that condition. Hence, by Kelley's 1959 there is a strictly positive measure on $\mathbb{D}$. We transfer the measure to the open-closed sets of the Stone space $S(\mathbb{D})$, extend the measure to the countably complete field of sets generated by them, and define the integral Π on the measurable functions. Π is defined on all of $C(S(\mathbb{D}), \overline{R})$. Transfer back, now, Π through the isomorphism to $\overline{\mathfrak{A}}$ and then, using 16.11 of Tarski 1949, to $\mathbb{F}$. This measure on $\mathbb{F}$ has the desired properties.

The notion of a strictly positive measure should be modified when we want the measure G-invariant. A measure that is G-invariant must vanish on all G-negligible sets. By a result of Bradford 1971, p. 194, we can see that most group of transformations of interest produce non-empty G-negligible sets. This is the case, for instance, when for every $n \in \omega$ there are disjoint $Y_0, \ldots, Y_{n-1}$ such that $X = \Sigma_{i<n} Y_i$ and $Y_i \equiv_G Y_j$ for every $i, j < n$. To take into account this fact we introduce the following notion:

We say that a measure μ on $\mathbb{F}$ is G-strictly positive if for any $A \in F$, $\mu(A) = 0$ implies that A is G-negligible.

Let N_G be the set of G-negligible elements. In Chuaqui 1977, it is proved that N_G is a countably complete ideal in $\mathbb{F}$. Thus, from 4.5. we get:

THEOREM 4.6. *Let* $\mathbb{F}$ *be a countable complete field of subsets of a set* X, *and* G *a group of transformations of* $\mathbb{F}$. *Then the following conditions are necessary and jointly sufficient for the existence of a countably additive,* G*-invariant, and* G*-strictly positive measure on* $\mathbb{F}$:

(i) $X \notin N_G$

(ii) $\mathbb{F} / N_G$ *has the Kelley property and is weakly countably distributive.*

It should be pointed out that (i) is not enough for the existence of a G-invariant measure. For a couterexample see Chuaqui 1973.

Similarly as for finite BA's, the measures obtained by these theorems are not, in general, unique. We need the condition of strong invariance. A measure μ on $\mathbb{F}$ is strongly G-invariant if for every $A, B \in F$ with

$A, B \notin N_G$, we have:

$\mu(A) \leq \mu(B)$ if and only if there is a $C \in F$ such that

$$A \equiv_G C \subseteq B .$$

The following theorem can be obtained from 2.10 of Chuaqui 1969:

THEOREM 4.7. *Let* $\mathbb{F}$ *be a countably complete field of subsets of a set* X, *and* G *a group of transformations of* $\mathbb{F}$. *Then, for the existence of a countably additive, strongly* G*-invariant, and* G*-strictly positive measure* μ *on* $\mathbb{F}$, *the following conditions are necessary and sufficient:*

(i) $X \notin N_G$,

(ii) *for any* $A, B \in F$, *there is a* $C \in F$, *such that* $A \equiv_G C \subseteq B$ *or* $B \equiv_G C \subseteq A$.

Moreover, under these conditions the measure μ *is unique.* Condition (ii) in 4.7. can be replaced by:

(ii') $\mathbb{F} / N_G$ *satisfies the countable chain condition (i.e. every set of disjoint elements of* F / N_G *is at most countable), and for any* $a \in F / N_G$, $a \neq 0$, *there is a sequence of disjoint elements* $x \in {}^{\omega}F / N_G$ *and a sequence* $y \in {}^{\omega}F / N_G$ *such that* $X / N_G = \Sigma_{i<\omega} x_i$ *and* $x_i \equiv_G y_i \leq a$ *for every* $i < \omega$.

There are several other alternatives to (ii'), that are obtained, just as is (ii'), from properties of simple cardinal algebras (see Tarski 1949, p. 117-120 and Chuaqui 197+).

The main problem that needs further study is the relation between the logical theory and the measure. I believe that the methods developed in Scott and Krauss 1966, might be useful in this connection, because once we have a measure μ on $\mathbb{F}$ we can put some of our simple probability structures in the framework of their probability systems with strict identity.

Unfortunately, the condition on their probability systems that requires the measure on the BA to be strictly positive, excludes some of the most natural simple probability structures from their framework. For instance, let us take the simple probability structure $\mathbb{K}$ formed by relational systems $\mathfrak{A}_0 = \langle C, +, a, 0\rangle_{a \in C}$, where $\langle C, +\rangle$ is a circle in the plane and 0 chooses one element of C. $\langle C, +, a\rangle_{a \in C}$ is the constant part of $\mathbb{K}$. G_K contains the translations of the circle. A natural probability measure P_K is obtained from the usual Lebesgue measure. It is easy to see that this measure does not satisfy Gaifman Condition, namely:

$$P_K(\exists v\ \overline{O}(v)) = 1 \quad \text{and} \quad sup_{F \in C^{(\omega)}}\ P_K(\bigvee_{a \in F}\ \overline{O}(t_a)) = 0,$$

where $C^{(\omega)}$ is the set of all finite subsets of C, $\overline{O}$ is the symbol representing O, and t_a the individual constant that denotes a.

5. Compound probability structures.

In this section I shall formalize the structures that appear when there are sequences of outcomes, as in the urn example of section one. The order of succesion of outcomes plays a fundamental role in these cases. In the urn model, there were only two succesive choices, so that the basic order was the system $<\{0, 1\}, \leq>$. In the general case, we may have any arbitrary partially ordered system $\mathbb{T} = <T, \leq>$. For this system, let $\overline{t} = \{s : s < t\}$ and $P = \{\overline{t} : t \in T\}$, where $s < t$ means $s \leq t$ and $s \neq t$. The possible succesive outcomes are determined by a function $\mathbb{K}$ with the following properties:

(i) $\mathcal{D}o\ \mathbb{K} = \{f : f$ is a function, $\mathcal{D}o f \in P$, and for each $t \in \mathcal{D}o f$, $f(t) \in \mathbb{K}_{f \restriction \overline{t}}\}$

(ii) For each $f \in \mathcal{D}o\ \mathbb{K}$, $\mathbb{K}_f$ is a simple probability structure with similarity type $s_f = <<m_i^f : i \in I_f>, J_f>$, and distinguished elements $<a_j^f : j \in J_f>$.

(iii) If $i \in I_f \cap I_g$ for $f, g \in \mathcal{D}o\ \mathbb{K}$, then $m_i^f = m_i^g$.

(iv) If $j \in J_f \cap J_g$ for $f, g \in \mathcal{D}o\ \mathbb{K}$, then $a_j^f = a_j^g$.

If $f \in \mathcal{D}o\ \mathbb{K}$, $\mathbb{K}_f$ is the set of possible outcomes provided that the succesive choices given by f are realized. The choice at t of an $f \in \mathcal{D}o\ \mathbb{K}$ should be in $\mathbb{K}_{f \restriction \overline{t}}$, the set of possible outcomes determined by the preceding choices. Requirements (iii) and (iv) are conditions of coherence: the same index should represent the same symbol, and thus the corresponding relation should have the same arity; a particular individual constant should always represent the same individual.

A compound probability structure is, then, a pair $<\mathbb{T}, \mathbb{K}>$ with the properties specified above. The set of possible (compound) outcomes is the set $\mathbb{H}_{<T, K>}$, defined by:

$$\mathbb{H}_{<T,K>} = \{F : F \in {}^{T} \cup \mathcal{D}o\mathbb{K}^{-1} \text{ and for each } t \in T,\ F(t) \in \mathbb{K}_{F \restriction \overline{t}}\}.$$

Intuitively, $\mathbb{T}$ represents the order of succesive outcomes (in general, it will be time) and each function $F \in \mathbb{H}_{<T,K>}$, a sequence of outcomes, such that each outcome depends on all preceding ones.

For each $F \in \mathbb{H}_{<T,K>}$, $F(t)$ is a relational system in $\mathbb{K}_{F|\overline{t}}$. The universe of $F(t)$ is denoted $|F(t)|$. The universe of F, in symbols $|F|$, is the union of $|F(t)|$ for $t \in T$. The universe of $<\mathbb{T}, \mathbb{K}>$, in symbols $|\mathbb{K}|$, is the union of $|F|$ for $F \in \mathbb{H}_{<T,K>}$.

The languages we have to consider for these compound probability structures must be more complicated than first-order languages. We must be able to talk about finite sequences of outcomes. In order to do this, we need finite sequences of elements of T. Thus, our language has four types of *variables* given by the denumerable sets Vl_1, Vl_2, Vl_3, and Vl_4. $Vl_1 = \{v_0, v_1, \ldots\}$ contains the variables used to refer to objects in $|\mathbb{K}|$; $Vl_2 = \{t_0, t_1, \ldots\}$ refer to elements of T; $Vl_3 = \{n_0, n_1, \ldots\}$ refer to natural numbers; and $Vl_4 = \{x_0, x_1, \ldots\}$ refer to finite sequences of elements of T (i.e. elements of ${}^{\omega}T$). The *logical constants* include (besides $\neg, \wedge, \vee, =, \forall$) the ternary relations S and P, which stand for addition and multiplication between natural numbers; the individual constants $\overline{0}$ and $\overline{1}$, which stand for the numbers 0 and 1; the binary relation L, which stands for $\leq$ between elements of T; and the ternary relation Val, such that $Val(x, n, t)$ is interpreted as: the natural number n is in the domain of the sequence x and x evaluated at n is t.

The *similarity type* for a language corresponding to $<\mathbb{T}, \mathbb{K}>$ is $s = <<m_i : i \in I>, J>$, where $I = \cup \{I_f : f \in Do\mathbb{K}\}$, $m_i = m_i^f$ for $i \in I_f$ and $J = \cup \{J_f : f \in Do\,\mathbb{K}\}$.

For each $i \in I$, we have an $m_i + 1$-ary predicate P_i, and for each $j \in J$, an individual constant c_j. The set of *individual constants* is denoted CN, $Vl_1 \cup CN$ is denoted IS_1, and $Vl_3 \cup \{\overline{0}, \overline{1}\}$ is denoted IS_3.

The notion of *atomic formula* is defined by:

(a) If $\alpha, \beta \in IS_1 \cup Vl_2 \cup IS_3 \cup Vl_4$, then $\alpha = \beta$ is an atomic formula.

(b) If $t \in Vl_2$ and $\alpha_0, \ldots, \alpha_{m_i - 1} \in IS_1$, then $P_i t \alpha_0 \ldots \alpha_{m_i - 1}$ is an atomic formula.

(c) If $n, m, p \in IS_3$, then $Snmp$ and $Pnmp$ are atomic formulas.

(d) If $x \in Vl_4$, $n \in IS_3$, and $t \in Vl_2$, then $Val(xnt)$ is an atomic formula.

(e) If $t_1, t_2 \in Vl_2$, then Lt_1t_2 is an atomic formula.

For the connectives we define, as usual, $\neg\phi$, $(\phi \wedge \psi)$, and $(\phi \vee \psi)$ as formulas if ϕ and ψ are formulas. For simplicity we consider only finitary languages. We allow *quantification* over any type of variable with the following restriction:

For $v \in V\ell_2$ and ϕ a formula, $\forall v \phi$ is a formula if and only if ϕ only contains individual symbols in IS_1 and at most one variable in $V\ell_2$ only appearing free in ϕ. For the other types of variables there are no restrictions.

We define *satisfaction of formulas in members of* $\mathbb{H}_{<T,K>}$. For each $F \in \mathbb{H}_{<T,K>}$, *an assignment in* F is a function g with domain $V\ell_1 \cup V\ell_2 \cup V\ell_3 \cup V\ell_4$ such that: if $v \in V\ell_1$, $g(v) \in |F|$; if $t \in V\ell_2$, $g(t) \in T$; if $n \in V\ell_3$, $g(n) \in \omega$; and if $x \in V\ell_4$, $g(x) \in {}^{\omega}T$. If g is an assignment in F and α a variable, $g(^{\alpha}_{\tau})$ stands for the assignment that coincides with g in every variable except, possibly, in α where it assigns τ. If α is an individual constant or a variable, we have: $\alpha^F_g = g(\alpha)$, if α is a variable; $\alpha^F_g = a_j$, if α is c_j; $\alpha^F_g = 0$, if α is $\overline{0}$; $\alpha^F_g = 1$, if α is $\overline{1}$.

If $t \in T$, and P_i. is a predicate, we define:

$$P_i^{F(t)} = R_i, \text{ where } R_i \text{ is the } i^{th} \text{ relation of } F(t), \text{ if } i \in I_{F|\overline{t}} ;$$

$$P_i^{F(t)} = \emptyset , \text{ otherwise.}$$

Now we can define by recursion g *satisfies* ϕ *in* F, for any formula ϕ, assignment g, and $F \in \mathbb{H}_{<T,K>}$; in symbols this will be $F \models \phi[g]$.

1) If ϕ is $\alpha = \beta$, then $F \models \phi[g]$ if and only if $\alpha^F_g = \beta^F_g$.

2) If ϕ is $P_i t \alpha_0 \ldots \alpha_{m_i - 1}$, then $F \models \phi[g]$ if and only if $P_i^{F(g(t))} \alpha^F_{0,g} \ldots \alpha^F_{m_i - 1, g}$.

3) If ϕ is $Smnp$ (or $Pmnp$), then $F \models \phi[g]$ if and only if $m^F_g + n^F_g = p^F_g$ (or $m^F_g \cdot n^F_g = p^F_g$).

4) If ϕ is $Val(xnt)$, then $F \models \phi[g]$ if and only if n^F_g is in the domain of x^F_g and x^F_g evaluated at n^F_g is t^F_g.

5) If ϕ is Lt_1t_2, then $F \models \phi[g]$ if and only if $t^F_{1_g} \leq t^F_{2_g}$.

6) If ϕ is $\neg\psi$, then $F \models \phi[g]$ if and only if not $F \models \psi[g]$.

7) If ϕ is $(\psi \wedge \theta)$, then $F \models \phi[g]$ if and only if $F \models \psi[g]$ and $F \models \theta[g]$.

8) If ϕ is $(\psi \vee \theta)$, then $F \models \phi[g]$ if and only if $F \models \psi[g]$ or $F \models \theta[g]$.

9) If ϕ is $\forall \upsilon \psi$ with $\upsilon \in V\ell_1$, then there are two cases:

9.1) There is a variable $t \in V\ell_2$ free in ψ; in this case we have, $F \models \phi[g]$ if and only if for every $\tau \in |F(g(t))|$, $F \models \psi[g(^{\upsilon}_{\tau})]$.

9.2) Otherwise we define, $F \models \phi[g]$ if and only if for every $\tau \in |F|$, $F \models \psi[g(^{\upsilon}_{\tau})]$.

10) If ϕ is $\forall \alpha \psi$ with $\alpha \in V\ell_2$ ($V\ell_3$ or $V\ell_4$), then $F \models \phi[g]$ if and only if for every $\tau \in T$ (ω or ${}^{\omega}T$) $F \models \psi[g(^{\alpha}_{\tau})]$.

For an assignment g and a formula ϕ, $\langle \mathbb{T}, \mathbb{K}\rangle \models \phi[g]$ if and only if for every $F \in \mathbb{H}_{\langle T,K\rangle}$, $F \models \phi[g]$.

In order to define $P_{\langle T,K\rangle}(\phi)$, the probability of a sentence ϕ (or, more, accurately, of the equivalence class of sentences logically equivalent to ϕ) in $\langle \mathbb{T}, \mathbb{K}\rangle$, we have to define a probability measure on the subsets of $\mathbb{H}_{\langle T,K\rangle}$ of the form:

$$Mod_{\langle T,K\rangle}(\phi) = \{F : F \in \mathbb{H}_{\langle T,K\rangle} \text{ and } F \models \phi\}.$$

With this measure, we define $P_{\langle T,K\rangle}(\phi)$ just as in section three. The problem, again, is to define a reasonable probability measure on these subsets of $\mathbb{H}_{\langle T,K\rangle}$.

For each $\delta \in \mathcal{D}o\, \mathbb{K}$, $\mathbb{K}_\delta$ is a simple probability structure and then, if possible, we can proceed as in section three to get a measure on an appropriate field of its subsets. This field of subsets of $\mathbb{K}_\delta$ is obtained as follows: We take all formulas that have just variables in $V\ell_1$, one variable α in $V\ell_2$, and symbols in the similarity type of $\mathbb{K}_\delta$. Omit the variable α and we get a sentence ϕ of a first-order language. Then the field of subsets of the form $Mod_{K_\delta}(\phi)$ for this type of sentence is the one needed. Let us call this field $\mathbb{M}_\delta$ and the corresponding measure μ_δ.

I do not have a general solution for passing from these measures μ_δ to a measure on the subsets of $\mathbb{H}_{\langle T,K\rangle}$. It is easy to do it, however, in two important cases:

CASE I. The order type of $\mathbb{T}$ is finite or ω, and for each $\delta \in \mathcal{D}o\, \mathbb{K}$, $\mathbb{K}_\delta$ is finite.

Consider $\mathcal{B}$ the family of subsets of $\mathbb{H}_{<T,K>}$ of the form $A=\{F : F \in \mathbb{H}_{<T,K>}$ and $F \mid \mathcal{D}o j = j\}$ for some $j \in \mathcal{D}o\, \mathbb{K}$. The field of subsets of $\mathbb{H}_{<T,K>}$ we are interested in is generated by $\mathcal{B}$, and if we define a measure on $\mathcal{B}$ it can be extended to this field. We define $\mu(A)$ as the product of the numbers $\mu_{j|\bar{t}}(\{j(t)\})$ for all $t \in \mathcal{D}o j$. Proceeding as in this case, we can define a probability measure for the urn model discussed in section one.

CASE II. $\mathbb{T}$ is arbitrary, but simply ordered, and for every $f, g \in \mathcal{D}o\, \mathbb{K}$, $\mathbb{K}_f = \mathbb{K}_g$. Let us call $\mathbb{K}_f$, P. Then $\mathbb{H}_{<T,K>}$ is the direct power of P, T times, i.e. $\mathbb{H}_{<T,K>} = {}^{T}P = \{F : \mathcal{D}oF = T$ and $\mathcal{D}oF^{-1} \subseteq P\}$. The desired measure is the product measure of the μ_f's. By the usual considerations in Probability Theory, we can obtain Bernouilli's law of large numbers.

6. EXAMPLES AND APPLICATIONS.

We shall first study a few probability models and analyze later how they may be applied. We have already seen the model for choosing a sample from a finite population. For further reference, I shall call this example, Example 1.

EXAMPLE 2. Distribution of r balls into n cells. There are several ways of setting up the simple probability structures depending on the statistics to be obtained. Which of these applies is determined by the evidence.

2a) Maxwell - Boltzmann statistics. In this case each partition of the (identifiable) balls into the r cells is equally likely. The set $\mathbb{K}$ can be considered as consisting of all structures $\mathfrak{A}_0 = \langle A, P_0, \ldots, P_{n-1}, O \rangle$ where A contains all pairs of numbers $\langle i, m \rangle$ with $i < r$ and $m < n$ (i.e. $A = r \times n$, if we adopt the usual set-theoretical conventions); P_m consists of all parts with m as second coordinate, and O is any subset of A that is a function with domain the set of all numbers less than r.

Each elements of A, $\langle i, m \rangle$ represents the fact that the ball i is in cell m. $P_0, \ldots P_{n-1}$ are the n cells and O is the particular partition chosen. It has to be a function, because each ball can be in only one cell.

The group G_K consists of all permutations f of A such that $f{*}O$ is a function for every function O with domain r. Thus, in this case G_K does not contain all permutations of A.

2b) Böse - Einstein statistics. In this case the balls are not identifiable.

The systems $\mathfrak{A}_0$ in $\mathbb{K}$, for this statistics, are of the same similarity type as before, i.e. $\mathfrak{A}_0 = \langle A, P_0, \ldots P_{n-1}, O \rangle$. The common part $\langle A, P_0, \ldots P_{n-1} \rangle$ is also the same, but O is a subset of A, such that if $\langle i,m \rangle \in O$, then every $\langle j,m \rangle$ for $j < i$ is also in O (i.e. for each $m < n, O^{-1*}\{m\} = i$ for some $i < r$).

G_K is not the group of all permutations of A, but contains only those that preserve this property of O.

2c) Fermi - Dirac statistics. In this case, partitions may have at most one ball in each cell. The systems here are similar to those above, with the condition that O should be a one - one function.

EXAMPLE 3. We shall analyze now a more complicated example. We shall not be able to give a detailed analysis because of the complicated physics involved, but, I hope, the discussion will be sufficient for understanding how to proceed. Suppose we have a circular roulette with infinitely many points. For simplicity, the roulette starts from a fixed position and a variable force with constant direction is applied. Each outcome results from the application of a particular force.

The systems in $\mathbb{K}$ may be taken to be of the form

$$\mathfrak{A}_I = \langle C \cup F, C, F, +_C, +_F, f, a, I \rangle_{a \in C}$$

where C is the set of points in the circle, F is the set of forces, $+_C$ represents translations in the circle, $+_F$, addition of forces, f is the continuous function that associates each initial force with a final position, and I is the set containing the initial force (I contains one elements of F).

In this example, if $g \in G_K$ and the distinguished elements are transformed by g to $g(a)(a \in C)$, then these new elements should satisfy the same sentences as the old ones (i.e. G_K satisfies conditions 1(a) and 1(b) in section 3). These transformations in G_K are isometries of the circle. Their effect on the semigroup of forces is more complicated. In particular, the functions g in G_K do not distinguish elements x, y of F such that $f(x) = f(y)$. Thus, g treats F as a circle by dealing with these two points alike. As f is continuous, g is also an isometry of this new "circle". Hence, if one takes two intervals A, B of the same length of this "circle" of forces, the class of structures that has I in A is equally likely to the class that has I in B. Which final positions are equally likely depends on the function f. If the roulette is balanced, we should have an f that sends equal intervals of forces into equal intervals of points, and then, equal intervals in the circle of points are equiprobable. If this is not the case, we might have point - intervals of different length equally likely.

When we add a variable initial position or a variable direction of force, the analysis is similar, though more complicated.

A similar method can be applied to the case of the throw of a coin or a die, because the outcome depends on the final position of the object. However, the situation in this case is even more complicated so we will not attempt a description. It is useful to notice, that in the case of a true coin or die, the models can be simplified thus: $\langle A, I \rangle$ where A is the set of faces and I contains the one that obtains.

We pass now to examples of compound structures. We have already discussed an instance of the urn model. I shall call it Example 4. All other cases of urn models can be represented by similar probability structures. For these examples, we define a probability measure by Case I of section 5.

EXAMPLE 5. An example of much interest is when we have the same experiment repeated several times. To be specific, let us take the tossing of a true coin n times. We take $\mathbb{T} = \langle n, \leq \rangle$. $\mathbb{K}_0$ will be the set of just two models $\mathfrak{A}_1 = \langle A, C_1 \rangle$ and $\mathfrak{A}_2 = \langle A, C_2 \rangle$ where $A = \{h, t\}$, $C_1 = \{h\}$ and $C_2 = \{t\}$ (h = heads, t = tails). For every $f \in Do\, \mathbb{K}$, $\mathbb{K}_f = \mathbb{K}_0$.

To define a probability measure for this example, we use the method discussed for Case II in section 5.

I shall now discuss briefly how these probability models can be applied. My views here are very much influenced by Lucas' views (see Lucas 1970) especially in Chapter 5. When applying Probability to particular cases, we have to distinguish two factors: the framework and the evidence. In my way of setting up things the framework is given by the particular probability structures chosen, and the evidence is what leads us to these structures.

We shall see first the case analyzed by Jeffreys 1961, p. 15., discussed also by Lucas 1970, p. 50.:

"Suppose that I know that Smith is an Englishman, but otherwise know nothing particular about him. He is very likely, on that evidence, to have a blue right eye. But suppose that I am informed that his left eye is brown - the probability is changed completely".

I agree with Lucas that this is not a probability of a singular proposition whose subjects is Smith. Instead, it is the probability of taking a sample of one element of a population. In the first case, the set $\mathbb{K}_E$ of possible outcomes consists of structures $\langle E,B,S \rangle$ where E is the set of Englishmen, B The blue - eyed Englishmen, and S a set of one element. In the second case, i.e. when we know that Smith has a brown left eye, the set $\mathbb{K}_B$ of possible outcomes consists of structures $\langle Br,B,S\rangle$ where Br is the class of Englishmen with a brown left eye. Let ϕ be the sentence "there is an x in S and x is in B". In the first case the probability of Smith having a blue eye is $P_{K_E}(\phi)$, while in the second it is $P_{K_B}(\phi)$, and these two probabilities are quite different.

The case of the throw of a coin can be analyzed as follows. From previous experiences, physical laws, etc.... we suppose at first that the coin used is a true coin with equal probabilities, as explained in Example 3. Further evidence might lead us to the conclusion that this is not the right model and we change it. Among the important items of further evidence is that provided by successive trials at throwing the coin. Here,using compound probability models, as those in Example 5, we may use Bernoulli's theorem and estimate the probability of a certain sequence. If a sequence obtains that would be very improbable according to our initial assumptions,we reject this initial model and look for another. My procedure agrees, in general terms, with that appearing in Lucas 1970 Chapter V which the reader may consult for further details.

The only statistical methods that I have not yet thought out carefully are some types of Bayesian statistics. Prima facie, they seem not to be justified, because they involve probabilities of probability hypotheses together with probabilities of events and this, in my system, would involve a mixture of object - language and metalanguage.However there might be methods of dealing with this mixture. A possibility is given by procedures similar to those appearing in Scott and Krauss 1966, section 6.

I believe that the methods I have presented in this paper provide an adequate definition of a probability measure taking into account the main characteristics of probability statements.

References.

Bradford, R.

1971, *Cardinal addition and the axiom of choice*, Ann. Math. Logic, vol.3, 111 - 196.

Carnap, R.

1950, Logical foundations of probability, Chicago University Press.

Carnap, R. and Jeffreys, R.

1971, Studies in inductive logic I, University of California Press, Berkeley and Los Angeles.

Chuaqui, R.

1965, A definition of probability based on equal likelihood, Ph. D. Dissertation, U. of California, Berkeley.

1969, *Cardinal algebras and measures invariant under equivalence relations*, Trans. Amer. Math. Soc., Vol. 142, 61 - 79.

1973, *The existence of an invariant measure and paradoxical decompositions*, Notices Amer. Math. Soc., Vol. 20, A - 636, Abstract 73T - B313.

1975, *A model - theoretical definition of probability*, Contributed papers, 5th Internacional Congress of Logic, Methodology and Philosophy of Science London, Ontario, Canada, VI 7-8.

1977, *Measures invariant under a group of transformations*, To appear in Pac. J. of Math.

197+, *Simple cardinal algebras and their applications to invariant measures*, To appear.

Fillmore, P. A.

1965, *The dimension theory of certain cardinal algebras*, Trans. Amer. Math. Soc., Vol. 117, 21 - 36.

Henkin, L., D. Monk and A. Tarski

1971, Cylindric algebras, Studies in Logic, North - Holland Pub. Co. Amsterdam.

Horn, A. and A. Tarski

1948, *Measures in Boolean algebras*, Trans. Amer. Math. Soc., Vol. 64, 467-497.

Jeffreys, H.

1961, Theory of Probability, 3rd. ed., Oxford University Press, Oxford.

Keisler, H. J.

1971, Model theory for infinitary logic, Studies in Logic., North-Holland Publishing Co. Amsterdam.

Kelley, J. L.

1959, *Measures on Boolean algebras* Pac. J. of Math., Vol. 9, 1165 - 1178.

Lucas, J. R.

1970, The concept of probability, Oxford University Press. Oxford.

Scott, D. and P. Krauss

1966, *Assigning probabilities to logical formulas*, Aspects of inductive logic, J. Hintikka and P. Suppes editors., Studies in Logic., North - Holland Pub. Co. Amsterdam.

Sikorski, R.

1969, Boolean algebras , 3rd. ed. Springer - Verlag, Berlin - Heidelberg.

Tarski, A.

1935, *Der Wahrheitsbegriff in den formalisierten Sprachen*, Studia Philos. (Warsaw), Vol. 1, 261 - 405 (English transl. in Logic, Semantics and Metamathematics, Oxford U. Press, 152-278).

1949, Cardinal Algebras , Oxford University Press, New York.

1954, *Contributions to the theory of models I,II*, Indagationes Mathematicae, Vol. 16, 572 - 588.

Instituto de Matemática
Universidad Católica de Chile
Santiago, Chile.

and

Departamento de Matemática
Universidade Estadual de Campinas
Campinas, São Paulo, Brazil.

Non-Classical Logics, Model Theory and Computability,
A.I. Arruda, N.C.A. da Costa and R. Chuaqui (eds.)

The Equivalence of Some Axioms of Strong Infinity

by *LUIZ PAULO DE ALCANTARA*

I. Introduction.

As pointed out by Mostowski 1967 the existential assumptions known as 'axioms of infinity' can be justified by two general principles:

(I) Principle of transition from potential to actual infinity.

(II) Principle of existence of singular sets.

An early application of the first principle is Dedekind's argument for the existence of infinite sets.

Also, the first principle allows us to formulate the axiom of inacces–sible numbers and Lévy's reflection schema for Zermelo - Fraenkel set theory.

The second principle allows us to formulate still stronger axioms.

Let us suppose that in constructing sets by means of the usual set-theoretic operations we obtain only sets with a property P. If there are no good reasons why all sets should have the property P we are free to add to the axioms a statement implying the existence of sets without the property P.

For example we apply this second principle in the formulation of the axiom stating the existence of measurable cardinals.

In this paper we study some statements related with the principle of existence of singular sets, which are equivalent to reflection principles.

We work in the system BG of Bernays - Gödel with the axioms of foundation and choice.

We employ the terminology and the notation of Drake 1974.

2. LÉVY'S AXIOM.

In this section we given an account of some results announced in de Alcantara 1974.

A. Lévy 1960 proposes an axiom schema (M) to be added to the Zermelo - Fraenkel axioms which postulates the existence of at least one inaccessible number in the range of every normal function defined for all ordinals (d.f.a.o.) and proves the

THEOREM 1. (Lévy)

(M) *is equivalent to the following schema* :

(M') *Every normal function d.f.a.o. has arbitrarily large fixed points which are inaccessible.*

THEOREM 2. (Lévy)

(M) *is equivalent to the conjunction of*:

(M") *Every normal function d.f.a.o. has at least one regular number in its range.*

(AI) *There exist arbitrarily large inaccessible numbers.*

Lévy's proof of theorem 2 involves a notion of inaccessible number whose equivalence with the usual one seems demonstrable only on the assumption of the axiom of choice (see Shepherdson 1952).

Assuming the axiom of choice it is possible to prove that (M) is equivalent to (M").

We need a lemma:

LEMMA 1.

(M") *implies*:

($\overline{\text{M}}$") *Every normal function d.f.a.o. has at least one regular fixed point.*

PROOF: Let f be a normal function d.f.a.o.; let f' be the derivative of f, i.e., the normal function which counts the fixed points of f.

Since f is d.f.a.o. then f' is also d.f.a.o.. By (M") there exists a ξ such that $f'(\xi) = \beta$ is regular.

Then,

$$f(\beta) = f(f'(\xi)) = f'(\xi) = \beta.$$

THEOREM 3.

(M) *is equivalent to* (M").

PROOF: Obviously (M) implies (M"). In order to prove the converse we give some preliminary definitions:

Let ξ be a cardinal, and t_ξ a sequence defined by

$$t_\xi(0) = \aleph_\xi$$
$$t_\xi(n+1) = 2^{t_\xi(n)}.$$

We put

$$\lambda_\xi = \bigcup_{n \in \omega} t(n) \qquad (\text{with GCH } \lambda_\xi = \aleph_{\xi+\omega}).$$

Let g be a function defined on the class of cardinal numbers by $g(\xi) = \lambda_\xi$.

We have $g(\xi) > \xi$ for all ξ.

Finally let $f: On \to On$ be a function defined by

$$f(0) = g(0)$$

$$f(\xi + 1) = g(f(\xi))$$

$$f(\lambda) = \sup_{\beta < \lambda} f(\beta), \text{ for } \lambda \text{ a limit ordinal.}$$

It is trivial to see that f is normal.

Now, we can prove (M).

Let h be a normal function. Then $h \circ f$ is normal and by (M") there exists a regular β such that

$$h(f(\beta)) = \beta$$

But,

$$\beta = h(f(\beta)) \geq f(\beta) \geq \beta$$

And we have $f(\beta) = \beta$.

Finally, we have to prove that β is inaccessible.

In fact, if

$$\mu < \beta = f(\beta) = \sup_{\xi < \beta} f(\xi),$$

then there exists $\xi (\xi < \beta)$ such that $f(\xi) > \mu$.

Thus,

$$\mu < g(\mu) < g(f(\xi)) = \lambda_{f(\xi)} = \bigcup_{n \in \omega} t_{f(\xi)}(n).$$

Then, there exists n $(n \in \omega)$ such that $\mu \leq t_{f(\xi)}(n)$.

And we have,

$$2^{\mu} \leq 2^{t_{f(\xi)}(n)} = t_{f(\xi)}(n + 1) < g(f(\xi)) = f(\xi + 1) \leq f(\beta) = \beta .$$

The proof is now complete.

Let us consider the following schemata:

(L) *For every increasing function* f *d.f.a.o. there exists an inaccessible number* α *such that*

$$(\forall \beta)\ (\beta < \alpha \rightarrow f(\beta) < \alpha).$$

(L') *For every increasing function* f *d.f.a.o. and every ordinal* γ , *there exists an inaccessible number* α , $\alpha > \gamma$ *such that*

$$(\forall \beta)\ (\beta < \alpha \rightarrow f(\beta) < \alpha).$$

(L") *For every increasing function* f *d.f.a.o. there exists a regular number* α *such that*

$$(\forall \beta)\ (\beta < \alpha \rightarrow f(\beta) < \alpha).$$

Arguments which are very similar to those of theorem 3 can be used in order to prove the equivalence between (L) and (L").

Assuming the axiom of choice we prove the

THEOREM 4.

(M) *is equivalent to each one of* (L) *and* (L').

PROOF: We show first that (L) and (L') are equivalent.

Let f be an increasing function. We define a function f_γ by

$$f_\gamma(\beta) = f(\gamma + \beta).$$

f_γ is increasing and then by (L)

$$(\forall \beta)\ (\beta < \alpha \rightarrow f(\beta) \leq f(\gamma + \beta) = f_\gamma(\beta) < \alpha).$$

Also,

$$\gamma \leq \gamma + \beta \leq f(\gamma + \beta) < \alpha .$$

We prove now the equivalence between (M) and (L).

Let f be an increasing function. We can associate to f a normal function f^* defined as follows:

$$f^*(0) = f(0);$$

$$f^*(\lambda) = \sup_{\beta<\lambda} f(\beta), \quad \text{for } \lambda \text{ a limit ordinal;}$$

$$f^*(\mu + 1) = \text{the least } \xi \text{ such that } \exists\eta(\xi = f(\eta)) \text{ and } \xi > f^*(\mu).$$

And we have,

$$f^*(\beta) \leq f(\beta) \leq f^*(\beta + 1) \quad \text{for all } \beta .$$

By (M) there exists an inaccessible α such that $f^*(\alpha) = \alpha$.
Let $\beta < \alpha$. Then,

$$f(\beta) \leq f^*(\beta + 1) < f^*(\alpha) = \alpha .$$

Conversely let f be a normal function. By (L) there exists an inaccessible α such that

$$(\forall \beta) \quad (\beta < \alpha \rightarrow f(\beta) < \alpha).$$

Therefore,

$$\alpha \leq f(\alpha) = \bigcup_{\xi<\alpha} f(\xi) \leq \alpha$$

and $f(\alpha) = \alpha$

3. Some new axioms of strong infinity.

In this section we introduce some new statements equivalent to (M).

DEFINITION. *Let f be a set valued function defined on the class of all ordinals.*

f is V - normal if and only if

(i) $\forall\alpha, \beta(\beta < \alpha \rightarrow f(\beta) \in f(\alpha))$;

(ii) $f(\lambda) = \bigcup_{\beta<\lambda} f(\beta)$, *if λ is a limit ordinal.*

In analogy with (L) we consider the following statement:

(L_V) *For every V - normal function d.f.a.o. there exists an inaccessible α such that*

$$(\forall\beta) \ (\beta < \alpha \rightarrow f(\beta) \in V_\alpha).$$

THEOREM 5. (L_V) *is equivalent to* (L).

PROOF: Let f be a V-normal function. We define $f_r : On \to On$ by

$f_r(\beta) = \rho(f(\beta))$, where $\rho(x)$ is the rank of x.

f_r is increasing, since if $\beta < \alpha$, $f(\beta) \in f(\alpha)$ and $\rho(f(\beta)) < \rho(f(\alpha))$.

By (L) there exists an inaccessible α such that

$$(\forall \beta)\ (\beta < \alpha \to f_r(\beta) < \alpha).$$

Then, if $\beta < \alpha$, $\rho(f(\beta)) < \alpha$ and $f(\beta) \in V_\alpha$.

Conversely, let $f : On \to On$ be an increasing function.

We can associate to f a normal function f^* defined as in the second part of theorem 4.

By (L_V) there exists an inaccessible α such that

$$(\forall \beta)\ (\beta < \alpha \to f^*(\beta) \in V_\alpha);$$

and we have $\beta + 1 < \alpha$ and $f^*(\beta + 1) \in V_\alpha$.

Therefore,

$$f(\beta) \in V_\alpha \quad \text{and} \quad f(\beta) < \alpha .$$

In order to give a new formulation of (L") we characterize the standard complete models of BG without the power set axiom. This will be done using some of the ideas developed in Kruse 1965 and Shepherdson 1952.

DEFINITION. *Let γ be a cardinal number.*

$$V(0, \gamma) = \emptyset ;$$

$$V(\alpha + 1, \gamma) = \mathcal{P}_\gamma V(\alpha, \gamma), \text{ where}$$

$$\mathcal{P}_\gamma(x) = \{ t \mid t \subset x \wedge card(x) < \gamma \} ;$$

$$V(\lambda, \gamma) = \bigcup_{\beta<\lambda} V(\beta, \gamma) \text{ for } \lambda \text{ a limit ordinal.}$$

DEFINITION. *Let γ be a cardinal number; x has power hereditarily less than γ if and only if*

(i) $Card(x) < \gamma$

(ii) $(\forall y)\ (y \in T\,C\,(x) \to card(y) < \gamma)$.

If x has power hereditarily less than γ, $card(TC(x)) \leq \gamma$. Thus $card(\rho(x)) \leq \gamma$ and the class of all sets whose power is hereditarily less than γ is a set denoted by $H(\gamma)$.

The following lemma can be easily proved.

LEMMA 2.

$$\mathcal{P}_{\gamma} H(\gamma) = H(\gamma).$$

By transfinite induction on α, we show

LEMMA 3.

$$V(\alpha, \gamma) = H(\gamma) \cap V_{\alpha}$$

$$V(\gamma, \gamma) = H(\gamma) \cap V_{\gamma}$$

LEMMA 4. *If* γ *is an infinite regular cardinal then* $H(\gamma) \subset V_{\gamma}$.

PROOF: If $x \in H(\gamma)$, then $x \subset V_{\gamma}$ and $card(x) < \gamma$. As γ is regular there exists $\gamma' < \gamma$ such that $x \subset V_{\gamma'}$. Thus $x \in V_{\gamma}$.

COROLLARY. *If* γ *is an infinite regular cardinal, then* $H(\gamma) = V(\gamma, \gamma)$.

LEMMA 5. *For* γ *a regular cardinal number we have*

(i) $(\forall x)\ (x \in U \rightarrow \bigcup x \in U)$

(ii) $(\forall x)\ (x \subset U \wedge (\exists y \in U)\ (card(x) \leq card(y)) \rightarrow x \in U)$ *if and only if*

$$U = V \quad \text{or} \quad U = H(\gamma).$$

For γ a regular cardinal number it is proved in Boffa 1970 that $card(H(\gamma)) = \sum_{\mu<\gamma} 2^{\mu}$.

We denote by BG* the system BG without the power set axiom.

From the lemmas 3, 4 and 5 we deduce

THEOREM 6. $M = \mathcal{P}U$ *is a standard complete model of* BG* *if and only if* $U = V$ *or* $U = V(\gamma, \gamma)$ *where* γ *is a regular cardinal greater than* $\aleph_0$.

In analogy with (L") we formulate the statements

(L''_V) *For every* V*-normal function* f, *d.f.a.o., there exists a regular cardinal* α, *such that* $\alpha > \aleph_0$ *and*

$$(\forall \beta)\ (\beta < \alpha \rightarrow f(\beta) \in V_{\alpha}).$$

($\mathcal{L}_V$) *For every V-normal function f d.f.a.o., there exists a regular cardinal α, such that $\alpha > \aleph_0$ and*

$$(\forall \beta)\ (\beta \in V(\alpha, \alpha) \rightarrow f(\beta) \in V(\alpha, \alpha)).$$

($\mathcal{L}''_V$) *For every V-normal function f, d.f.a.o., there exists u such that $S \subset M^{BG^*}(u)$ and*

$$(\forall \beta)\ (\beta \in u \rightarrow f(\beta) \in u).$$

THEOREM 7. ($\mathcal{L}_V$) *is equivalent to* ($\mathcal{L}''_V$).

The proof follows easily from theorem 6.

THEOREM 8. ($\mathcal{L}_V$) *implies* ($\mathcal{L}''_V$)

PROOF: Let $f : On \rightarrow V$ be a V-normal function.

By ($\mathcal{L}''_V$) there exists a standard complete model u of BG* such that

$$(\forall \beta)\ (\beta \in u \rightarrow f(\beta) \in u).$$

But then, there exists a regular $\gamma > \aleph_0$ and $u = V(\gamma, \gamma)$, i.e. for $\beta < \gamma$

$$\beta \in V(\gamma, \gamma) \rightarrow f(\beta) \in V(\gamma, \gamma).$$

As $V(\gamma, \gamma) \subset V_\gamma$ we have $f(\beta) \in V_\gamma$.

We can easily see that (L_V) implies ($\mathcal{L}_V$) because for an inaccessible α,

$$H(\alpha) = V_\alpha$$

Thus we have the following diagram :

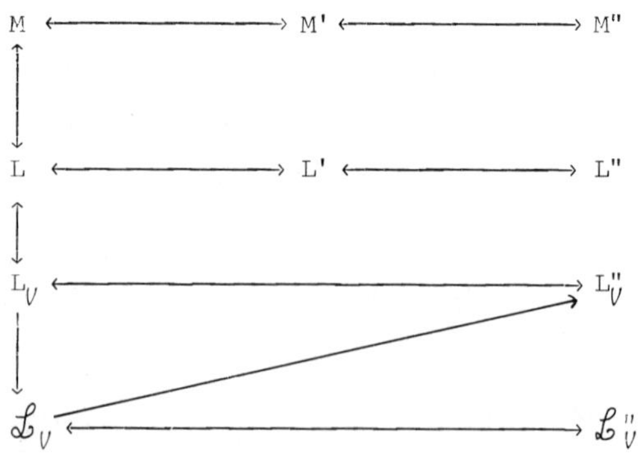

REFERENCES.

Boffa, M.

1970, *Sur l'ensemble des ensembles héréditairement de puissance inférieure à un cardinal infini donné*, Bulletin de la Société Mathématique de Belgique, 22, 389 - 392.

de Alcantara, L. P.

1974, *On new axiom schemata of strong infinity in axiomatic set theory* (abstract), The Journal of Symbolic Logic, 39, 410.

Drake, F.

1974, Set Theory, North - Holland, Amsterdam.

Kruse, A. H.

1965, *Grothendieck universes and the super - complete models of Shepherdson*, Compositio Mathematicae, 17, 96 - 101.

Lévy, A.

1960, *Axiom schemata of strong infinity in axiomatic set theory*, Pacific Journal of Mathematics, 10, 223 - 238.

Mostowski, A.

1967, *Recent results in set theory*, in Problems in the Philosophy of Mathematics, North - Holland, Amsterdam.

Shepherdson, J.

1952, *Inner Models for Set Theory*, The Journal of Symbolic Logic, 14, 225 - 237.

Departamento de Matemática
Universidade Estadual de Campinas
Campinas, São Paulo, Brazil.

Non-Classical Logics, Model Theory and Computability,
A.I. Arruda, N.C.A. da Costa and R. Chuaqui (eds.)

PRESENT PROBLEMS ABOUT INTERVALS IN RELATION-THEORY AND LOGIC.

by *ROLAND FRAÏSSÉ*.

0. INTRODUCTION AND SUMMARY.

Everybody knows the two definitions of an interval in a chain, or total ordering relation. Firstly, the *absolute interval*, which is a set closed under intermediarity: if x and $y > x$ belong to the interval, then every z such that $x < z < y$ belongs to it. Secondly the *relative interval*: given two elements a and $b > a$, whose set $\{a,b\}$ is called the *bound*, the relative interval is the set of x's such that $a < x < b$. Analogous definition of the relative interval of x's such that $x < a$, or such that $x > a$. More generally, given an arbitrary subset F of the base E of the ordering relation, a *relative interval with bound* F will be a subset of $E - F$, maximal by inclusion among sets D such that, if t belongs to F, then all x's of D are $< t$, or all are $> t$. Any absolute interval D is a relative interval, by taking the bound $F = E - D$; and conversely any relative interval is obviously closed by intermediarity, then is an absolute interval. But this equivalence between definitions will not subsist in generalizations to arbitrary relations.

In the years 1950, the author proposed the following generalizations (see, for instance, Fraïssé 1973, p. 107). Recall that, given two relations A and B of common arity n, a bijection f from a subset F of the base $|A|$, to a subset G of $|B|$, is called a *local isomorphism* of A towards B, when f is an isomorphism of the restriction A/F onto B/G. For being a local isomorphism, it is sufficient, and obviously necessary, that f restricted to every set of $< n$ elements of its domain F, be a local isomorphism. Recall that f is called a *local automorphism* of A, when f is a local isomorphism of A towards A.

These notions, as the following notions of intervals, are immediately extensible to any *multirelation*, which is a finite sequence of relations that are called its *components*, with a common base; then the *arity* of the multirelation will be the maximun of the arities of its components. Generally we will call them relations, by abbreviation, except when new components are explicitly added.

Given a relation A of base $E = |A|$, a subset D of E is called an *A-absolute interval*, or simply an A-interval, when any local automorphism of the restriction A/D, extended by the identity on $E - D$, gives a local automorphism of A: we shall say simply that it is *extensible* by the identity on $E - D$. Note that it is sufficient to consider local automorphism of A/D on domains of $p \leq n-1$ elements, and to extend them by identity on $n-p$ elements of $E-D$. If A is a chain, we refind the usual interval. If A is a partial ordering relation of base E, then D is an A-interval iff, for any element t of $E-D$, either all elements of D are $< t$, or all are $> t$, or all are incomparable to t.

Given a relation A of base E, take a subset F of E. Then a subset D of $E-F$ is called an (A,F)-*interval*, or an A-*relative interval* with F as a *bound*, when D is a maximal set, by inclusion, among sets D' such that any local automorphism of A/D', extended by identity on F, gives a local automorphism of A. If A is a chain, the possible sets D' have, for any element t of F, all their elements $< t$ or all $> t$; and by taking for D a maximal D', we get a set closed for intermediarity, and so a usual interval.

Any absolute interval D is obviouly a relative interval, by taking the bound $E-D$. However the converse is false, already for a partial ordering relation A. For instance, given a and $b > a$ (modulo A) the set of all x's such that $a < x < b$ is a relative interval D, with $\{a,b\}$ as a bound. Now suppose there exist x,y in D, and t out of D, with $a < x < t$ and t incomparable to y and b. Then the local automorphism from the singleton $\{x\}$ to $\{y\}$, is extensible by the identity on $\{a,b\}$, but not by the identity on $\{t\}$.

During twenty years, our knowledge about intervals, extervals (which are their complementary sets), and related topics, was pratically reduced to the statement of the exercise 6, page 107, in Fraïssé 1973. Some recent researches of Foldes, Gillam, and Pouzet, seriously increased this knowledge. Moreover the author recently became conscious of several logical problems involving interval theory. For instance, logical problems about intervallary extensions of a relation, which are natural generalizations of the classical extension of the rational ordering by the real ordering relation. In connection to that, the notion of a compact set of ultrafilters, and of a compact relation (see 5.1 bellow), leads to logical problems. Finally we shall recall Ehrenfeucht's results about the comparison, sum, and product theory among ordinals, reported in Feferman 1957, where the notion of interval is deeply connected with logic; and we shall propose several possible generalizations of these results to comparison, sum, product theory among ordering relation, or among arbitrary relations.

1. ELEMENTARY PROPERTIES OF ABSOLUTE INTERVALS AND EXTERVALS.

It is obvious that the empty set, the total base, and the singleton of any element, are intervals. There exist relations in which these are the only ones; for instance the consecutivity C on natural integers: $C(x,y) = +$ when $y = x + 1$, and $C(x,y) = -$ in other cases. Indeed if a set D of integers has at least two elements, and is not the whole base, take $x \in D$, $x' \in D$ and $y \in E - D$ (where E is the base), but $y = x + 1$ or $y = x - 1$; then the transformation of x into x' is a local automorphism of C, unextensible by the identity on y.

A *multirelation* is a finite sequence of relations on the same base, called the *composant* relations. The notions of local isomorphism, automorphism, interval are immediately extended to multirelations. A *unary* multirelation is a multirelation whose composant relations are unary ones (taking value + or - for each element in the base).

If A *is a unary multirelation, then any subset of the base is an* A-*interval.*

1.1 Recall that a relation B is *free-interpretable* in a relation A on the same base, when any local automorphism of A is an automorphism of B. It is equivalent to say that there exists a free logical formula, e.g., a formula without quantifier, with a predicate substituable by A , and a number n of free variables equal to the arity of B, taking the value $B(x_1, \ldots, x_n)$ when the predicate is A and $x_1,\ldots,x_n$ are elements of the common base.

If A *and* B *are each free-interpretable in the other, then the* A *-intervals and the* B *-intervals are the same.*

Consequently, *if there exists a unary multirelation B such that* A *and* B *are mutually free-interpretable, then all subsets of the base* |A| *are* A-*intervals*. Example: concatenate a unary relation U and the binary relation of equivalence with two classes, defined by the values + and - taken by U. The converse is not true: take the chain on two elements; see also 1.9 bellow .

Given a relation A, consider as equivalent any two elements u,v of |A| when the transposition (u,v) extended by the identity on all other elements, is an automorphism for A. *There are finitely many equivalence classes so defined for* A, *iff there is a unary multirelation in which* A *is free-interpretable*. Indeed for a unary multirelation U, local automorphisms are exactly those bijections which transform any element into an other giving the the same value to each component of U; each local automorphism of finite domain is generated by some tranpositions between elements giving same value to U, in the sense that it is obtained by composition of these transpositions and then by restriction to its domain. Coming back to A, the same argument proves that A is free-interpretable by the unary multirelation whose components are defined from each equivalence class of elements of |A|.

Now note that, if a relation B is free-interpretable in A, it is possible either there are more A-intervals than B-intervals, or there are more B-intervals than A-intervals, or that none of both sets of intervals is included in the other. First example: A is a unary relation taking at least twice the value + and once the value - , and B is the corresponding equivalence relation. If $A(x) = A(x') = +$ and $A(y) = -$, the pair $\{x,y\}$ is an A-interval, as all the subsets of the base, and is not a B-interval, the transformation of x into y being an automorphism of B, unextensible by the identity on $\{x'\}$. Second example: A is the chain of natural integers, and B is the unary relation taking always the value +. Third example: $A = (N,O)$ where N is the chain of natural integers, and O is the unary relation, singleton of zero; B is the equivalence relation with two classes, the singleton of zero and the set of positive integers. Then B is free-interpretable in O, then in A. But the set $\{0,1\}$ is an A-interval without being a B-interval, and the set $\{1,3\}$ is a B-interval without being an A-interval.

1.2. *Given a relation* A, *any intersection of* A-*intervals is an* A-*interval.*

□ Consider a set of A-intervals D_i, and their intersection U. Let f be a local automorphism of the restriction A/U. In order f be extensible by the identity on $|A| - U$, it is sufficient that f be extensible by the identity on any finite subset of $|A| - U$. For such a finite set H, we can consider an arbitray sequence of the elements $u_1,\ldots,u_h$ of H, and extend f by the identity on $\{u_1\}$, then on $\{u_1,u_2\}$, and so on. For each element u_i of H, there exists a D_i such that $u_i \in |A| - D_i$, so that the addition of the identity on $\{u_i\}$ transforms the local automorphism of A into another local automorphism of A; note that it is irrelevant that previous u_j $(j < i)$ be elements of D_i or elements of $|A| - D_i$. □

1.3. *Given a set of A-intervals, filtering by inclusion, then their union is an A-interval.*

□ Let D_i be these intervals, and U their union. Let f be a local automorphism of the restriction A/U, with a domain and a codomain which can be supposed finite. Then there exists atleast a D_i including the domain and the codomain of f. Consequently, f is extensible by the identity on $|A| - D_i$, thus by the identity on $|A| - U$. □

1.4. *Let n be the arity of A (the maximun arity of components if A is a multirelation); let U, V be two A-intervals; suppose that any restriction of $A/(U \cup V)$ of cardinal $\leq n-1$, admits an isomorphic relation among the restrictions of $A/(U \cap V)$. Then the union $U \cup V$ is an A-interval.* In the case of a chain A, or more generally of a reflexive binary relation we refind that the union of two intervals with a common element is an interval.

□ Let f be a local automorphism of A, with its domain and codomain included in $U \cup V$. It is sufficient to suppose these domains are each of cardinality $m \leq n-1$, and to prove that f is extensible by the identity on n-m arbitrary elements out of $U \cup V$. Let us note $u_1,\ldots,u_n$ the elements of the domain of f, and $v_1, \ldots, v_m$ their transformed elements by f. By hypothesis, there exist $w_1,\ldots,w_m$ belonging to the intersection $U \cap V$, the transformation of u_i into w_i $(i = 1,\ldots,m)$ and the transformation of v_i into w_i being local automorphisms of A. It is sufficient to prove that the first transformation is extensible by the identity on $n-m$ arbitrary elements out of $U \cup V$. The analogous proof works for the second transformation.

We may always suppose that there is a $p \leq m$ with $u_1,\ldots,u_p$ belonging to U, and $u_{p+1},\ldots,u_m$ belonging to V and out of U. The transformation of u_i into w_i $(i = 1,\ldots,p)$ is a local automorphism of A, with domain and codomain included in U. Thus it is extensible by the identity on u_j $(j = p+1, \ldots, m)$, and moreover by the identity on $n-m$ arbitrary elements out of

$U \cup V$. On another side, the transformation of $w_i, \ldots, w_p, u_{p+1}, \ldots, u_m$ into the w_i $(i = 1, \ldots, m)$ is a local automorphism of A, since it is composed of two local automorphisms, as it is seen by going through the u_i $(i = 1, \ldots, m)$. Its domain and codomain are included in V; thus it is extensible by the identity on $n - m$ arbitrary elements out of $U \cup V$. Finally the transformation of u_i into w_i $(i = 1, \ldots, m)$ is so extensible.□

1.5. Given a relation A, let us call an A-*exterval* the complementary set of an A-interval.

A subset D of the base $|A|$ is an A-exterval iff, for any local automorphism f of A/D, and any subset G *of $D^* = |A| - D$ either f is not extensible by the identity on G, or f is extensible by any local automorphism of A/D^* with domain G.*

□ Suppose D is an exterval, then $D^* = |A| - D$ is an interval. Let g be a local automorphism of A/D^* with domain G . Now suppose that the local automorphism f of A/D is extensible by the identity I_G on G, so the union $f \cup I_G$ is a local automorphism of A. The set D^* being an interval, g is extensible by the identity $I_{F'}$ on the codomain F' of f. Thus the union $f \cup g$ is a local automorphism, as obtained by composition of $f \cup I_G$ and $I_{F'} \cup g$.

Suppose D is not an exterval, then D^* is not an interval. Thus there exists a local automorphism g of A/D^* and a subset F of D, such that $I_F \cup g$ is not a local automorphism. Then the identity I_F on F is obviously extensible by the identity on G, without being extensible by g; against our condition.□

1.6. *Given a relation* A *and two disjoint* A-*intervals* F *and* G, *and a local automorphism f of the restriction* A/F *and another g of* A/G, *then the union $f \cup g$ is a local automorphism of* A. It is sufficient to extend f by the identity on the domain of g, and g by the identity on the domain of f, and to compose.

Let A be a relation, E its base, n its arity. Let us share E into disjoint A -interval D_i. To each D_i, let us associate a finite subset d_i of D_i that we call the *representant* of D_i, such that each restriction of A/D_i with cardinality $\leq n - 1$ admits an isomorphic restriction of A/d_i. Let us call *intervallary image*, or more simply image of A, the restriction of A to the union of the representants d_i.

There exists only one relation A *of given base* E, *shared into disjoint* A-

intervals D_i, *of given restrictions* A/D_i *and with a given intervallary image* $A/\cup d_i$ *where each* d_i *is a representant of* D_i.

□ Suppose there exist two relations A and A' satisfying our conditions; it is sufficient to prove that, for any subset F of E, with cardinality $\leq n$, we have $A/F = A'/F$. This is obvious if F is included in a D_i or in the union $\cup d_i$. Suppose the contrary: each intersection $F_i = F \cap D_i$ is of cardinality $\leq n-1$. By hypothesis, there exists an isomorphism of A/F_i onto a restriction of A/d_i. These isomorphisms have disjoint domains since each domain is included in a D_i; same remark for codomains; by the preceding statement, their union is a local automorphism of A, and also of A'. So $A/\cup d_i$ being identical to $A'/\cup d_i$, this automorphism gives $A/F = A'/F$. □

1.7. *Given a relation* A, *a subset* D *of its base* $|A|$ *and an A-interval* U, *the intersection* $D \cap U$ *is an* (A/D)*-interval.*

Note that, given an (A/D)-interval V, there does not necessarily exist an A-interval U such that $V = D \cap U$. For instance, take for A the partial ordering relation, Boolean lattice represented by a usual 3-dimensional cube whose minimun u and maximun v are two opposite vertices; take for D the set of the three vertices x, y, z immediately preceding v, and take $V = \{x, y\}$. Then an A-interval including V has necessarily the element immediately anterior to x and z (and incomparable to y), and then necessarily z itself: contradiction.

Given A *and an A-interval* D, *the* (A/D) *intervals are exactly the A-intervals included in* D, *or equivalently the intersections with* D *of all A-intervals.*

1.8. *Let* E *be the set of natural integers, and* N *the usual chain on* E. *If a relation* A *on the same base* E *is free-interpretable in* N *and if any pair of integers is an A-interval, then all permutations of* E *are automorphisms for* A. (This statement and the following are communicated by M. Pouzet.)

□ Let $u < v < w$ be three integers. As A is free-interpretable in N, the transformation of u into w is a local automorphism of A; as the pair $\{u, w\}$ is an A-interval, this local automorphism is extensible by the identity on v. Always by free-interpretability of A in N, any transformation which saves the order of integers is a local automorphism of A: by composition, the transposition between any two integers is a local automorphism of A. It follows that any permutation of E is an automorphism of A. □

1.9. *Given a relation A, if any pair of elements of the base |A| is an A-interval, then:*

(1) there is a partition of the base into finitely many classes, the transposition between any two elements of a same class, extended by the identity, giving an automorphism for A;

(2) there is a unary multirelation in which A is free-interpretable; moreover if there is at most one class in conclusion (1), which reduces to a singleton, then there is a unary multirelation B such that A and B are mutually free-interpretable.

□ Consider as equivalent any two elements u,v of the base |A| when the transposition (u,v) is a local automorphism for A; since the pair $\{u,v\}$ is an A-interval, this transposition extended by the identity on other elements, gives an automorphism for A. In order to get our conclusion (1) suppose there are infinitely many classes in our equivalence. Using the axiom of choice, take a denumerable sequence of elements a_i (i natural integer), mutually not equivalent. Call N the ω-chain of a_i's, with $N(a_i,a_j) = +$ when $i \leq j$. Let n be the arity of A; consider as equivalent any two sets with the same cardinal $p \leq n$, when the transformation of the one into the other, which preserves the ordering N, is a local automorphism of A. Using Ramsey's theorem, take a denumerable set U of elements a_i, that we re-numerate, so that any local automorphism of N/U, on $\leq n$ elements, and consequently any local automorphism of N/U, is a local automorphism of A/U: in other words A/U is free interpretable in N/U. By 1.7 any pair of elements of U is an (A/U)-interval. By 1.8 all transpositions in U are automorphisms for A/U, and then for A since any pair of elements is an A-interval: contradiction, proving our conclusion (1).

We know from 1.1 and our conclusion (1), that A is free-interpretable in the unary multirelation B whose components are defined each by an equivalence class of elements of |A|. To prove that B is free-interpretable in A, it is sufficient to take any two elements u,v in different classes, and prove that the transformation of u into v is not a local automorphism for A. Suppose the contrary for u and v, and suppose there exists $u' \neq u$ and equivalent to u. The transformation of u into v being extensible, since $\{u,v\}$ is an A-interval, the transposition (u',v) is a local automorphism for A: so u,u' and v are equivalent: contradiction. □

2. The finite-val, a Boolean notion.

Consider a relation A and a subset D of its base |A|. For each positive integer p, consider as equivalent two p-uples $u_1,\ldots,u_p$ and $v_1,\ldots,v_p$, where the u's and v's are elements of D, when the transformation of u_i into v_i $(i = 1,\ldots,p)$ is a local automorphism of A, extensible by the identity on $|A| - D$. Then we call D an A-*finite-val*, when there are finitely many classes in this equivalence.

If n is the arity of A, it is sufficient to limit the length of sequences to $p \leq n-1$ and the identities to $n-p$ elements.

For any relation, any finite union of disjoint intervals is a finite-val (the condition of disjointness will be cancelled after 2.3 below).

□ Two p-uples are equivalent iff their first terms are both in a same interval, their second terms both in a same interval, and so on, and obviously if the transformation of the one into the other is a local automorphism. This gives a finite number of equivalence classes.□

For a chain, the finite-vals are exactly all finite unions of intervals. However this is not the general case. For instance,consider the cyclic ordering, obtained by starting from a chain A, and giving to any 3-uple x,y,z the value + iff $x \leq y \leq z$ or $y \leq z \leq x$ or $z \leq x \leq y$ (modulo A). Take for instance A = usual chain of natural integers. Then the only intervals of the cyclic ordering are the empty set, the base and singletons. Indeed for any other set D, the transposition between two distinct elements x,y of D, is a local automorphism, and is not extensible by the identity on any element out of D. However for any integer u,the set of integers $\geq u$ is a finite-val, as it is seen by considering as equivalent any two p-uples of such integers, when the transformation of the first p-uple into the second is a bijection saving the ordering.

As a consequence of the previous statement, *any finite subset of the base is a finite-val.*

2.1. *The complementary set of any finite-val (in reference to the base) is a finite-val.*

□ Let E be the base and D a finite-val; let n be the arity of the relation. For each positive integer $p \leq n-1$, consider any two p-uples in D as equivalent when the transformation of the one into the other is a local automorphism, extensible by the identity on $E - D$: by hypothesis, there are finitely many classes of this equivalence. Now for each p-uple ($p \leq n-1$) in D, let us take a unique representant belonging to the same class, and call H the finite subset of D, union of all these representants. For each positive integer q, let us consider as equivalent any two q-uples in $E - D$, when the transformation of the one into the other is a local automorphism extensible by the identity on H: as H is finite, there is only a finite number of equivalence classes.

It is now sufficient to prove that, if g is such a local automorphism in $E - D$, extensible by the identity on H, then g is extensible by the identity on D. Let F be any subset of D, of cardinality $\leq n - 1$: it is sufficient to prove that g is extensible by the identity on F, since in this case, g extended by the identity on D, and then restricted to any subset with cardinality $\leq n$, will give a local automorphism. By hypothesis, there exists a subset of F' of H and a local automorphism f with domain F and codomain F', extensible by the identity on $E - D$, thus extensible by the identity I_G on G. As $f \cup I_G$ and $g \cup I_{F'}$ (identity on F') are local automorphisms, so is $f \cup g$ by composition. As f is extensible by the identity on $E - D$, thus by the identity $I_{G'}$ on G', the union $f^{-1} \cup I_{G'}$, is a local automorphism, and by composition $I_F \cup g$ is a local automorphism. □

Consequently, *the complementary set of any finite subset of the base is a finite-val; any finite intersection of extervals is a finite-val.*

2.2. *If a relation B is free-interpretable in A, then any A-finite-val is a B-finite-val.* Indeed, if two finite sequences are transformed one into the other by a local automorphism of A, then this transformation is a local automorphism of B. So if they are equivalent for an A-finite-val D , this equivalence subsists for D and B.

Consequently if there exists a unary multirelation in which A is free-interpretable, then all subsets of the base are A-finite-vals; for the converse statement, see 2.5 below.

Given a relation B and a B-finite-val D, there exists a relation A in which B is free-interpretable, D being an A-interval.

□ Let n be the arity of B. Consider the equivalence among the n-uples in D, putting two n-uples in the same class when the transformation of one into the other is a local automorphism of B, extensible by the identity on $|B| - D$. For each class U, take the n-ary relation with base $|B|$ having the value + for n-uples of U and the value - otherwise. Then the finite sequences of these relations and of all the components of B, constitute an n-ary multirelation A in which B is obviously free-interpretable. Moreover, if two n-uples in D are transformed by a local automorphism of A, then they belong to the same class precedently defined from B and D, then this local automorphism is extensible by the identity on $|B| - D$, giving another automorphism of A .□

Consequently, *given a multirelation B, a subset D of the base is a B-finite-val iff there exists a multirelation A in which B is free-interpretable, D being an A-interval* (M. Pouzet, 1975, not published).

If D is a finite-val, then D increased or diminished by a finite subset of the base, is a finite-val. For each element to add or to suppress, take its singleton relation (unary relation taking the value + for the considered element) and add it to the multirelation A in the previous statement.

2.3. *The union and the intersection of two finite-vals is a finite-val* (M. Pouzet, 1976, not published).

□It is sufficient, in view of 2.1, to prove it for the union. Suppose the contrary; let U and V be two finite-vals such that the union $U \cup V$ is not a finite-val. Let p be the least positive integer for which there exists an infinite sequence of p-uples in $U \cup V$, mutually not equivalent , in this sense that the transformation of any one p-uple into another, extended by the identity out of $U \cup V$, is not a local automorphism. For each p-uple, let us call u the terms belonging to U and not to V, call v the terms belonging to V and not to U, and w the terms belonging to $U \cap V$. We can always suppose that, for a given index $r \leq p$, the term of index r is always a term u, or always v, or always w; so that, simplifying the presentation without loss of generality, we can suppose $p = 3$ with one term u, one v, and one w in each 3-uple. For any integer i, let us call u_i, v_i, w_i these three terms. (Since U is a finite-val, each sequence $u_i\, w_i$ is in U, these sequences belong to a finite number of classes, for the equivalence defined by local automorphisms extensible by the identity out of U. Thus we can suppose that all sequences $u_i\, w_i$ belong to the same class: consequently for any i, the transformation of $u_i\, v_i\, w_i$ into $u_0\, v_i\, w_0$ is a local automorphism. On another side, since V is a finite-val, this set V diminished of the finite set of terms w_0 is a finite-val (see 2.2). So the terms v belonging to this finite-val, and the terms u_0 and w_0 being out of it, we can suppose that, for any integers i, j, the transformation of $u_0\, v_i\, w_0$ into $u_0\, v_j\, w_0$ is a local automorphism. Thus, for any integers i, j, we have local automorphisms from $u_i\, v_i\, w_i$ to $u_0\, v_i\, w_0$, then $u_0\, v_j\, w_0$, then to $u_j\, v_j\, w_j$: contradiction.□

2.4. *Let* E *be the set of natural integers,* N *the usual chain on* E. *If* A *is free-interpretable in* N *and is any set of integers is an* A-*finite-val, then all permutations of* E *are automorphisms for* A (this statement and the following are due to M. Pouzet).

□ Suppose there exists a permutation of E which changes A; then if n is the arity of A, there exists a permutation on at most $2n$ integers, which changes A, and so a transposition between two integers, which changes A . By the free-interpretability of A in N, there exists an integer u such that, for any integer x, the transposition $(x, x+u)$ changes A. For the same reason, the transformation of any x into $x+2u$ is a local automorphism of A not extensible by the identity on the singleton $\{x+u\}$, but extensible by the identity on integers $< x$ or $> x+2u$. Consequently, the set of all multiples of $2u$ is not a finite-val. □

2.5. *Given a relation* A, *if any subset of the base* |A| *is an* A-*finite-val, then the conclusion* (1) *of* 1.8 *is valid; so by* 1.1, *there is a unary multirelation in* A *is free-interpretable*. Note that (2) of 1.8 is not necessarily valid: example, a binary equivalence relation with finitely many classes.

□ Consider as equivalent any two elements u, v of the base |A| when the transposition (u,v), extended by the identity on all other elements of |A|, is an automorphism for A. Suppose there are infinitely many equivalence classes. Using the axiom of choice, take a denumerable subset E_0 of the base, such that the transposition between any two elements of E_0, extended by the identity, changes A. Moreover, take an ω-chain N_0 on the base E_0, and by Ramsey's theorem, suppose that the restriction A/E_0 is free-interpretable in N_0. Let n be the arity of A; for each $p \le n$, consider as equivalent any two subsets of E_0 with cardinal p, when the unique local automorphism of N_0 which transforms the one into the other, being a local automorphism of A, is still extensible by the identity out of E_0. Since E_0 is a a finite-val, there are only finitely many equivalence classes: using Ramsey's theorem, we get a denumerable subset E_1 of E_0 in which all subsets are equivalent, and this for each $p \le n$. Iterating n times, we get the successive sets $E_0 \supseteq E_1 \supseteq E_2 \supseteq \ldots \supseteq E_n$.

Let u, v be two distinct elements of E_n. By our hypothesis, the transposition (u,v) changes A; so there exist elements $x_1, \ldots, x_{n-1}$ in the base |A|, such that (u,v), extended by the identity on all x's, changes the restriction $A/\{u, v, x_1, \ldots, x_{n-1}\}$. There exists at least one integer h $(1 \le h \le n)$ such that $E_{h-1} - E_h$ does not have any x among its elements. By the preceding statement 2.4, the transposition (u,v) extended by identity, is an automorphism for $A_h = A/E_h$ but not for the restriction of A to the union $E_h \cup (|A| - E_{h-1})$. However in this restriction, any local automorphism of $N_h = N_0/E_h$ is a local automorphism of A_h and is extensible by the identity on $|A| - E_{h-1}$. Consequently, given any three elements $u < w < v$ (modulo N_h) in E_h, the transformation of u into v is a local automorphism of A,extensible by the identity on all elements of E_h which are $< u$ or $> v$, and by the identity on $|A| - E_{h-1}$; but not extensible by the identity on $|A| - E_{h-1}$ increased by the element w. So that the set of elements having an even rank in E_h, for instance, is not an A-finite-val: contradiction proving our conclusion (1). The conclusion (2) follows from 1.1. □

3. THE SUBVAL, AN INTERMEDIARY NOTION BETWEEN INTERVAL AND FINITE-VAL.

Given a relation A, a subset D of the base $|A|$ will be called an A-*subval* when, for any local automorphism f of A/D, and any subset G of $D^* = |A| - D$, either f is not extensible by the identity on G, or f is extensible by the identity on any subset of D^*, obtained from G by any local automorphism of A/D^*.

3.1. *Any interval or exterval is a subval.*

□ This is obvious for an interval. Let D be an exterval and E the base, so that $E - D$ is an interval. Let f be a local automorphism of A/D, with domain F and codomain F', and let g be a local automorphism of $A/(E-D)$, with domain G and codomain G'. Suppose that f is extensible by the identity I_G on G: we have to prove that f is again extensible by the identity $I_{G'}$ on G'. The union $f \cup I_G$ being a local automorphism, and $E - D$ an interval,the union $g \cup I_{F'}$ is a local automorphism and so $f \cup g$ by composition (see also 1.5). Moreover $g \cup I_F$ and thus $g^{-1} \cup I_F$ are local automorphisms: by composition with $f \cup g$, we get $f \cup I_{G'}$. □

There exists a subval which is neither an interval,nor an exterval. Take the cyclic ordering relation alredy defined in 2,from the total ordering of natural integers: for any integer u, the set of integers $\geq u$ is a subval. Because if $u = 0$, this is the whole base; if $u \geq 1$, any local automorphism among integers $\geq u$, is extensible by the identity on the singleton $\{0\}$, iff it preserves the usual total ordering between integers; and in such a case, it is extensible by the identity on all the integers $< u$. Now it is sufficient to take $u \geq 2$, in order that the set of integers $\geq u$ be neither an interval nor an exterval: we already noticed in 2 that the only intervals of the cyclic ordering are the empty set, the base and the singletons.

3.2. *Any subval is a finite-val.*

□ Let n be the arity of our relation A; we know that, in order for D to be a finite-val, it is sufficient that sequences of length $p \leq n$ fall into a finite number of equivalence classes, each of these classes being so defined, that equivalent sequences give a local automorphism extensible by identity on any set of $n - p$ elements out of D; and all that for each $p \leq n$. Now suppose D be a subval. For each $q \leq n$, consider as equivalent any two subsets G and G' of $E - D$ (where E is the base) with cardinal q, when the restrictions A/G and A/G' are isomorphic. Take a representant in each of these equivalent classes, and let H be the finite set, union of these representants. Now consider any two sequences u, v in D, with a same length $\leq n$, as equivalent when the transformation of u into v is a local automorphism of A, extensible by the identity on H. As H is a finite set, there are finitely many classes of equivalent sequences. It is sufficient now to note that, D being a subval, the transformation of u into an equivalent sequence v of length $p \leq n$, if it is extensible by the identity on H, is also extensible by the identity on any subset of $E - D$ with cardinal $n - p$, since for such a subset G there exist a subset G' of H with A/G' isomorphic to A/G. □

There exists a finite-val which is neither a subval nor the complementary set of a subval. To see it, note that, for a total ordering relation A, the A-subvals are exactly the A-intervals and the A-extervals. Indeed for any set D which is neither an interval nor an exterval, there exists x, y, in D and z, t out of D with $x < z < y < t$ or $t < x < z < y$. In any case the transformation of x into y is extensible by the identity on the singleton $\{t\}$ but not by the identity on $\{z\}$. Now the union of two intervals admitting another interval between them and another after, for a total ordering, is a finite-val which is neither a subval nor the complementary of a subval.

There exists a subval whose complementary set is not a subval. Indeed, take the partial ordering relation obtained from two total orderings A and B, with any element of $|A|$ uncomparable to any element of $|B|$. Take a median interval B' of B, and call D the union of $|A|$ and $|B'|$. Now D is a subval, since any local automorphism in D, either transforms an element of $|A|$ into an element of $|B'|$, or conversely, and then is not extensible to any identity out of $\mathcal{D}$; or it is the union of a local automorphism of A and of one of B', and then is extensible to the identity on $E-\mathcal{D}$ (where E is the base). However $E-\mathcal{D}$ is not a subval, since the transformation of an element of $|B|$, anterior to B', into an element posterior to B', is extensible by the identity on the singleton of any element of $|A|$, but not on the singleton of any element of $|B'|$.

Finally, note that the notion of subval is not preserved by free interpretability, as is the notion of finite-val. Let C be a chain, or total ordering, A a unary relation on the base $|C|$, with value + for elements of a C-interval, and B another similar unary relation, corresponding to a C-interval disjoint from the former, and even letting at least one element between them, and at least one element after them. Now the union of these C-intervals A and B is neither an interval nor an exterval for C, thus it is not a C-subval. However it is an $(A\ B\ C)$-subval, where $A\ B\ C$ is the multirelation of components A, B, C. Indeed if f is a local automorphism of $A\ B\ C$ inside the union of C-intervals A and B, then f cannot transform an element of the C-interval A into an element of the C-interval B, because A takes value + in the former and - in the latter. So f is the union of a local automorphism of C inside A and a local automorphism inside B, and then f is extensible by any identity out of the union of intervals A and B.

PROBLEM: If A is free-interpretable in a chain, the complementary set of an A-subval is it an A-subval?

4. The relative interval.

Note that the empty set is a relative interval, with the whole base as its bound; the whole base is a relative interval, the unique one with the empty set as its bound; the singleton of any element u, with the base minus u as a bound (other bounds can exist).

Note that for a given bound F, the condition about sets $\mathcal{D}'$ in our introduction: "any local automorphism of $A/\mathcal{D}'$ is extensible by the identity on F", is an inductive condition. Precisely, if for a same F, we have an ascending chain of sets $\mathcal{D}'$ (for inclusion), their union is still a $\mathcal{D}'$. Indeed if n is the arity of A, for being a local automorphism of $A/\mathcal{D}'$ and being extensible by the identity on F, it is sufficient that a bijection f have all its restricted bijections to any set with cardinal $p \leq n$, satisfying the same property, reduced itself to subsets of F with cardinal $n-p$, and that for $p = 1,2,\ldots,n$.

As an immediate consequence, given a subset F of the base, any singleton out of F being obviously a set D' with the above mentionned property, there exists an (A,F)-interval which includes this singleton. In other words, *the union of all* (A,F)*-intervals is* $|A| - F$.

Other consequence: *if* $G \subseteq F$, *any* (A,F)*-interval is included in an* (A,G)*-interval.*

4.1. *For a binary relation* A, *and a finite subset* F *of its base, there are finitely many* (A,F)*-intervals.*

□ Consider as equivalent any two elements u,v of $|A| - F$, when the transformation of u into v is a local automorphism of A, extensible by the identity on F. Since F is finite, there are finitely many classes in this equivalence. Then any interval with F as a bound , as soon as it contains an element u, has to contain all the class of u; as it is seen because a bijection f is a local automorphism of A iff its restricted bijections to any pair of elements of the domain, are themselves local automorphisms. □

Note that the statement does not extend to ternary relations. For instance, take the cyçlic ordering defined in 2, from the usual chain of natural integers, and take for F the singleton of 0. Each singleton of an integer $u \neq 0$ satisfies the condition of extensibility to F of any local automorphism (necessarily reduced to the identity on u); and the addition of another integer $v \neq 0$ and $\neq u$ is impossible, the transposition (u,v) being a local automorphism of the cyclic ordering, inextensible by the identity on 0. So any singleton of an integer $\neq 0$ is an interval with bound F.

4.2. *For a binary relation* A, *and a subset* F *of its base, if* U, V *are each an* (A,F)*-interval, and if for any element* x *of the union* $U \cup V$, *there exist a* y *of the intersection* $U \cap V$, *the transformation of* x *into* y *being a local automorphism of* A, *then* $U = V$.

□ According to the maximality of relative intervals, for inclusion , it is sufficient to prove that the union $U \cup V$ is an (A,F)-interval, or simply satisfies the extensibility condition. It is even sufficient to prove it from a local automorphism of $A/(U \cup V)$, with a domain of only one element . Let x and z be two elements of the union $U \cup V$, the transformation of x into z being a local automorphism. By hypothesis there exists an element y of the intersection, the transformation of x into y being a local automorphism, and so the transformation of y into z. Since x and y belong both to U, or both to V, the first local automorphism is extensible by the identity on F; same result for the second, and finally for the transformation of x into z. □

PROBLEM: For a ternary relation, existence of two distinct intervals, with a same bound, each restriction in the union, with cardinal 2, having an isomorphic restriction in the intersection. More strongly, each finite restriction in the union, having an isomorphic restriction in the intersection.

Note that 1.7 does not subsist for relative intervals. For instance, take a base E of four elements u, x, y, z; the set $F = \{u\}$; the unary relation A with value + for x, z, u and - for y; the binary symmetrical relation B with value + for (u,x), (u,z) and their converses, value - in other cases. Now the pair $\{x,y\}$ is an (AB,F)-interval: the element z cannot be added since the transformation of x into z is a local automorphism inextensible to the element u. Take $D = \{y,z,u\}$ and consider the restriction to D of A B, the bound F being unchanged. Now the intersection of D and the interval $\{x,y\}$ reduces to the singleton $\{y\}$: it is not an $(AB/D,F)$-interval, being not maximal for inclusion, since in the pair $\{y,z\}$, the transformation of y into z is not a local automorphism.

5. Closures of a relation; compact relation.

These closures, already defined in Fraïssé 1974, p. 185, extend to relations the closure procedure whereby the chain of real numbers is derived from the chain of rationals.

Given a relation A of base E, we define an A-*filter* as a set $\mathfrak{F}$ of nonempty A-intervals (absolute intervals) satisfying the following conditions: (1) any A-interval including an element of $\mathfrak{F}$ is also an element of $\mathfrak{F}$; (2) the intersection of two elements of $\mathfrak{F}$ is an element of $\mathfrak{F}$ (it is an A-interval, by 1.2).

A maximal A-filter will be called an A-*ultrafilter*. Any A-filter may be extended to an A-ultrafilter. An A-ultrafilter is said to be *trivial* if it consists of all A-intervals including a singleton. If the A-ultrafilter is not trivial, the intersection of all its elements is empty, and each element is infinite. For a given A-interval D and A-ultrafilter $\mathfrak{F}$, either D intersects each element of $\mathfrak{F}$, and is therefore an element of $\mathfrak{F}$, or there exists an element of $\mathfrak{J}$ disjoint from D. Consequently, if $\mathfrak{J}$ and $\mathfrak{J}'$ are distinct A-ultrafilters, there exist elements D of $\mathfrak{J}$ and D' of $\mathfrak{J}'$ with D and D' disjoint.

Let us complete the base E by embedding it in the set E^* of all A-ultrafilters: by identifying each trivial A-ultrafilter with the element of E generating it. With each non-trivial A-ultrafilter $\mathfrak{J}$ let us associate a relation $A(\mathfrak{J})$ with base $E(\mathfrak{J})$ and the same arity as A. The bases $E(\mathfrak{J})$ are assumed to be disjoint from E and from one another. The relations $A(\mathfrak{J})$ are subjected to the following condition, which may always be satisfied: for any finite subset F of $E(\mathfrak{J})$ and any element D of $\mathfrak{J}$, there exists at least one isomorphism of the restriction $A(\mathfrak{J})/F$ onto a restriction of A/D.

Once the relations $A(\mathfrak{J})$ have been chosen (some of them may have empty bases), the *closure* A^+ of A is unambiguously defined as follows on the union E^+ of E and the sets $E(\mathfrak{J})$ for all $\mathfrak{J}$. Let n be the arity of A and $x_1, \ldots, x_n$ elements of E^+. If some x_i belongs to E, we replace it by $x'_i = x_i$ and say that it is fixed. Now group all x_i lying in a same set $E(\mathfrak{J})$ together, and consider their images under a local isomorphism of $A(\mathfrak{J})$ towards A/D, where D is an element of $\mathfrak{J}$. Denote these images by x'_i; we stipulate that different members D of different ultrafilters $\mathfrak{J}$ be pairwise disjoint and contain no fixed elements x_i. We then set $A^+(x_1,\ldots,x_n) = A(x'_1,\ldots, x'_n)$; this value is independent of the specific intervals D and isomorphisms chosen (by 1.6).

Let us recall two problems of the above reference:

PROBLEM 1. Given a relation A and a nontrivial A-ultrafilter $\mathfrak{J}$, associate an empty relation with every other nontrivial A-ultrafilter. Does there exist a nonempty relation $A(\mathfrak{J})$ giving as closure a logical (elementary) extension of A ?

PROBLEM 2. Let $\mathfrak{J}$ and $\mathfrak{G}$ be two nontrivial A-ultrafilters, $A(\mathfrak{J})$ and $A(\mathfrak{G})$ relations each of which yields a logical extension of A via closure (the relations associated with all other A-ultrafilters being empty). Is the extension obtained by considering $A(\mathfrak{J})$ and $A(\mathfrak{G})$ a logical extension of A ?

5.1. Starting from a set E, consider a set of filters $\mathfrak{J}$ on E, each element of such a filter $\mathfrak{J}$ being a subset of E. Call this set of filters *compact* if, for any choice function f such that $f(\mathfrak{J})$ is an element of $\mathfrak{J}$ for each $\mathfrak{J}$, there exist finitely many $\mathfrak{J}$'s such that the union of the sets

$f(\mathfrak{F})$ is E. *For any set E, the set of all ultrafilters on E is a compact set.* Indeed suppose the contrary: there exists a choice function f such that, for any finite set U of ultrafilters, $E_U = E$ minus the union of $f(\mathfrak{F})$ for all $\mathfrak{F}$ of U, is nonempty. So the supersets of the E_U's constitute a filter on E; take a finer ultrafilter: it would be an ultrafilter on E, distinct from all ultrafilters (communicated by M. Jean).

Coming back to relations, we say that a relation A is *compact* when the set of all A-ultrafilters is compact. By the preceding remark, any unary relation or multirelation is compact. By the same argument, it is seen that any chain is a compact relation.

A relation A is compact iff, for any finite union of A-intervals, the complementary set (referently to the base) is a finite union of A-intervals.

Let us prove firstly the following lemma.

Call E the base of A; suppose that, for any A-interval D, the complementary set E - D is a finite union of A-intervals. Then any ultrafilter on E, once reduced to A-intervals, gives an A-ultrafilter.

Note that, for the consecutivity C on natural integers, the only C-intervals being the empty set, the whole base and singletons, any non trivial ultrafilter, once reduced to C-intervals, gives only the base: this is not a C-ultrafilter.

□ Let $\mathfrak{F}$ be an ultrafilter on E, and $\mathbf{A}$ the set of all A-intervals belonging to $\mathfrak{F}$. Suppose $\mathbf{A}$ is not an A-ultrafilter: there exists an A-interval D with $D \notin \mathbf{A}$ and D intersects any element of $\mathbf{A}$. The complementary set $E - D \in \mathfrak{F}$. By hypothesis $E - D$ is a finite union of A-intervals: so one of them $D^* \in \mathfrak{F}$ and consequently $D^* \in \mathbf{A}$. Finally D and D^* are disjoint sets; on another side D intersects any element of $\mathbf{A}$, and so D intersects D^*: contradiction. □

Proof of the proposition:

□ Suppose that, for any finite union of A-intervals, the complementary set is still a finite union of A-intervals. If A is not compact, there is a choice function f such that, for any finite set U of A-ultrafilters, E_U = E minus union of $f(\mathfrak{F})$'s for all $\mathfrak{F}$ of U, is nonempty. Then the supersets of E_U's constitute a filter on E. Take a finer ultrafilter, and reduce it to A-intervals: this gives an A-ultrafilter, by the above lemma. Call it V and note that $f(V) \in V$. For each finite set U of A-ultrafilters, E_U is by hypothesis a finite union of A-intervals: one of them belongs to V. Taking $U = \{V\}$, we get $E_U = E - f(V)$; there exists an element of V which is included in $E - f(V)$, thus disjoint from $f(V)$: contradiction; so A is compact.

Conversely suppose A compact; suppose there exists a finite set of A-intervals D, with E minus union of D's being not a finite union of A-intervals. To each A-ultrafilter $\mathfrak{J}$, associate $f(\mathfrak{J})$ being a D belonging to $\mathfrak{J}$; or $f(\mathfrak{J})$ element of $\mathfrak{J}$ and disjoint from each D. With this choice function, no finite union of $f(\mathfrak{J})$ can give E: contradiction. □

6. Problems about intervals, sums, products, in connection with Ehrenfeucht's logical study of ordinals.

In Feferman 1957, several important results due to Ehrenfeucht are related. For instance, the class of all ordinals, with the usual comparison $<$, admits as a logical (elementary) restriction, the relation $<$ on ordinals less than ω^ω. Analogous result with the comparison $<$ and sum, the smallest logical restriction being $(<,+)$ on ordinals less than $\omega^{(\omega^\omega)}$. Analogous result with the comparison, sum and product, the smallest logical restriction being $(<, +, \cdot)$ on ordinals less than ω power $(\omega^{(\omega^\omega)})$, or ω superpower 4. Moreover Ehrenfeucht proved the decidability of theories in $<$ and in $(<, +)$; the theory in $(<, +, \cdot)$ being obviously undecidable since it gives easily a representation of the arithmetic with $+$ and $\cdot$ on natural integers.

In a general manner, given a class **E** and a 'relation' **A** defined on **E**, the readjusted Löwenheim-Skolem theorem always gives a denumerable logical restriction. But it is not at all obvious to see what restriction is convenient, and the elements of the base of the denumerable logical restriction are not themselves always denumerable structures.

For instance, let us start from the class of all chains, with the comparison $\leq$, defining $A \leq B$ when there exists a restriction of B which is isomorphical to A. This comparison is reflexive, transitive, but not antisymmetric, even if we identify isomorphic chains: if ω^- is the symmetric chain of the ordinal ω, then $\omega(\omega^-)$, where each element of ω is replaced by a chain isomorphic to ω^-, is $\leq$ and $\geq$ to $1 + \omega(\omega^-)$, obtained by addition of a first element. We call *equimorphic* such chains $\leq$ and $\geq$ each to the other. Also the chain Q of rationals is equimorphic to $Q + 1$ or to $Q + \omega$.

Now let us consider the theory of comparison between chains, analogous to Ehrenfeucht's theory among ordinals. There are several theories, that are precised if we interpret the identical symbol = as the identity between chains, or as isomorphy, or as equimorphy between chains. A first problem is to know if these theories are decidable. A second problem is to find denumerable logical restrictions. It is excluded that the class of all denumerable chains gives a logical restriction: indeed the chain Q of rationals is $\geq$ each denumerable chain; so the model of all denumerable chains satisfies the formula $\exists_x \forall_y \, y \leq x$, obviously unacceptable for the class of all chains. It is also excluded to take only finite chains, because we would have the formula saying that any chain admits an immediately inferior chain. Furthermore it is excluded to take only finite and some denumerable chains. Indeed by a known theorem due to Dushnik and Miller 1940, for any chain X strictly superior to Q (rationals) and inferior to R (chain of real numbers), there exists another chain $< X$ and still $> Q$. This implies an infinite strictly decreasing sequence, that is false for denumerable chains , by Laver 1971.

An analogous problem is asked by considering the class of all relations of a given arity, with the same comparison: existence of a restriction of B isomorphical to A. Even the problems about addition and multiplication, which are obviously defined among chains, can be asked for the class of all relations. Indeed we can say that a relation C is a sum of A and B if the base C is the union of A and B, and if any union of a local automorphism of A and a local automorphism of B, is a local automorphism of C (see Fraïssé 1973, p. 108).

References.

Dushnik, B. and E. W. Miller

1940, *Concerning similarity transformations of linearly ordered sets*, Bull. Amer. Math. Soc., vol. 46 , 322-326.

Feferman, S.

1957, *Some recent works of Ehrenfeucht and Fraïssé*, Summer Institute of Symbolic Logic, mimeographied, 201-209.

Foldès, I.

1973, *Relations denses et dispersées; extension d'un théorème de Hausdorff*, C. R. Acad. Sc. Paris, 277 (A), 269-271.

Fraïssé, R.

1973, Course of Mathematical Logic, vol. 1, Reidel, Dordrecht.

1974, Course of Mathematical Logic, vol. 2, Reidel, Dordrecht.

Laver, R.

1971, *On Fraïssé's order type conjecture*, Thesis, Berkeley 1969, and Annals Math. vol. 93, 89-111.

Département de Mathématiques
Université de Provence
Marseille, France.

Non-Classical Logics, Model Theory and Computability,
A.I. Arruda, N.C.A. da Costa and R. Chuaqui (eds.)

PROJECTIVE LOGICS AND PROJECTIVE BOOLEAN ALGEBRAS (*)

by *RICARDO MORAIS*

I. INTRODUCTION.

Lusin and Sierpinski started, in 1925, the development of the theory of projective sets (cf. Lusin 1925 and Sierpinski 1925) but soon afterwards this research practically ended due to the complexity of the problems and the lack of better tools to work with.

It was not until the late sixties that some new results were obtained (cf. Fenstad 1971, Moschovakis 1970, and Kechris 1973) using the recently proposed, but still questionable, axiom of "Projective Determinancy" (cf. Addison and Moskovakis 1968).

Recently A. Nerode thought of develop a logic L_A that would be suitable for the study of the analytic sets, or projective sets of level 1, as $L_{\omega_1\omega}$ was to Borel sets. This was done by his student P. J. Campbell, and further strengthened in E.Ellentuck 1975 using different methods. This paper follows a sugestion of Ellentuck of trying the generalization of this approach to all levels of the projective hierarchy.

(*) The author is indebted to Professor Ellentuck for his orientation, and to Rutgers University (U.S.A.) and CAPES (Brazil) for their financial support.

II. PROJECTIVE LOGICS.

0. PRELIMINARIES.

Let $[seq]^n$ denote the set of all n-tuples $(v_1,\ldots,v_n)$ of finite sequences of natural numbers satisfying $\ell(v_1) = \ldots = \ell(v_n)$, where $\ell(v)$ denotes the length of v.

Throughout this text f and g will denote elements of ${}^{\omega}\omega$; n, m, and k will be natural numbers and v will stand both for elements of $[seq]^n$, and for elements of seq the set of finite sequences of natural numbers.

The most important tool developped in this work is a pair of families of subsets of $[seq]^n$, denoted respectively by $Full_n$ and $Full_n^*$.

DEFINITION 1. a) *$F \in Full_n$ (F is a Full-n set) if and only if $F \subset [seq]^n$ and F satisfies the following condition given by a string of quantifiers with n alternations:*

$$(\exists f_1)(\forall f_2)\ldots(Qf_n)(Q'k)((f_1 \mid k,\ldots,f_n \mid k) \in F)$$

where Q and Q' are distinct quantifiers.

Similarly we define

b) *$G \in Full_n^*$ (G is a Full-n-star set) if and only if $G \subset [seq]^n$ and*

$$(\forall f_1)(\exists f_2)\ldots(Q'f_n)(Qk)((f_1 \mid k,\ldots,f_n \mid k) \in G).$$

F and G will denote elements of $Full$ and $Full^*$ respectively, and the word countable will always mean either finite or denumerable.

Finally, if $\Phi = \{\phi_i \mid i \in I\}$ is a countable set of formulas of a formal language, the conjunction $\wedge\Phi$ will also be written $\bigwedge_{i\in I} \phi_i$ or simply $\wedge_I \phi_i$ if no confusion arises.

1. THE LANGUAGE L_{pn}

We start with a first order logic L with countably many relation, function and constant symbols, and ω_1 variables. Let $L_{\omega_1\omega}$ be the infinitary logic over L as defined in Keisler 1971, p. 6. We obtaion L_{pn} from $L_{\omega_1\omega}$ by adjoining two new operators P_n and P_n^* in the following way:

Formulas of L_{pn} are those of $L_{\omega_1\omega}$ with the addition: if ϕ is a map from $[seq]^n$ into formulas of L_{pn} then $P_n(\phi)$ and $P_n^*(\phi)$ are formulas of L_{pn}.

Extend the notion of *satisfaction* by defining:

(1) $\mathfrak{A} \models P_n(\phi)[h]$ iff $(\exists F \in Full_n)(\forall v \in F)\ \mathfrak{A} \models \phi(v)[h]$

$\mathfrak{A} \models P_n^*(\phi)[h]$ iff $(\exists G \in Full_n^*)(\forall v \in G)\ \mathfrak{A} \models \phi(v)[h]$

Occurrences of variables and constants in $P_n(\phi)$ and $P_n^*(\phi)$ are defined to be those in the collection of all $\phi(v)$ for $v \in [seq]^n$.

Define $\mathfrak{A}$-*validity* and *validity* in the usual way.

The notion of *subformula* of Keisler 1971, p. 11 is extended by

(2) $Sub(P_n(\phi)) = \{P_n(\phi)\} \cup \bigcup_v Sub(\phi(v))$.

$Sub(P_n^*(\phi)) = \{P_n^*(\phi)\} \cup \bigcup_v Sub(\phi(v))$.

Extend the notion of *moving the negation inside* (cf. Keisler 1971, p.11) as follows:

(3) $(P_n(\phi))\neg$ is $P_n^*(\neg\phi)$

$(P_n^*(\phi))\neg$ is $P_n(\neg\phi)$,

where $\neg\phi$ is the map that takes v into $\neg\phi(v)$.

Axioms of L_{pn} will consist of the nine axioms for propositional logic as presented in Bell and Slomson 1969, p. 36, plus the six axioms for $L_{\omega_1\omega}$ as in Keisler 1971, p. 15.

The *Rules of Inference* will be

(R1) Modus Ponens

(R2) if $\vdash \psi \to \phi$ then $\vdash \psi \to \forall x\phi$ if x is not free in ψ .

(R3) if for every $\phi \in \Phi \vdash \psi \to \phi$ then $\vdash \psi \to \bigwedge\Phi$.

(R4) if for every $F \in Full_n \vdash \bigwedge_F \phi(v) \to \psi$ then $\vdash P_n(\phi) \to \psi$.

(R5) if for every $G \in Full_n^* \vdash \psi \to \bigvee_G \phi(v)$ then $\vdash \psi \to P_n(\phi)$.

We only have to worry about introducing $P_n(\phi)$ because the facts about $P_n^*(\phi)$ are easily gotten from those about $P_n(\phi)$ using the following instance of the axiom $\vdash \psi\neg \longleftrightarrow \neg\psi$.

PROPOSITION 2. $\vdash P_n^*(\neg\phi) \leftrightarrow \neg P_n(\phi)$

Now before we proceed with the studies of L_{pn} let us collect some important properties of the families $Full_n$ and $Full_n^*$.

PROPOSITION 3. *Suppose that for every* $v \in [seq]^n$, $\lambda(v)$ *is a meta-statement about* v . *Let* Q *and* Q' *be as in Definition 1. Then the following are true meta-statements.*

a) $(\exists f_1)\cdots(Qf_n)(Q'k)\lambda(f_1 \mid k,\ldots,f_n \mid k)$ *iff* $(\exists F \in Full_n)(\forall v \in F)\lambda(v)$

b) $(\forall f_1)\cdots(Q'f_n)(Qk)\lambda(f_1 \mid k,\ldots,f_n \mid k)$ *iff* $(\exists G \in Full^*_n)(\forall v \in G)\lambda(v)$

c) $(\exists F \in Full_n)(\forall v \in F)\lambda(v)$ *iff* $(\forall G \in Full^*_n)(\exists v \in G)\lambda(v)$

d) $(\forall F \in Full_n)(\exists v \in F)\lambda(v)$ *iff* $(\exists G \in Full^*_n)(\forall v \in G)\lambda(v)$

It is important here to note the complete symmetry between these two families of sets. It is tnis symmetry that will make possible the majority of our proofs, besides elegantly reducing all the work in half.

Now, in order to better understand the behavior of these families, we have:

PROPOSITION 4. *If* $n > 1$, $f \in {}^{\omega}\omega$ *and* $H \subset [seq]^n$, *let*

$$H_f = \{(v_2,\ldots,v_n) \mid (f \mid \ell(v_2),\ v_2,\ldots,v_n) \in H\}.$$

Then

a) $H \in Full_n$ *iff* $(\exists f)\ (H_f \in Full^*_{n-1})$

b) $H \in Full^*_n$ *iff* $(\forall f)\ (H_f \in Full_{n-1})$

Finally the next proposition shows that the families $Full_n$ and $Full^*_n$ are well mixed together. The proof is by induction on n.

PROPOSITION 5. $(\forall F \in Full_n)(\forall G \in Full^*_n)(F \cap G \neq \emptyset)$

2. THE CONSISTENCY PROPERTY.

Let C be a countable set of constant symbols not appearing in L. Let M be the first order logic obtained by adding each $c \in C$ to L, and from M construct the logic M_{pn}.

A *basic term* is either a constant symbol of M_{pn} or a term of the form $f(t_1,\ldots,t_k)$ where $t_1,\ldots,t_k$ are basic terms and f is a k-ary function symbol of L.

The definition that follows was taken from Keisler 1971, p. 11, but here, besides adapting it to the present situation (namely adding clauses C8 and C9) we also modified, to simplify the proofs, the notion of basic term, and clauses C10 and C11.

DEFINITION 6. *A Consistency Property is a set* S *satisfying for each* $s \in S$:

(C1) *s is a countable set of sentences of* M_{pn} ;

(C2) *if* ϕ *is a sentence of* M_{pn} *then either* $\phi \notin s$ *or* $\neg\phi \notin s$;

(C3) *if* $\neg\phi \in s$ *then* $s \cup \{\phi\neg\} \in S$;

(C4) *if* $\wedge\Phi \in s$ *then* $s \cup \{\phi\} \in S$ *for each* $\phi \in \Phi$;

(C5) *if* $\vee\Phi \in s$ *then* $s \cup \{\phi\} \in S$ *for some* $\phi \in \Phi$;

(C6) *if* $\forall x\phi \in s$ *then* $s \cup \{\phi(c)\} \in S$ *for each* $c \in C$;

(C7) *if* $\exists x\phi \in s$ *then* $s \cup \{\phi(c)\} \in S$ *for some* $c \in C$;

(C8) *if* $P_n(\phi) \in s$ *then* $s \cup \{\wedge_F\phi(v)\} \in S$ *for some* $F \in Full_n$;

(C9) *if* $P^*_n(\phi) \in s$ *then* $s \cup \{\wedge_G\phi(v)\} \in S$ *for some* $G \in Full^*_n$;

(C10) *if* t *and* t' *are basic terms and* $t = t' \in s$ *then* $s \cup \{t' = t\} \in S$;

(C11) *if* t *and* t' *are basic terms,* $t = t' \in s$ *and* $\phi(t) \in s$ *then* $s \cup \{\phi(t')\} \in S$;

(C12) *if* t *is a basic term then there is a* $c \in C$ *such that* $s \cup \{t = c\} \in S$.

The definition of Consistency Property is this long because we want to have:

THEOREM 7. (Model Existence Theorem). *If* S *is a Consistency Property and* $s_0 \in S$, *then* s_0 *has a model.*

PROOF: Without loss of generality we way assume that each subset of an element of S is again in S. To construct the model satisfying s_0 we start with the smallest set Y of formulas of M_{pn} for which:

(i) $s_0 \subset Y$

(ii) Y is closed under subformulas.

(iii) If t is a term, t' a basic term and $\phi(t) \in Y$ then $\phi(t') \in Y$.

(iv) if $\neg\phi \in Y$ then $\phi\neg \in Y$.

(v) If $c \in C$ and t is a basic term then $c = t \in Y$.

Let $X = \{\phi_0, \phi_1, \ldots\}$ be the countably infinite set of sentences of Y, and $T = \{t_0, t_1, \ldots\}$ be the set of basic terms. Starting with s_0 construct an increasing sequence of elements of S as follows. Suppose we have s_m, and we build $s_{m+1} \supset s_m$:

(1) if $s_m \cup \{\phi_m\} \notin S$ let $s'_{m+1} = s_m$;

(2) if $s_m \cup \{\phi_m\} \in S$ we consider the following cases:

(2.1) ϕ_m is $\bigvee\Phi$ then for some $\phi \in \Phi$, $s'_{m+1} = s_m \cup \{\phi_m\} \cup \{\phi\} \in S$,

(2.2) ϕ_m is $\exists x\phi$ then for some $c \in C$, $s'_{m+1} = s_m \cup \{\phi_m\} \cup \{\phi(c)\} \in S$,

(2.3) ϕ_m is $P_n(\phi)$ then for some $F \in Full_n$, $s'_{m+1} =$
$= s_m \cup \{\phi_m\} \cup \{\bigwedge_F \phi(v)\} \in S$,

(2.4) ϕ_m is $P^*_n(\phi)$ then for some $G \in Full^*_n$, $s'_{m+1} =$
$= s_m \cup \{\phi_m\} \cup \{\bigwedge_G \phi(v)\} \in S$,

(2.5) ϕ_m is any other formula, $s'_{m+1} = s_m \cup \{\phi_m\} \in S$;

(3) finally, since in any case $s'_{m+1} \in S$, there is $c \in C$ such that
$s_{m+1} = s'_{m+1} \cup \{c = t_m\} \in S$.

Next define $s_\omega = \bigcup_m s_m$ and define an equivalence relation on C by: $c \sim d$ iff $c = d \in s_\omega$.

Let $[c]$ be the equivalence class of $c \in C$ and let $A = \{[c] \mid c \in C\}$. This is the universe of the model that will satisfy s_0.

Now for each k-ary relation symbol P_m of L and each k-ary function symbol f_m of L define a relation R_m on A^k and a function F_m from A^k into A by:

(a) $([c_1],\ldots,[c_k]) \in R_m$ iff $P_m(c_1,\ldots,c_k) \in s_\omega$

(b) $F_m([c_1],\ldots,[c_k]) = [c_0]$ iff $c_0 = f_m(c_1,\ldots,c_k) \in s_\omega$.

Note now that if $\phi \in s_\omega$ and

(a) ϕ is $\bigwedge\Phi$, then $\theta \in y$ for each $\theta \in \Phi$;

(b) ϕ is any other formula, then $\phi \in y$.

Then use this fact to show that the structure

$$\mathfrak{A} = \langle A, \{R_m \mid m \in \omega\}, \{F_m \mid m \in \omega\}, A \rangle$$

satisfies s_0.

Theorem 7 is a nice tool to use in the proof of

THEOREM 8. (The Completeness Theorem for L_{pn}) *If ϕ is a sentence of*

L_{pn} *then* $\vdash \phi$ *iff* $\models \phi$

PROOF: To show that every theorem is valid we prove that the rules of inference (R4) and (R5) preserve validity.

(a) Rule (R4).

Suppose $\forall F \in Full_n$, $\mathfrak{A} \models \Lambda_F \phi(v) \to \psi$ then it is not the case that $\exists F \in Full_n$, $\mathfrak{A} \models \Lambda_F \phi(v) \wedge \neg\psi$.

But by Definition 1, $\mathfrak{A} \models P_n(\phi)$ implies that $\exists F \in Full_n$, $\mathfrak{A} \models \Lambda_F \phi(v)$, and therefore implies not $\mathfrak{A} \models P_n(\phi) \wedge \neg\psi$ or equivalently, $\mathfrak{A} \models P_n(\phi) \to \psi$.

(b) Rule (R5).

Suppose $\forall G \in Full_n^*$, $\mathfrak{A} \models \psi \to V_G\, \phi(v)$. Then $\mathfrak{A} \models \neg\psi$ or $(\forall G \in Full_n^*)$ $(\exists v \in G)$ $\mathfrak{A} \models \phi(v)$ and hence by Proposition 3, $\mathfrak{A} \models \neg\psi$ or $(\exists F \in Full_n)(\forall v \in F)$ $\mathfrak{A} \models \phi(v)$, which implies $\mathfrak{A} \models \psi \to P_n(\phi)$.

Now we have to show that every valid sentence is a theorem. In order to do that we let S be the set of finite sets of sentences s of M_{pn} such that only finitely many $c \in C$ occur in s and not $\vdash \neg\Lambda s$.

We then show S is a Consistency Property and the result follows because if ϕ is not a theorem in L_{pn} then ϕ is not a theorem in M_{pn} and hence $\{\neg\phi\} \in S$. By the Model Existence Theorem $\neg\phi$ has a model and therefore ϕ is not valid.

We exemplify the proof that S is a Consistency Property be proving (C8) and (C9).

(C8) Suppose $P_n(\phi) \in s$ but $(\forall F \in Full_n)(s \cup \{\Lambda_F \phi(v)\} \notin S$.

Since $P_n(\phi) \in s$, we have $\forall F \in Full_n$ $\vdash \neg\Lambda(s \cup \{\Lambda_F \phi(v)\})$, and so $(\forall F \in Full_n)$ $(\vdash \Lambda_F \phi(v) \to \neg\Lambda s)$; then by (R4) $\vdash P_n(\phi) \to \neg\Lambda s$ and, since $P_n(\phi) \in s$, $\vdash \neg\Lambda s$, a contradiction.

(C9) Suppose $P_n^*(\phi) \in s$ but $(\forall G \in Full_n^*)$ $(s \cup \{\Lambda_G \phi(v)\} \notin S)$; then again $\vdash \neg\Lambda(s \cup \{\Lambda_G \phi(v)\})$ for every $G \in Full_n^*$ and so $(\forall G \in Full_n^*)$ $(\vdash \Lambda s \to V_G \neg\phi(v))$, which implies, by (R5), $\vdash \Lambda s \to P_n(\neg\phi)$.

Therefore, by Proposition 2, $\vdash \Lambda s \to \neg P_n^*(\phi)$ or, equivalently, $\vdash \neg\Lambda s$, a contradiction.

There is another projective logic of interest to us, namely:

DEFINITION 9. *The logic* L_p *is defined to be the union of all* L_{pn} *as* n *runs over* ω, *or in other words:*

(a) *In* L_p, $P_n(\phi)$ *is a formula for every* n.

(b) *The rules of inference* (R4) *and* (R5) *now read*

(R4) $(\forall n \in \omega)$ *if* $(\forall F \in Full_n) \vdash \Lambda_F \phi(v) \to \psi$ *then* $\vdash P_n(\phi) \to \psi$,

(R5) $(\forall n \in \omega)$ *if* $(\forall G \in Full^*_n) \vdash \psi \to V_G \phi(v)$ *then* $\vdash \psi \to P_n(\phi)$.

Obviously L_p is complete since all L_{pn} are.

There is one important theorem particular to L_p:

THEOREM 10. *In* L_p *the following is a rule of inference:*

(R4') (a) (*for* $n = 1$) *if* $\forall f \vdash \Lambda_k \phi(f \mid k) \to \psi$ *then* $\vdash P_1(\phi) \to \psi$;

(b) (*for* $n > 1$) *if* $\forall f \vdash P^*_{n-1}(\phi_f) \to \psi$ *then* $\vdash P_n(\phi) \to \psi$;

where $\phi_f(v_2,\dots,v_n) = \phi(f \mid \ell(v_2), v_2,\dots,v_n)$.

We conclude this section with the remark that the downward Löwenheim-Skolem-Tarski theorem holds for both L_p and L_{pn}.

III. n-PROJECTIVE BOOLEAN ALGEBRAS.

1. INTRODUCTION.

In this section we define a new kind of Boolean algebras, called n-projective Boolean algebras, which are generalizations of the Suslin algebras introduced by L. Rieger in 1955 (cf. Rieger 1955).

Our work, however, is patterned after a recent paper by E. Ellentuck (Ellentuck 197+) in which he studies the S-algebras of Rieger based on his previous paper on Suslin logic (Ellentuck 1975).

Rieger's idea with the Suslin algebras was to provide a structure in which one could model Π^1_1 analysis. Our algebras are intended to help model Π^1_n analysis.

1. BASIC RESULTS.

Let B be a Boolean algebra.

The *joins* and *meets* of B will be denoted respectively by *Sup* and *Inf*. The *infinite join* of the family $\{b_i \mid i \in I\}$ is denoted by $\mathit{Sup}_{i \in I}\, b_i$ or simply by $\mathit{Sup}_i\, b_i$ if it clear which set I is.

If ϕ is a map from $[seq]^n$ into B we shall use the notation $P_n(\phi)$ for the following element of B, provided it exists:

$$P_n(\phi) = \mathit{Sup}_F \ \mathit{Inf}_{v \in F}\, \phi(v),$$

and, similarly,

$$P_n^*(\phi) = \mathit{Sup}_G \ \mathit{Inf}_{v \in G}\, \phi(v),$$

where, as usual, F runs over Full_n and G over Full_n^* .

The symbols $P_n(\phi)$ and $P_n^*(\phi)$ will be used both for the above suprema and the previously defined formulas of L_{pn}, but this should lead to no confusion.

DEFINITION 11. *A σ-Boolean algebra B is an n-projective Boolean algebra (n-PBA for short) if $P_n(\phi) \in B$ for every map ϕ from $[seq]^n$ into B and if in adition B satisfies the following distributive law:*

$$(4) \qquad \mathit{Sup}_F \ \mathit{Inf}_{v \in F}\, \phi(v) = \mathit{Inf}_G \ \mathit{Sup}_{v \in G}\, \phi(v).$$

DEFINITION 12. *A σ-Boolean algebra is an ω-projective Boolean algebra (ω-PBA for short) if it is n-PBA for every $n \in \omega$.*

Formula (4) is a very powerful distributive law and not all algebras closed under P_n and P_n^* satisfy it. In fact, there are complete Boolean algebras in which (4) fails. In our work, however, we need this distributivity to tie things up (see for example condition (6) below), and we are thus forced to introduce it as part of the definition.

To get an example in which (4) fails see Morais 1976.

Another way to see the importance of (4) is the next proposition which presents three equivalent formulations of (4).

PROPOSITION 13. *If B is a Boolean algebra and ϕ is a map from $[seq]^n$ into B, let $-\phi$ be the map defined by*

$$(-\phi)(v) = -\phi(v)$$

where $-$ is the symbol for complementation in B. Then the following are equivalent:

(4) $\underset{F}{Sup}\ \underset{v\in F}{Inf}\ \phi(v) = \underset{G}{Inf}\ \underset{v\in G}{Sup}\ \phi(v)$ *for all* ϕ.

(5) $\underset{F}{Inf}\ \underset{v\in F}{Sup}\ \phi(v) = \underset{G}{Sup}\ \underset{v\in G}{Inf}\ \phi(v)$ *for all* ϕ.

(6) $P_n(\phi) = -P^*_n(-\phi)$ *for all* ϕ.

(7) $P^*_n(\phi) = -P_n(-\phi)$ *for all* ϕ.

Now, using these equivalences, we can get several properties of projective Boolean Algebras, namely:

PROPOSITION 14. *A* σ*-Boolean algebra* B *is* n-PBA *if and only if* B *is closed under the* P^*_n *operator and* (4) *holds.*

PROPOSITION 15. *If* $n > 1$ *and* B *is* n-PBA, *then* B *is* $(n-1)$-PBA.

PROPOSITION 16. *If* $n > 1$ *then every* n-PBA *satisfies:*

(a) $P_n(\phi) = \underset{f}{Sup}\ P^*_{n-1}(\phi_f)$,

(b) $P^*_n(\phi) = \underset{f}{Inf}\ P_{n-1}(\phi_f)$,

where f *runs in* ${}^{\omega}\omega$ *and* ϕ_f *is defined by*

$$\phi_f(v_2,\ldots,v_n) = \phi(f \mid \ell(v_2),\ v_2,\ldots,v_n).$$

PROPOSITION 17. *Every* n-PBA *satisfies:*

(a) $P_n(\phi) = \underset{f_1}{Sup}\ \underset{f_2}{Inf} \ldots \underset{f_n}{Sup}\ \underset{k}{Inf}\ \phi(f_1 \mid k,\ldots,f_n \mid k)$,

(b) $P^*_n(\phi) = \underset{f_1}{Inf}\ \underset{f_2}{Sup} \ldots \underset{f_n}{Inf}\ \underset{k}{Sup}\ \phi(f_1 \mid k,\ldots,f_n \mid k)$,

where we took n *odd as an example.*

PROPOSITION 18. *The complete Boolean algebra* $2 = \{0,1\}$ *is* ω-PBA.

PROOF: Since 2 is complete we have just to show (4) holds in 2.

Now $P_n(\phi) = 0$ iff $\underset{F}{Sup}\ \underset{v\in F}{Inf}\ \phi(v) = 0$ iff $(\forall F \in Full_n)(\exists v \in F)(\phi(v)=0)$ iff (by Proposition 3) $(\exists G \in Full^*_n)(\forall v \in G)(\phi(v) = 0)$ iff $\underset{G}{Inf}\ \underset{v\in G}{Sup}\ \phi(v) = 0$ iff $\underset{G}{Sup}\ \underset{v\in G}{Inf}\ -\phi(v) = 1$ iff $P^*_n(-\phi) = 1$ iff $-P^*_n(-\phi) = 0$ and therefore (6) holds.

The most important example of an ω - PBA however is given by the following:

THEOREM 19. *The Lindenbaum algebra L_p of the ω - projective logic L_p is* ω - PBA.

PROOF: Let $|\phi|$ denote the equivalence class of the formula ϕ in L_p.

Let Φ be a map from $[seq]^n$ into L_p and define a map ϕ from $[seq]^n$ into formulas of L_p by choosing for each $v \in [seq]^n$ a representative formula $\phi(v)$ of the equivalence class $\Phi(v)$. We then show

(8) $P_n(\Phi) = |P_n(\phi)|$,

and hence L_p is closed under the P_n operator.

First we have to prove that the formula $P_n(\phi)$ does not depend on the choice of the map ϕ.

Its enough to show that for any other map ψ : if

$\forall v \in [seq]^n \vdash \phi(v) \rightarrow \psi(v)$ then $\vdash P_n(\phi) \rightarrow P_n(\psi)$.

By (R4) this follows from

(9) $\forall F \in Full_n \vdash \bigwedge_F \phi(v) \rightarrow P_n(\psi)$,

which in turn follows from, (by R5),

$(\forall G \in Full_n^*)(\forall F \in Full_n) \vdash \bigwedge_F \phi(v) \rightarrow \bigvee_G \psi(v)$.

But this is trivial since by Proposition 5, given any F and G, $F \cap G \neq \emptyset$. Therefore (9) holds.

Now to finish the proof of Theorem 19 we have to show that the distributive law (4) holds in L_p.

We shall need,

(10) $P_n^*(\Phi) = |P_n^*(\phi)|$

which is equivalent to,

$$\sup_G |\bigwedge_G \phi(v)| = |P_n^*(\phi)| ,$$

and so we have to prove:

(i) $(\forall G \in Full_n^*) \vdash \bigwedge_G \phi(v) \rightarrow P_n^*(\phi)$

and

(ii) If $(\forall G \in Full_n^*) \vdash \bigwedge_G \phi(v) \rightarrow \psi$ then $\vdash P_n^*(\phi) \rightarrow \psi$.

PROOF of (i): From proposition 5 get

$$(\forall G \in Full_n^*)(\forall F \in Full_n) \vdash \wedge_F \neg\phi(v) \to \vee_G \neg\phi(v)$$

now apply (R4) and use Proposition 2.

PROOF of (ii): Start with the hypothesis

$$(\forall G \in Full_n^*) \vdash \neg\psi \to \vee_G \neg\phi(v),$$

then apply (R5), and use Proposition 2. Finally (8), (10) and Proposition 2 give

$$P_n(\Phi) = - P_n^*(-\Phi)$$

and therefore (6) holds, which is equivalent to (4).

NOTE: Evidently exactly the same proof shows that the Lindenbaum algebra of L_{pn} (denoted L_{pn}) is n-PBA.

3. Free n-projective Boolean algebras.

DEFINITION 20. *An nP-homomorphism between Boolean algebras is a σ-homomorphism that preserves the P_n operator. An ωP-homomorphism is a σ-homomorphism that preserves P_n for every $n \in \omega$.*

DEFINITION 21. *Let B be an n-PBA and $G \subset B$. Then:*

(a) *G nP-generates B if B is the smallest n-PBA containing G.*

(b) *G freely nP-generates B if G nP-generates B and in addition given any other n-PBA B' and any map $h : G \to B'$ there is an nP-homomorphism $H : B \to B'$ which extends h.*

(c) *Similarly define ωP-set of generators and free ωP-set of generators.*

DEFINITION 22. *An n-PBA is a free nP-algebra if it contains a free nP-set of generators. Similarly, define a free ωP-algebra.*

If is a common practice in any text about "free" structures to first talk about its uniqueness and afterwards to prove its existence. The following two propositions are proven in the same way it is usually done for general Boolean algebras. See for example Halmos 1963, p. 42.

PROPOSITION 23. *If B is a free nP-algebra, G the set of free nP-generators and h the given map from G into the n-PBA B', then the nP-homomorphism $H : B \to B'$ that extends h is unique.*

PROPOSITION 24. *Any two free nP-algebras whose set of generators have the same cardinality are nP-isomorphic.*

Now to present an example of a free ωP-algebra (the existence of a free nP-algebra is proved similarly) we proceed as follows.

First define a *propositional logic* L_κ for each cardinal κ and then we show that the Lindenbaum algebra $\mathcal{L}_\kappa$ of L_κ is a free ωP-algebra with κ generators.

L_κ is going to have a set of κ variables

$$\{p_\alpha \mid \alpha < \kappa\},$$

and the propositional connectives $\neg$ and $\bigwedge$. As in L_{pn}, we introduce the operators P_n and $P_n^\star$ and let the set of formulas be the least set such that

(a) p_α is a formula for each ordinal $\alpha < \kappa$.

(b) if ϕ is a formula then so is $\neg\phi$

(c) if Φ is a countable set of formulas then $\bigwedge\Phi$ is a formula.

(d) if ϕ is a map from $[seq]^n$ into formulas then $P_n(\phi)$ and $P_n^\star(\phi)$ are formulas (for every $n \in \omega$).

Define "*moving the negation inside*" for formulas of L_κ as we did for L_{pn} with the addition:

$$\text{"}p_\alpha\neg \quad \text{is} \quad \neg p_\alpha\text{"}.$$

For axioms take the nine axioms of propositional logic as in Bell and Slomson 1969, p. 36, plus $\vdash \phi\neg \leftrightarrow \neg\phi$ and $\vdash \bigwedge\Phi \to \phi$, where Φ is a countable set of formulas and $\phi \in \Phi$.

For rules of inference take those of L_{pn} with the exception of (R2).

A realization of L_κ is a map f from the set of variables into the ω-PBA $2 = \{0, 1\}$, which is inductively extended to all formulas as follows:

(a) $f(\neg\phi) = -f(\phi)$,

(b) $f(\bigwedge\Phi) = \operatorname{Inf}_{\phi \in \Phi} f(\phi)$,

(c) $f(P_n(\phi)) = P_n(f(\phi))$ and $f(P_n^\star(\phi)) = P_n^\star(f(\phi))$, where $f(\phi)$ is the map defined by $f(\phi)(v) = f(\phi(v))$ for $v \in [seq]^n$.

We say that a formula ϕ is valid if $f(\phi) = 1$ in all realizations f .

Now, before we prove that L_κ is an ωP-algebra on κ generators, we need:

PROPOSITION 25. *Let B be an ωP-algebra and f any map from the variables of L_κ into B. Extend f to all formulas by rules* (a) *through* (c). *Then*

$$\vdash \phi \quad \textit{implies} \quad f(\phi) = 1.$$

In particular, by Proposition 18, *every theorem of L_κ is valid.*

PROOF: First note that because of properties (a) and (b) f satisfies:

(11) $f(\phi \to \psi) = 1$ if and only if $f(\phi) \leq f(\psi)$.

It is routine to show that the axioms are mapped into 1 , but we check, as an example, that the axiom $\phi\neg \leftrightarrow \neg\phi$ is mapped into 1 for the case ϕ is $P_n(\psi)$.

By (11) we have to show,

$$f(P_n(\psi)\neg) = f(\neg P_n(\psi)) .$$

But

$$f(P_n(\psi)\neg) = f(P_n^*(\neg \psi)) = P_n^*(f(\neg \psi) =$$

$$P_n^*(- f(\psi)) = -P_n(f(\psi)) = - f(P_n(\psi)) = f(\neg P_n(\psi)).$$

where the fourth equality follows from (6).

Similarly, using (11) it is easy to prove that the rules of inference preserve the property of being mapped into 1. As an example we check for (R4).

Suppose $\forall F \in Full_n$, $f(\wedge_F \phi(v) \to \psi) = 1$ and we have to show $f(P_n(\phi) \to \psi) = 1$.

By (11) and property (b) we have

$$(\forall F \in Full_n) \operatorname{Inf}_{v \in F} f(\phi(v)) \leq f(\psi) .$$

Therefore

$$\operatorname{Sup}_F \operatorname{Inf}_{v \in F} f(\phi(v)) \leq f(\psi).$$

But by definition this is $P_n(f(\phi)) \leq f(\psi)$, and hence $f(P_n(\phi)) \leq f(\psi)$. Thus by (11), $f(P_n(\phi) \to \psi) = 1$.

We therefore conclude that every theorem of L_κ is mapped into 1.

We are now in position to show

THEOREM 26. *L_κ is a free ωP-algebra on exactly κ generators.*

PROOF: First it is clear that the same proof used to show that L_p was ω-PBA (Theorem 19) can be repeated here to show L_κ is ω-PBA.

Next let $G = \{|p_\alpha| \mid \alpha < \kappa\}$ and let there be given an arbitrary ωP-algebra B together with a map $h : G \to B$.

Now using h define a map f from the variables of L_κ into B by

$$f(p_\alpha) = h(|p_\alpha|),$$

and extend f inductively to all formulas of L_κ.

By Proposition 25 and (11) it is easy to show that f is constant in every equivalence class $|\phi|$, and so the following is a well defined map from L_κ into B:

$$H(|\phi|) = f(\phi).$$

This H is the desired ωP-homomorphism extending h, and hence it only remains to show that the cardinality of G is κ. But this is easy, for given $\alpha, \beta < \kappa$ with $\alpha \neq \beta$, Proposition 25 can help to show that $p_\alpha \leftrightarrow p_\beta$ is not a theorem and hence $|p_\alpha| \neq |p_\beta|$.

4. A REPRESENTATION THEOREM FOR FREE nP-BOOLEAN ALGEBRAS.

We start this section with a completeness theorem for L_κ. This is done the same way we did for L_{pn} and so we omit the proof, although we point out the basic points. First we define:

DEFINITION 27. A *κ-Consistency Property, where κ is a cardinal, is a set S satisfying for all $s \in S$*

(K1) *s is a countable set of formulas of L_κ;*

(K2) *if ϕ is a formula of L_κ then either $\phi \notin s$ or $\neg\phi \notin s$;*

(K3) *if $\neg\phi \in s$ then $s \cup \{\phi\neg\} \in S$;*

(K4) *if $\bigwedge\Phi \in s$ then $s \cup \{\phi\} \in S$ for all $\phi \in \Phi$;*

(K5) *if $\bigvee\Phi \in s$ then $s \cup \{\phi\} \in S$ for some $\phi \in \Phi$;*

(K6) *if $P_n(\phi) \in s$ then $s \cup \{\bigwedge_F \phi(v)\} \in S$ for some $F \in Full_n$;*

(K7) *if* $P^*_n(\phi) \in s$ *then* $s \cup \{\Lambda_G \phi(v)\} \in S$ *for some* $G \in Full^*_n$.

Next we have:

PROPOSITION 28. *If* S *is a* κ *- Consistency Property and* $s_o \in S$ *then there is a realization* f *of* L_κ *for which* $f(\phi) = 1$ *for all* $\phi \in s_o$.

PROOF: This proof is patterned after the one for the Model Existence Theorem (Theorem 7). We start s_o and construct a sequence (s_m) of elements of of S with the desired closure properties. Then let $s_\omega = \bigcup_m s_m$ and define a map from the variables of L_κ into 2 by

$$f(p_\alpha) = 1 \quad \text{iff} \quad p_\alpha \in s_\omega$$

Then extend f inductively to all formulas and because of the way the sequence (s_m) was constructed we have

$$f(\phi) = 1 \quad \text{for all} \quad \phi \in s_\omega .$$

Finally, we have:

PROPOSITION 29. *If* ϕ *is not a theorem of* L_κ *then for some realization,* $f(\phi) = 0$.

PROOF: Just like we did for L_{pn} we show that the set of all finite sets s of formulas of L_κ for which not $\vdash \neg \Lambda s$ is a κ - Consistency Property. Then use Proposition 28 to get the result.

DEFINITION 30. *An* nP *- field of sets is a* σ *- field of sets closed under the operator* P_n. *If a* σ*- field of sets is closed under* P_n *for every* $n \in \omega$ *we call it an* ωP *- field of sets.*

Notice that we did not mention any distributive law here. This however is no surprise because we have:

PROPOSITION 31. *Every* nP *- field of sets is* n - PBA.

THEOREM 32. *For each cardinal* κ *there is an* nP *- field of sets (respectively* ωP *- field of sets) that is* nP *- generated (*ωP *- generated) by* κ *of its elements.*

PROOF: Let $X = 2^\kappa$ be the set of maps from κ into $2 = \{0, 1\}$ and define for each $\alpha < \kappa$

$$g_\alpha = \{f \in 2^\kappa \mid f(\alpha) = 1\} .$$

Next, let $Q = \{g_\alpha \mid \alpha < \kappa\}$ which is a subset of the power set of X, and let $B_{\kappa n}$ (respectively B_κ) be the smallest nP-algebra (ωP-algebra) containing Q.

Since the power set of X is a complete field of sets, $B_{\kappa n}$ and B_κ are well defined.

Finally, to show that the cardinality of Q is κ take $\alpha \neq \beta$ and choose any map $f \in 2^\kappa$ for which $f(\alpha) \neq f(\beta)$. Hence if, say, $f(\alpha) = 1$ then $f \in g_\alpha$ but $f \notin g_\beta$, and therefore $g_\alpha \neq g_\beta$.

Now copying what we did for L_κ we construct a propositional logic $L_{\kappa n}$ for each $n \in \omega$ in such a way that their corresponding Lindenbaum algebras $L_{\kappa n}$ are free nP-algebras.

Our representation theorem for nP-algebras is an immediate consequence of the next very important proposition.

PROPOSITION 33. *L_κ is ωP-isomorphic to B_κ; and for every $n \in \omega$, $L_{\kappa n}$ is nP-isomorphic to $B_{\kappa n}$.*

PROOF: We prove only that L_κ is ωP-isomorphic to B_κ.

The ωP-isomorphism $H : L_\kappa \to B_\kappa$ we are looking for is defined inductively by:

(a) For every ordinal $\alpha < \kappa$, $H(|p_\alpha|) = g_\alpha$.

(b) $H(|\neg\phi|) = H(|\phi|)'$, where A' denotes the complement of A.

(c) $H(|\wedge\Phi|) = \bigcap_{\phi \in \Phi} H(|\phi|)$.

(d) $H(|P_n(\phi)|) = \bigcup_F \bigcap_{v \in F} H(|\phi(v)|)$.

This definition makes H an ωP-homomorphism, and we have to show it is one-to-one and onto.

To show H is one-to-one we define for each $f \in 2^\kappa$ a realization f' of L_κ by

$$f'(p_\alpha) = f(\alpha)$$

(of course extending f' inductively to all formulas).

Next, by induction on the complexity of ϕ, we show $H(|\phi|) = \{f \in 2^\kappa \mid f'(\phi) = 1\}$.

Finally, we have to prove that if $H(|\phi|)$ is the 1 of B_κ then $|\phi|$ is the 1 of L_κ.

But if $H(|\phi|) = 2^\kappa$ then for every $f \in 2^\kappa$, $f'(\phi) = 1$ which means that every realization of L_κ satisfies ϕ. Therefore ϕ is a theorem by Proposition 29, and hence $|\phi|$ is the 1 of L_κ.

Lastly, since the image of L_κ under H is an ωP-algebra which contains Q, the ωP-set of generators of B_κ, we have that H is onto.

Now given any ωP-algebra B, let κ be the cardinality of the set B. Since L_κ is a free ωP-algebra we can get an ωP-homomorphism from L_κ onto B.

Therefore the previons proposition gives:

THEOREM 34. (The Representation Theorem for Projective Algebras).

(a) *Any nP-algebra is an nP-homomorphic image of an nP-field of sets.*

(b) *Any P-algebra is an ωP-homomorphic image of an ωP-field of sets.*

IV. CONCLUSION.

Our representation theorem for free projective Boolean algebras provided us with a "bridge" from logic to set theory, but so far nothing was specifically shown so as to give a relationship between the projective field of sets and the projective sets of Lusin and Sierpinski. Our terminology therefore lacks some justification, which is however given by the following and last theorem:

THEOREM 35. *For $n > 0$, $\underset{\sim}{\Delta}^1_{n+1}$ is an n-projective field of sets, where $\underset{\sim}{A}$ stands for "boldface A".*

PROOF: (for a detailed proof please see Morais 1976).

We will show that $\underset{\sim}{\Delta}^1_{n+1}$ is closed under the P_n, but this is not enough, however, to prove that it is n-projective because the definition of an n-projective algebra starts with a σ-algebra. But it is easy to see that the same argument used below can be repeated to show that $\underset{\sim}{\Delta}^1_{n+1}$ is closed under the P_1 operator, and this in turn is a generalization of countable unions and intersections (cf. Kuratowski and Mostowski 1968, p. 341).

Let now ϕ be any map from $[seq]^n$ into Δ^1_{n+1}, and we show that $x \in P_n(\phi)$ can be given both by a Σ^1_{n+1} and a Π^1_{n+1} predicate.

By Proposition 3 $x \in P_n(\phi)$ has two equivalent formulations, namely:

(a) $(\exists F \in Full_n)(\forall v \in F)(x \in \phi(v))$

and

(b) $(\forall G \in Full^*_n)(\exists v \in G)(x \in \phi(v))$.

We are going to use (a) (respectively (b)) to show that $x \in P_n(\phi)$ is given by a Σ^1_{n+1} (respectively Π^1_{n+1}) predicate.

First, since $[seq]^n$ is countable, there is a 1-1 recursive map λ from ω onto $[seq]^n$ and so we can think that the domain of ϕ is ω. Let $\underline{v}$ be the integer associated with $v \in [seq]^n$ by means of λ. In addition, we substitute the sets F and G by their respective characteristic functions $g : [seq]^n \to \{0, 1\}$, and as we did for ϕ we think the domain of g as ω.

Therefore (a) is equivalent to $(\exists g \in {}^\omega\omega)(\forall k \in \omega)\ [(range\ g = \{0, 1\}$ *and* $\{v \in [seq]^n \mid g(v) = 1\} \in Full_n$ *and* $g(k) = 1) \to x \in \phi(k)]$.

The expression range $g = \{0, 1\}$ is written

$$(\forall m \in \omega)(g(m) = 0 \quad or \quad g(m) = 1).$$

The expression $\{v \in [seq]^n \mid g(v) = 1\} \in Full_n$ is equivalent to

$$(\forall f_1)(\exists f_2)\ldots(Qf_n)(Q'm)(g(f_1 \mid m, \ldots, f_n \mid m) = 1)$$

which is Σ^1_n.

Finally since $\phi(k) \in \Delta^1_{n+1} \subset \Sigma^1_{n+1}$ for all k we can make $x \in \phi(k)$ a Σ^1_{n+1} predicate.

Therefore if we write $\forall 1$ and $\exists 1$ for quantification over reals and $\forall 0$ and $\exists 0$ for quantification over numbers, the statement (a) now reads:

$$\exists 1\ \forall 0\ [\ (\forall 0 \wedge \underbrace{\exists 1\ \ldots\ Q1}_{n}\ \ Q'0) \to \underbrace{\exists 1\ \ldots\ Q'1}_{n+1}\ Q0\ ]\ .$$

Then using the Tarski - Kuratowski algorithms (cf. Rogers 1967, p. 307) we simplify the above to

$$\underbrace{\exists 1\ \ldots\ \exists 1}_{n+1}\ \forall 0\ [\ (\wedge\ ) \to\ ]$$

which is a $\underset{\sim}{\Sigma}^1_{n+1}$ predicate.

Now using (b), since $G \in Full^*_n$ is $\underset{\sim}{\Pi}^1_n$ and since $\phi(k) \in \underset{\sim}{\Delta}^1_{n+1} \subset \underset{\sim}{\Pi}^1_{n+1}$, we, start with

$$\forall 1 \exists 0 \left[(\forall 0 \wedge \underbrace{\forall 1 \ldots Q'1}_{n} Q0) \rightarrow \underbrace{\forall 1 \ldots Q1}_{n+1} Q'0 \right]$$

and end up with a $\underset{\sim}{\Pi}^1_{n+1}$ predicate.

It is clear by now that one of the most interesting notions that came up along this work was that of $Full^*_n$ generalizing Ellentuck's Full sets, and its counterpart $Full_n$. The symmetry between these two classes of sets not only helped cutting all our proofs in half but also, and more significantly, without this symmetry most of our proofs could not have come through, specially our last theorem in which the simultaneous use of $Full_n$ and $Full^*_n$ was fundamental.

For these reasons we foresee an increasing use of these notions in the future studies of projective sets.

To conclude this work, among several interesting questions for which all this machinery is applicable, we selected two that we are particularly interested in investigating, namely:

(1) What kind of interpolation theorem holds in L_{pn} or L_p, if any?

(2) If M is a universe of sets and B is an n-projective Boolean algebra, what can be accomplished inside the Boolean valued model M^B?

REFERENCES.

Addison, J. A. and Y. Moskovakis

1968, *Some consequences of the axiom of definable determinateness*, Proc. Nat. Acad. Sci., Vol. 59, 708 - 712.

Bell, J. L. and A. B. Slomson

1969, Models and Ultraproducts, North - Holland, Amsterdam.

Ellentuck, E.

1975, *The foundations of Suslin Logic*, The Journal of symbolic Logic, vol. 40, 567-575

197+, *Free Suslin algebras,* Submitted to Czech. Math. Journal.

Fenstad, F.

1971, *The axiom of determinateness,* Proceedings of the Second Scandinavian Logic Symposium, Ed., J. E. Fenstad, North - Holland, Amsterdam,41-61.

Halmos, P.

1963, Lectures on Boolean Algebras, Van Nostrand Reinhold Company, London.

Kechris, A.

1973, *Measure and category in effective descriptive set theory,* Annals of Math. Logic, Vol. 5, 337 - 384.

Keisler, H. J.

1971, Model Theory for Infinitary Logic, North - Holland, Amsterdam.

Kuratowski, K. and A. Mostowski

1968, Set Theory, North - Holland, Amsterdam.

Lusin, N.

1925, *Sur les ensembles projectives de M. Henri Lebesgue,* C. R. Acad. Sci. de Paris.

Morais, R.

1976, Projective Logic, Ph. D. Thesis, Rutgers University, U.S.A.

Moschovakis, Y.

1970, *Determinancy and prewellorderings of the continuum,* in Mathematical Logic and Foundations of Set Theory, Ed.Y.Bar-Hillel, North-Holland, Amsterdam, 24 - 62.

Rieger, L.

1955, *Concerning Suslin algebras (S - Algebras) and their representation* (Russian), Czech. Math. Journal, Vol. 5. 99 - 142.

Rogers, H.

1967, Theory of Recursive Functions and Effective Computability, MacGraw - Hill.

Sierpinski, W.

1925, *Sur une classe d'ensembles,* Fund. Math. Vol. 7, 237 - 243.

Instituto de Matemática
Universidade Federal do Rio de Janeiro
Rio de Janeiro, RJ., Brazil

Non-Classical Logics, Model Theory and Computability,
A.I. Arruda, N.C.A. da Costa and R. Chuaqui (eds.)

Some Theorems on Omitting Types, With Applications to Model Completeness, Amalgamation, and Related Properties.

by *CHARLES C. PINTER*

1. Introduction.

Many topics of current interest in model theory involve models which omit designated sets of types. For example, the existentially closed models of a theory T are precisely those which omit a certain set of types. The same is true for the generic models of T for finite forcing, the completing models of T, the models which are amalgamation bases for T, and so on. In fact, important properties of a theory T hold iff all the models of T omit certain given types: this is the case of a theory being model complete, finitely forcing complete, having the amalgamation property, the congruence extension property, and many others.

One of the objects of this paper is to show that many superficially unrelated results of model theory are, in fact, consequences of the same simple theorems on omitting types. By systematically developing certain rather elementary observations on omitting types, we can recapture - and unify-many known results, find a number of new ones, and almost trivialize some classical results. As an example of the latter, we show that Lindstrom's theorem on model completeness is a consequence of a very simple observation on omitting types in theories which are categorical in some infinite power.

In Section 2 of this paper we develop several properties of omitting types, and in Section 3 we illustrate their uses with a variety of examples.

Our standard reference to model theory will be Chang and Keisler 1973.

We assume throughout that L is a countable, finitary, first-order language; a *theory* T is a deductively closed set of sentences of L. To simplify notation, we will let $\bar{v}$ designate any finite sequence $(v_1,\ldots,v_n)$ of variables, and, if $\mathfrak{A}$ is a structure, we will let $\bar{a}$ designate any finite sequence $(a_1,\ldots,a_n)$ of members of $\mathfrak{A}$. For any formula ϕ, we will use the symbols $\phi(\bar{v})$ and $\phi[\bar{a}]$ with their obvious meanings, provided that the free variables of ϕ occur among $\bar{v}$ and that the sequence $\bar{a}$ matches the sequence $\bar{v}$.

2. SOME THEOREMS ON OMITTING TYPES.

Let $\sigma(\bar{v}) = \{\sigma_n(\bar{v}) : n \in \omega\}$ be a sequence of formulas of L. If T is a theory of L, we say that T *locally omits* σ iff for every formula $\phi(\bar{v})$ which is consistent with T, $\phi(\bar{v}) \wedge \neg\sigma_n(\bar{v})$ is consistent with T for some $n \in \omega$. Let Σ be a countable set of sequences σ of formulas; if T locally omits every $\sigma \in \Sigma$, we will say that T is Σ-*complete*. A model $\mathfrak{A} \models$ T will be said to *omit* Σ iff $\mathfrak{A}$ omits each $\sigma \in \Sigma$. The Omitting-types Theorem asserts:

(2.1) *If* T *is* Σ-*complete, then* T *has a model which omits* Σ. *If* T *is a complete theory:* T *is* Σ-*complete iff* T *has a model which omits* Σ.

DEFINITION. We say that T is Σ-*consistent* iff $T \subseteq T'$, where T' is consistent and Σ-complete.

Clearly, T *has a model omitting* Σ *iff* T *is* Σ-*consistent.* Furthermore, if T is a complete theory T *is* Σ-*consistent iff* T *is* Σ-*complete.*

(2.2) THEOREM. *For any theory* T, *the following are equivalent:*

(i) *Every extension of* T *is* Σ-*consistent.*

(ii) *Every extension of* T *is* Σ-*complete.*

(iii) *For every formula* $\phi(\bar{v})$ *and every* $\sigma \in \Sigma$, *there are integers* $i_1,\ldots,i_q$ *such that*

$$T \vdash (\exists\bar{v})\,\phi(\bar{v}) \leftrightarrow \bigvee_{j=i_1,\ldots,i_q} (\exists\bar{v})[\phi(\bar{v}) \wedge \neg\sigma_j(\bar{v})]$$

PROOF:

(iii) $\Longrightarrow$ (ii) $\Longrightarrow$ (i) is immediate.

(i) $\Longrightarrow$ (iii): Every model of T is a model of a complete extension of T, that is, of a complete, Σ-complete theory. Thus, for each formula $\phi(\bar{v})$, every model $\mathfrak{A} \models T$ satisfies the sentence

$$(\exists\bar{v})\phi(\bar{v}) \leftrightarrow \bigvee_{i \in \omega}(\exists\bar{v})[\phi(\bar{v}) \wedge \neg\sigma_i(\bar{v})] .$$

By compactness, we have (iii). □

EXAMPLE. Let T be a consistent theory in the language L^{ω} of ω-logic. Then, every extension of T has an ω-model iff every extension of T is ω-complete iff for each formula $\phi(\bar{v})$ there are integers $i_1,\ldots,i_q$ such that

$$T \vdash (\exists v)[N(v) \wedge \phi(v)] \rightarrow \phi(i_1) \vee \ldots \vee \phi(i_q).$$

The next two theorems have a great many applications, which will be developed in Section 3 of this paper.

(2.3) THEOREM. (i) *Every intersection of* Σ - *complete theories is* Σ-*complete.*

(ii) *Every* Σ - *complete theory is the intersection of all its complete,* Σ - *complete extensions.*

PROOF:

(i) is immediate, using the contrapositive of the definition of "T locally omits σ".

(ii) : We will show that if T is Σ - complete and ϕ is any sentence which is consistent with T , then $T \cup \{\phi\}$ is Σ - complete; (ii) will follow immediately from this. Well, suppose T is Σ - complete and $\psi(\bar{v})$ is consistent with $T \cup \{\phi\}$. Then $\psi(\bar{v}) \wedge \phi$ is consistent with T , so for some $n \in \omega$, $\psi(\bar{v}) \wedge \phi \wedge \neg\sigma_n(\bar{v})$ is consistent with T . Thus $\psi(\bar{v}) \wedge \neg\sigma_n(\bar{v})$ is consistent with $T \cup \{\phi\}$. □

(2.4) THEOREM. *The following are equivalent:*

(i) T *is* Σ - *complete*

(ii) *For all sentences* ϕ, $T \vdash \phi$ *iff* ϕ *is true in every model of* T *which omits* Σ .

PROOF: We will show that (ii) holds iff T is the intersection of all its complete, Σ - complete extensions; our result will then follow by (2.3). Suppose (ii) holds, and let $\{T_i : i \in I\}$ be the set of all the complete, Σ - complete theories containing T . Now, $T \subseteq \bigcap_i T$; furthermore, if $\phi \in \bigcap_i T_i$, then ϕ is true in every model of T omitting Σ, hence $\phi \in T$. Conversely, suppose $T = \bigcap_i T_i$; certainly $\phi \in T$ implies that ϕ is true in every model of T omitting Σ. On the other hand, if ϕ is true in every model of T omitting Σ, then $\phi \in T_i$ for each $i \in I$, so $\phi \in \bigcap_i T_i$. □

DEFINITION. Let T be Σ-consistent. The intersection of all the complete, Σ-complete extensions of T is called the Σ-*completion* of T, and is denoted by T^Σ.

T^Σ is the smallest Σ-complete theory containing T. We will see in the next Section that forcing-companions, model-companions, and other familiar constructions of model theory are specific examples of Σ-completions.

It is obvious that *any model of* T *which omits* Σ *has to be a model of* T^Σ. From this fact, we immediately deduce:

(2.5) THEOREM. $T^\Sigma \vdash \phi$ *iff* ϕ *is true in every model of* T *which omits* Σ.

In many cases which we are led to consider, the class $\mathfrak{J}$ of all the models of T which omit Σ turns out to be an elementary class. For such cases, the following is useful:

(2.6) THEOREM. *Suppose the class* $\mathfrak{J}$ *of all the models of* T *which omit* Σ *is an elementary class. Then* $T^\Sigma = \mathrm{Th}(\mathfrak{J})$, *and* T^Σ *is axiomatized by*

$$T \cup \{(\forall\bar{v}) \bigvee_i \neg\sigma_i(\bar{v}) : \sigma \in \Sigma\}$$

where each $\bigvee_i$ *is a finite disjunction.*

PROOF: $T^\Sigma = \mathrm{Th}(\mathfrak{J})$ by (2.5). Thus, each model of T^Σ omits Σ, and our result follows from a simple application of compactness. □

In the sequel, it will be useful to note that, if T is Σ-consistent, there is an obvious inductive procedure for constructing the Σ-completion T^Σ of T.

To construct T^Σ, we define a sequence $\mathcal{U}_0 \subseteq \mathcal{U}_1 \subseteq \cdots$ of consistent theories as follows:

(2.7) (i) $\mathcal{U}_0 = T$.

(ii) Given $\mathcal{U}_\kappa$, define $\mathcal{U}_{\kappa+1}$ by: for each $\sigma \in \Sigma$ and each formula $\psi(\bar{v})$, if $\mathcal{U}_\kappa \vdash \psi(\bar{v}) \to \sigma_n(\bar{v})$ for each $n \in \omega$, then $\neg(\exists\bar{v})\psi(\bar{v}) \in \mathcal{U}_{\kappa+1}$.

(iii) If α is a limit ordinal, $\mathcal{U}_\alpha = \bigcup_{\gamma<\alpha} \mathcal{U}_\gamma$.

Finally, let T^Σ be the deductive closure of $\bigcup_{\kappa<\omega_1} \mathcal{U}_\kappa$. Our assumption that

T is Σ - consistent guarantees that each $\mathcal{U}_\kappa$ is consitent, and clearly, for some countable ordinal μ, $\nu \geq \mu$ implies $\mathcal{U}_\nu = \mathcal{U}_\mu$.

The preceding construction may be interpreted as a completeness theorem for truth in the class of models of T which omit Σ, with the closure condition (ii) taken as an infinitary rule of proof. Specifically, we have the following generalization of the ω - rule, which we shall call the Σ-*rule*: For each $\sigma \in \Sigma$, from $\psi(\bar{v}) \to \sigma_n(\bar{v})$ for each $n \in \omega$ infer $\neg(\exists\bar{v})\psi(\bar{v})$, where $\psi(\bar{v})$ is any formula of L. Σ- *logic* is formed by adding the Σ - rule to the axioms and rules of inference of the first - order logic L and allowing infinitely long proofs. We have the following completeness theorem for Σ-logic:

(2.8) *A theory* T *in* L *is consistent in* Σ - *logic iff* T *has a model omitting* Σ. *Furthermore, if* ϕ *is any sentence of* L, ϕ *can be deduced from* T *in* Σ - *logic iff* ϕ *is true in every model of* T *which omits* Σ.

We say that T *cofinally omits* Σ iff every infinite model of T can be extended to a model of T which omits Σ. The following simple observations have many applications:

If every denumerable model of T omits Σ, then every model of T omits Σ. Indeed, if $\mathfrak{A} \models T$, $\bar{a}$ is in $\mathfrak{A}$, and $\bar{a}$ realizes σ, then by the downward Löwenheim - Skolem Theorem, T has a denumerable model $\mathfrak{A}' \prec \mathfrak{A}$ such that $\bar{a}$ is in $\mathfrak{A}'$, and clearly $\bar{a}$ realizes σ in $\mathfrak{A}'$. From this observation and the upward Löwenheim - Skolem Theorem, we infer the following for every infinite cardinal α:

$(2.9)^\alpha$ *If every model of* T *of power* α *omits* Σ, *then every model of* T *omits* Σ.

Furthermore, the downward Löwenheim - Skolem Theorem yields:

(2.10) *If* T *cofinally omits* Σ, *then every model* $\mathfrak{A} \models T$ *can be extended to a model* $\mathfrak{B} \models T$ *such that* $\mathfrak{B}$ *omits* Σ *and card* $\mathfrak{A}$ = *card* $\mathfrak{B}$.

$(2.9)^\alpha$ has manifold applications, among which are the following: a theory T has the amalgamation property iff every model of T of a given infinite power α is an amalgamation base of T. (This improves Yasuhara 1974, Theorem 1.13.1). T is model - complete iff every model of T of a given infinite power α is existentially closed over T. Analogous statements may be inferred for the strong amalgamation property, the congruence extension property (see Bacsich and Rowlands Hughes 1974), and many others. Thus, to know if a theory T has one of the above - named properties, it suffices to watch only the models of a given infinite power α.

We end this section with some observations on *categoricity in power and omitting types*.

Our first proposition generalizes (and, in a sense, trivializes) Lindstrom's Theorem on model - completeness.

(2.11) PROPOSITION. *Let* T *be a theory which is* α *- categorical for some* $\alpha \geq \omega$. *Let* Σ *be any countable set of types such that* T *cofinally omits* Σ . *Then every model of* T *omits* Σ .

PROOF: Take any $\sigma \in \Sigma$, and suppose σ is consistent with T . Then T has a model of power α which realizes σ . Because T is α - categorical, no model of T of power α omits σ : in view of (2.10), this contradicts our assumption that T cofinally omits σ . We must conclude that σ is inconsistent with T , hence every model of T omits Σ .□

(To obtain Lindstrom's Theorem, we note: T is model - complete iff every model of T is existentially closed over T ; $\mathfrak{A}$ is existentially closed over T iff $\mathfrak{A}$ omits a certain set of types Σ ; if T is inductive, then every model of T can be extended to an existentially closed model of T .)

Now, let T be a theory which is ω_1-categorical but not ω-categorical. If a type σ is consistent with T and T locally omits σ , then σ is omitted in some countable models of T , but realized in all its uncountable models. If every countable model of T omitting σ were to admit a countable proper elementary extension omitting σ , an ω_1 - chain of such models could be formed, and their union would be an uncountable model of T omitting σ ,which is impossible. Thus, there are maximal countable models of T omitting σ , (*maximal* in the sense that they admit no proper elementary extension omitting σ). The same argument holds for a countable set of types Σ . We conclude:

(2.12) PROPOSITION. *Let* T *be* ω_1 *- categorical but not* ω *- categorical. Let* Σ *be a countable set of types which are consistent in* T *and such that* T *is* Σ*-consistent. Then there are models of* T *which are maximal with respect to omitting* Σ . *These models are countable; every elementary chain of countable models of* T *omitting* Σ *must end with one of these maximal models.*

A great many consequences may be drawn from this simple observation. A canonical application is to atomic models: every theory T which is ω_1-categorical but not ω - categorical has a countable maximal atomic model, which is therefore also minimal. Applying this result to T, as well as to the theories $\mathrm{Th}\,\mathfrak{A}_A$ where $\mathfrak{A}$ is any countable model of T, we recover many of the results of Morley 1967.

3. SOME APPLICATIONS.

In this section, we select a few typical applications of the results developed in Section 2. In particular, we will illustrate the uses of the notion of Σ-completion.

(I) EXISTENTIALLY CLOSED MODELS.

Let T be a theory; we recall that if $\mathfrak{A} \models T_\forall$, $\mathfrak{A}$ is said to be *existentially closed over* T iff for any $\mathfrak{B} \models T$, $\mathfrak{A} \subseteq \mathfrak{B}$ implies $\mathfrak{A} \prec_1 \mathfrak{B}$. We will abbreviate "existentially closed over T" by e.c. over T.

For each formula $\alpha \in \forall_1$, let σ^α designate the type $\{\alpha, \neg\varepsilon : \varepsilon \in \exists_1$ and $T \vdash \varepsilon \to \alpha\}$. Let Σ be the set of all the types σ^α as α ranges over all the formulas of $\forall_1$. It is well known that if $\mathfrak{A} \models T_\forall$, $\mathfrak{A}$ is e.c.over T iff $\mathfrak{A}$ omits Σ.

It is important to note that $\mathfrak{A}$ is e.c. over T iff $\mathfrak{A}$ is e.c. over $T_\forall$; indeed, the sentences $(\forall \bar{v})\,[\varepsilon \to \alpha]$ which determine the composition of each type σ^α are sentences of $T_\forall$.

By Theorems 2.4 and 2.5, we immediately conclude:

(3.1) PROPOSITION. *T is Σ-complete iff T is the set of all the sentences which are true in every e.c. model of T. Furthermore, $T_\forall^\Sigma$ is the theory of all the models which are e.c. over T.*
(in other words, $\phi \in T_\forall^\Sigma$ *iff ϕ is true in every model which is e.c.over* T).

It is worth remarking that $T_\forall^\Sigma$ may be constructed from $T_\forall$ by the procedure outlined in (2.7), and its theorems may be obtained from $T_\forall$ by deduction in Σ-logic.

We note also that

(3.2) $T_\forall^\Sigma$ *is mutually model-consistent which $T_\forall$, hence with* T.
Indeed, every model of $T_\forall$ can be extended to an e.c. model of $T_\forall$, that is, to a model of $T_\forall^\Sigma$.

Now, suppose that the class of all the models which are e.c. over T is elementary. By Theorem 2.6, this class is axiomatized by $T_\forall^\Sigma$ and

$$T_{\forall}^{\Sigma} = T_{\forall} \cup \{\alpha \rightarrow \varepsilon : \alpha \in \forall_1\}$$

where, for each $\alpha \in \forall_1$, ε is an existential formula such that $T \vdash \varepsilon \rightarrow \alpha$. Obvioulsy $T_{\forall}^{\Sigma}$ is model - complete, hence $T_{\forall}^{\Sigma}$ is the model - companion of $T_{\forall}$, and therefore also the model - companion of T. We conclude:

(3.3) PROPOSITION. *If* T *has a model - companion, the latter is axiomatized by*

$$T_{\forall}^{\Sigma} = T_{\forall} \cup \{\alpha \rightarrow \varepsilon : \alpha \in \forall_1\},$$

where, for each $\alpha \in \forall_1$, ε *is an existential formula such that* $T \vdash \varepsilon \rightarrow \alpha$.

It is easily verified from the relevant definitions (see, e.g.,Robinson 1971) that the infinitely T - generic models are exactly the infinitely $T_{\forall}$-generic models. Thus, by Fisher and Robinson 1972, Theorem 3.4 .

(3.4) PROPOSITION. *For any theory* T, *the following are equivalent:*

(a) *The infinitely* T - *generic models form an elementary class.*

(b) *The models which are e.c. over* T *form an elementary class.*

(c) T *has a model companion.*

If one of these conditions hold, then $T_{\forall}^{\Sigma}$ *is the model companion of* T *and axiomatized the class of infinitely* T - *generic models.*

(This is a variant of Fisher and Robinson 1972, Theorem 3.4. In particular, the latter applies to inductive theories only, whereas the present (3.6) holds for arbitrary theories).

In one of the major results of their paper, Fisher and Robinson demonstrate that the forcing companion of any inductive theory T (that is, the theory of the generic models of T) can be decomposed into the forcing companions of the components of $T_{\forall}$. (A *component* of a universal theory U is a minimal irreducible extension of U). We show now that an analogous result holds for $T_{\forall}^{\Sigma}$, the theory of the models which are e.c. over T.

(3.5) PROPOSITION. *Let* $\{T_i : i \in I\}$ *be the set of all the components of* $T_{\forall}$. *The class of models which are e.c. over* T *is the union of the classes of e.c. models of the* T_i, *and these are mutually disjoint. In particular,* $T_{\forall}^{\Sigma} = \bigcap_{i \in I} T_i^{\Sigma}$.

PROOF: If $\mathfrak{A}$ is e.c. over $T_{\forall}$, $\mathfrak{A} \models T_i$ for some $i \in I$. Now if $\mathfrak{A} \subseteq \mathfrak{B} \models T_i$, then $\mathfrak{B} \models T_{\forall}$, hence $\mathfrak{A} \prec_1 \mathfrak{B}$; this shows that $\mathfrak{A}$ is e.c. over T_i. Conversely, if $\mathfrak{A}$ is e.c. over T_i for some $i \in I$, then $\mathfrak{A}$ can be embedded in a model of $T'_i = T_i \cup \{\neg\phi \in \exists_1 : \phi \in T_j\}$, hence $\mathfrak{A} \models T'_i$. Thus, $\mathfrak{A}$ cannot be embedded in a model of any T_j, $j \neq i$. So clearly, $\mathfrak{A}$ is e.c. over $T_\forall$. □

(Note that if T is inductive, then $T^\Sigma = \bigcap_{i \in I} T_i^\Sigma$.)

(II) THE AMALGAMATION PROPERTY.

A model $\mathfrak{A} \models T_\forall$ is called T-*amalgamative* (or a T-*amalgamation base*), if each diagram

(3.6) 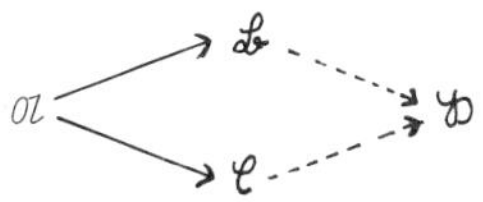 $\mathfrak{B}, \mathfrak{C}, \mathfrak{D} \models T$

can be completed. (See, e.g., Bacsich and Rowlands Hughes 1974, and Yasuhara 1974).

For every pair (α, β) of universal formulas, let $\sigma^{(\alpha,\beta)}$ designate the type $\{\neg\varepsilon, \neg\delta : \varepsilon, \delta \in \exists_1,\ T \vdash \varepsilon \to \alpha \text{ and } T \vdash \delta \to \beta\}$.

Let Σ be the set of all the types $\sigma^{(\alpha,\beta)}$, as (α, β) ranges over all the pairs of universal formulas such that $T \vdash \alpha \vee \beta$. It is known (Bacsich and Rowlands Hughes 1974, Theorem 2.4) that if $\mathfrak{A} \models T_\forall$, then $\mathfrak{A}$ is T-amalgamative iff $\mathfrak{A}$ omits Σ. We note, as in the preceding example, that a model is T-amalgamative iff it is $T_\forall$-amalgamative.

By Theorems 2.4 and 2.5, we immediately conclude:

(3.7) PROPOSITION. *T is Σ-complete iff T is the set of all the sentences which are true in every T-amalgamative model of T. Furthermore, $T_\forall^\Sigma$ is the theory of all the T-amalgamative models.*

(That is, $T_\forall^\Sigma \vdash \phi$ iff ϕ is true in every T-amalgamative model).

We remark that $T_\forall^\Sigma$ may be constructed from $T_\forall$ by the procedure of (2.7), and its theorems may be obtained from $T_\forall$ by deduction in Σ-logic.

(3.8) *$T_\forall^\Sigma$ is mutually model-consistent with $T_\forall$, hence with T.*

For every model $\mathfrak{A} \models T_\forall$ can be extended to a T-amalgamative model $\mathfrak{B} \models T_\forall^\Sigma$.

If the class of all the T-amalgamative models is elementary, then by Theorem 2.6, it is axiomatized by $T_\forall^\Sigma$ and

(3.9) $$T_\forall^\Sigma = T_\forall \cup \{\varepsilon_z \vee \delta_z : z \in Z\},$$

where Z is the set of all pairs (α, β) of universal formulas such *that* $T \vdash \alpha \vee \beta$, $\varepsilon_z, \delta_z \in \exists_1$, *and for each* $z = (\alpha, \beta) \in Z$, $T \vdash \varepsilon_z \to \alpha$ *and* $T \vdash \delta_z \to \beta$.

By compactness, T has the amalgamation property (AP) iff T includes a set of formulas $T_\forall \cup \{\varepsilon_z \vee \delta_z : z \in Z\}$ as described under (3.9). But then the theory $T_\forall \cup \{\varepsilon_z \vee \delta_z : z \in Z\}$ already has the AP. Thus, we are led to the following scheme for classifying theories having the AP:

(3.10) *Every theory* $T_\forall \cup \{\varepsilon_z \vee \delta_z : z \in Z\}$ *which is mutually model-consistent with* T *has the* AP. *Among the theories which are mutually model-consistent with* T, *a theory* U *has the* AP *iff* U *contains a set of sentences* $T_\forall \cup \{\varepsilon_z \vee \delta_z : z \in Z\}$.

A minor generalization of the AP is very useful and occurs increasingly in the literature: If S and T are theories, we say that S *has the* T*-amalgamation property* (S *has the* T-AP) iff the diagram (3.6) can be completed for every model $\mathfrak{A} \models S$; that is, iff every model $\mathfrak{A} \models S$ is T-amalgamative. One verifies, by an easy argument such as (Yasuhara 1974, 1.9.3), that S *has the* T*-amalgamation property iff* S_I *has the* $T_\forall$*-amalgamation property*. Thus, there is no loss of generality if we now confine our attention to the case where S is an inductive theory and T is a universal theory. We proceed now to give a decomposition result for the AP in the spirit of Proposition (3.5); we will show that the relation "S has the T-AP" is determined by relations "S' has the T'-AP" among the components S' of S and T' of T.

(3.11) LEMMA. (i) *If* T *is a universal theory, two structures* $\mathfrak{A}$ *and* $\mathfrak{B}$ *are models of the same component of* T *iff they have a joint extension which is a model of* T.

(ii) *If* S *is an inductive theory, two structures* $\mathfrak{A}$ *and* $\mathfrak{B}$ *are models of the same component of* S *iff there is a model* $\mathfrak{C}$ *of* S *such that* $\mathfrak{A} \prec_1 \mathfrak{C}$ *and* $\mathfrak{B} \prec_1 \mathfrak{C}$.

(Part (i) is well-known. Part (ii) may be established by using Part (i) after expanding the language so as to make existential formulas atomic).

(3.12) COROLLARY. *Let* T *be a universal theory, and suppose* T *is model-consistent with* $\mathfrak{A}$. *Then all the models of* T *which are extensions of* $\mathfrak{A}$ *are models of the same component* T_1 *of* T.

(3.13) COROLLARY. *Let* T *be a universal theory,* S *an inductive theory, and suppose* T *is model-consistent with* S. *If* S *is irreducible, then exactly one component* T_1 *of* T *is* $\subseteq S$, *and every other component of* T *is inconsistent with* S.

($T \subseteq S_\forall$, so if $\mathfrak{A}, \mathfrak{B} \models S$, then by 3.11 (ii), there is a $\mathfrak{C} \models S$ such that $\mathfrak{A} \prec_1 \mathfrak{C}$ and $\mathfrak{B} \prec_1 \mathfrak{C}$; by 3.11(i), $\mathfrak{A}$ and $\mathfrak{B}$ are models of the same component of T, and all the other components of T are inconsistent with S).

(3.14) THEOREM. *Let* T *be a universal theory,* S *an inductive theory, and suppose* T *is model-consistent with* S. *Then*

(i) S *has the* T-AP *iff every component of* S *has the* T-AP.

Next, assume that S *is irreducible:*

(ii) S *has the* T-AP *iff* S *has the* T'-AP *for some component* T' *of* T.

((i) is immediate and (ii) follows from Corollaries 3.12 and 3.13).

In particular, S *has the* T-AP *iff every component* S' *of* S *has the* T'-AP *for some component* T' *of* T.

(III) COUNTABLE FRAGMENTS OF $L_{\omega_1\omega}$.

Throughout this discussion, we let L_A be an arbitrary but fixed countable fragment of $L_{\omega_1\omega}$. The logical symbols of L_A are $\neg$, $\bigvee$, $\forall$, $\exists$, and $=$, and we take $\bigwedge$ to be defined by $\bigwedge\Phi = \neg\bigvee\{\neg\phi : \phi \in \Phi\}$. Let L be the finitary language whose relation, function and constant symbols are those of L_A. We form a new first-order language L' by adding to L a new relation symbol $R_{\bigvee\Phi}$ for every infinite disjunction $\bigvee\Phi$ in L_A. Every formula ϕ in L_A is associated with a formula ϕ' in L', obtained from ϕ by replacing each occurrence of any infinite disjunction $\bigvee\Phi$ by $R_{\bigvee\Phi}$.

Let V' be the theory of L' whose axioms are all the formulas $\phi' \to R_{\bigvee\Phi}$, where $\phi \in \Phi$ and $\bigvee\Phi$ ranges over all the infinite disjunctions in L_A. We call $\mathfrak{A}$ a *standard model* of L' if $\mathfrak{A} \models V'$ and, for each infinite disjunction $\bigvee\Phi$ in L_A, $\mathfrak{A}$ omits the type $\{R_{\bigvee\Phi}, \neg\phi' : \phi \in \Phi\}$. Let Σ denote the set of these types, with $\bigvee\Phi$ ranging over all the infinite disjunctions in L_A. Then a model $\mathfrak{A}$ of L' is standard iff $\mathfrak{A} \models V'$ and $\mathfrak{A}$ omits Σ.

The relationship between standard models of L' and models of L_A is given by:

(3.15) (i) *If* $\mathfrak{A}$ *is a standard model of* L', *then* $\mathfrak{A}$ *has a reduct* $\mathfrak{A}'$ *which is a model of* L_A, *and* $\mathfrak{A} \models \phi \leftrightarrow \phi'$ *for every formula* ϕ *in* L_A.

(ii) *If* $\mathfrak{B}$ *is a model of* L_A, $\mathfrak{B}$ *has a unique expansion to a standard model* $\mathfrak{B}'$ *of* L' *such that* $\mathfrak{B}' \models \phi \leftrightarrow \phi'$ *for every formula* ϕ *in* L_A.

(This result is essentially known, and routine to verify).

Let V_A designate the Σ - completion of V'. Obviously, every standard model of L' is a model of V_A. If ϕ is any sentence of L_A, then ϕ is true in every model of L_A iff ϕ' is true in every standard model of L' iff $V_A \vdash \phi'$. Thus,

(3.16) *For every formula* ϕ *in* L_A, $\vdash_{L_A} \phi$ *iff* $V_A \vdash \phi'$.

The following is a characterization of Σ- complete theories in L'.

(3.17) PROPOSITION. *Let* T *be a theory in* L' *such that* $V_A \subseteq$ T. T *is* Σ - *complete iff* T *satisfies the following condition: for each infinite disjunction* $\bigvee\Phi$ *in* L_A, *if* $T \vdash (\forall\bar{v})\,[\phi' \rightarrow \psi]$ *for every* $\phi \in \Phi$, *then* $T \vdash (\forall\bar{v})\,[R_{\bigvee\Phi} \rightarrow \psi]$.

PROOF: Suppose T is Σ - complete; this means that if $T \vdash \xi \rightarrow \neg\phi'$ for every $\phi \in \Phi$ and $T \vdash \xi \rightarrow R_{\bigvee\Phi}$, then ξ is inconsistent with T. Now suppose that $T \vdash \phi' \rightarrow \psi$, that is, $T \vdash \neg\psi \rightarrow \neg\phi'$ for every $\phi \in \Phi$. We claim that $T \vdash \neg\psi \rightarrow \neg R_{\bigvee\Phi}$; otherwise, $\neg\psi \wedge R_{\bigvee\Phi}$ is consistent with T, yet $T \vdash \neg\psi \wedge R_{\bigvee\Phi} \rightarrow \neg\phi'$ for each $\phi \in \Phi$ and $\vdash \neg\psi \wedge R_{\bigvee\Phi} \rightarrow R_{\bigvee\Phi}$, which is impossible. Thus, $T \vdash R_{\bigvee\Phi} \rightarrow \psi$. The converse is immediate. □

If T is any set of sentences in L_A, let $T' = V_A \cup \{\phi' : \phi \in T\}$. A set of sentences T in L_A is said to be *deductively closed in* L_A iff T includes the axioms of L_A and is closed under the rules of inference for L_A. By (3.16) and (3.17) we conclude that for any set T of sentences in L_A,

(3.18) T *is deductively closed in* L_A *iff* T' *is* Σ - *complete.*

If T is a set of sentences of L_A, the *deductive closure* of T is the smallest deductively closed set of sentences containing T. One easily verifies that,

(3.19) U *is the deductive closure of* T *iff* U' *is the* Σ - *completion of* T'.

We say that T is a *complete* set of sentences in L_A iff for every sentence

ϕ in L_A , $\phi \in T$ or $\neg\phi \in T$ (but not both). T is *consistent in* L_A iff the deductive closure of T is a proper subset of L_A . Thus, a complete set of sentences of L_A is consistent iff it is deductively closed.

Combining (3.18) - (3.19) with (2.3) - (2.5), we get the following, where T is any set of sentences of L_A:

(3.20) *T is deductively closed iff T is the intersection of all its complete, consistent extensions.*

(3.21) *T is consistent in L_A iff T has a model.*

(3.22) *If T is consistent in L_A , $T \vdash_{L_A} \phi$ iff ϕ is true in every model of T.*

The last statement immediately yields the *completeness Theorem for* $L_{\omega_1\omega}$. The *Omitting - types Theorem for* L_A follows easily from (3.18) and the observation that if T' is Σ - complete and Φ - complete, it is also $(\Sigma \cup \Phi)$-complete.

Although (2.1), the Omitting - types Theorem, does not hold for uncountable languages, the construction of L' outlined above is possible even when L_A is an uncountable fragment and when $L_{\omega_1\omega}$ has uncountably many relation, function or constant symbols; (3.15) remains valid in this case. From the observation that *every elementary substructure (in L') of a standard model is a standard model,* together with the downward Löwenheim - Skolem Theorem for finitary languages, we get the following strong form of the Löwenheim-Skolem Theorem for $L_{\omega_1\omega}$: (If L_A is any fragment of $L_{\omega_1\omega}$, we let $\# L_A$ designate the cardinal $\|L\|$ + the cardinal number of the set of sentences of L_A which are infinite disjunctions $\bigvee\Phi$; we note that $\# L_A = \|L'\|$) ;

(3.23) *Let L_A be a fragment of $L_{\omega_1\omega}$ (where $L_{\omega_1\omega}$ may have uncountably many non - logical symbols). Let $\mathfrak{A}$ be a model of L_A such that card $\mathfrak{A} = \alpha$ and $\alpha \geq \beta \geq = L_A$. Given any set $X \subseteq A$ of power $\leq \beta$, there is a $\mathfrak{B} \prec_{L_A} \mathfrak{A}$ of power β containing X.*

Many other properties of $L_{\omega_1\omega}$ follow similarly from their finitary counterparts and (3.15) - in particular, known statements relating to elementary chains, homogeneous models and indiscernibles. Related logical systems, such as weak second - order logic and logic with the added quantifier "there exist infinitely many", may be treated in a similar manner.

4. FINAL REMARKS.

In Section 2 we gave the definition: T *cofinally omits* Σ iff every infinite model of T can be extended to a model of T which omits Σ. It is shown in Bacsich and Rowlands Hughes 1974 that this notion has several important applications in the special case where T is an inductive theory and each σ in Σ consists of universal formulas. A semantic characterization of "T cofinally omits σ" is given in Bacsich and Rowlands Hughes 1974 for this special case, but it has the disadvantage of involving a rather unwieldy sentence of $L_{\omega_1\omega}$ which seriously restricts its usefulness. The following is a simple characterizarion of the same notion:

(4.1) THEOREM. *Let* T *be an inductive theory, and* Σ *a countable set of types* σ *consisting of universal formulas.* T *cofinally omits* Σ *iff for each* σ *in* Σ, *the following condition holds:*

(4.2) *For any choice* $\{\varepsilon_n : n \in \omega\}$ *of existential formulas such that* $T \vdash \varepsilon_n \to \sigma_n$ *for each* $n \in \omega$, $\{\varepsilon_n : n \in \omega\}$ *is inconsistentent with* T.

REMARK: (4.2) is equivalent to: σ *is not in the deductive closure of any consistent existential type, with respect to* T.

PROOF of the Theorem: Our condition is necessary: for if there is a $\sigma \in \Sigma$ and $\{\varepsilon_n : n \in \omega\}$ as in (2.9) such that $\{\varepsilon_n : n \in \omega\}$ is *consistent* with T, then there is a model $\mathfrak{A} \models T$ and an $\bar{a}$ in $\mathfrak{A}$ such that $\mathfrak{A} \models \varepsilon_n[\bar{a}]$ for each $n \in \omega$; clearly, no extension of $\mathfrak{A}$ to a model of T can omit σ. To prove the sufficiency of our condition, we will construct an extension of $\mathfrak{A} \models T$ which will omit Σ. Let Φ be a set of sentences of $L(\mathfrak{A})$ which is maximal with respect to the two conditions: (i) $\mathcal{D}(\mathfrak{A}) \subseteq \Phi \subseteq \exists_1(\mathfrak{A})$, and (ii) $T \cup \Phi$ is consistent. Let $\mathfrak{A}' \models T \cup \Phi$; by the maximality of Φ, if $\mathfrak{A}' \models \phi[\bar{a}]$ for any $\phi \in \forall_1$ and $\bar{a}$ in $\mathfrak{A}$, then $T \cup \Phi \vdash \phi(\bar{a})$, hence $T \vdash \varepsilon \to \phi$ for some $\varepsilon \in \exists_1$ and $\mathfrak{A}' \models \varepsilon[\bar{a}]$. Thus, for each $\sigma \in \Sigma$ and $\bar{a}$ in $\mathfrak{A}$, if $\bar{a}$ satisfies σ in $\mathfrak{A}'$, then $\bar{a}$ satisfies $\{\varepsilon_n : n \in \omega\}$ in $\mathfrak{A}'$, where $T \vdash \varepsilon_n \to \sigma_n$ for each $n \in \omega$. By assumption, this is impossible. It follows that for each $\bar{a}$ in $\mathfrak{A}$, $\bar{a}$ does not satisfy any σ in $\mathfrak{A}'$. Using this last observation ω times, we construct a chain $\mathfrak{A} = \mathfrak{A}_0 \subseteq \mathfrak{A}_1 \subseteq \cdots \subseteq \mathfrak{A}_n \subseteq \cdots (n \in \omega)$ of models of T, with $\mathfrak{A}_{n+1} = \mathfrak{A}'_n$, and let $\mathfrak{B} = \bigcup_{n\in\omega} \mathfrak{A}_n$. Clearly $\mathfrak{B} \models T$ and $\mathfrak{B}$ omits each $\sigma \in \Sigma$. □

One easily verifies directly that if T and Σ are as in (4.1), then T *cofinally omits* Σ *iff* Σ *is omitted in every existentially closed model of* T.

A final observation relating to categoricity is of some interest. We will say that T is α - *categorical with respect to* Σ iff T is Σ-consistent and all the models of T of power α which omit Σ are isomorphic. Now suppose T is ω - categorical with respect to Σ ; models of T which omit Σ are models of T^{Σ}, and T^{Σ} is the intersection of all its complete extensions having countable models omitting Σ ; we deduce that T^{Σ} *is complete*. Let τ be any type such that T^{Σ} locally omits τ ; if a model of T omitting Σ were to realize τ, so would a countable model omitting Σ ; but another countable model omitting Σ would omit τ, which is impossible. Thus, every model of T omitting Σ is atomic:

(4.3) T *is* ω - *categorical with respect to* Σ *iff the models of* T *which omits* Σ *are exactly the atomic models of* T^{Σ} . *In particular,* T^{Σ} *is a complete, atomic theory.*

The notion of α - categoricity with respect to Σ yields to the same methods, roughly, as are used in $L_{\omega_1\omega}$, and similar results can be obtained.

REFERENCES.

Bacsich, P. and D. Rowlands Hughes

1974, *Syntactic characterizations of amalgamation, convexity and related properties,* The Journal of Symbolic Logic, vol. 39, 433 - 451.

Chang, C. C. and H. J. Keisler

1973, Model Theory, North - Holland, Amsterdam.

Fisher, E. and A. Robinson

1972, *Inductive theories and their forcing companions,* Israel J. Math., vol. 12, 95 - 107.

Morley, M.

1967, *Countable models of* $\aleph_1$ *- categorical theories,* Israel J. Math., vol. 5, 65 - 72.

Robinson, A.

1971, *Infinite forcing in model theory,* Proceedings of the Second Scandinavian Symposium in Logic (Oslo 1971), North-Holland, Amsterdam, 317-340.

Yasuhara, M.
1974, *The amalgamation property, the universal - homogeneous models, and the generic models*, Math. Scand., vol. 34, 5 - 36.

Department of Mathematics
Bucknell University
Lewisburg, Pennsylvania, U.S.A.

Non-Classical Logics, Model Theory and Computability,
A.I. Arruda, N.C.A. da Costa and R. Chuaqui (eds.)

Semi-Formal Beth Tableaux

by *ANDRÉS R. RAGGIO*

Beth Tableaux are one of the most elegant systematizations of quantification theory. But they have a serious drawback: formulas with functions cannot be analyzed directly but only indirectly eliminating the functions with the help of new predicates. Hilbert-type axiomatizations are in this respect simpler; in order to handle functions only the morphology must be changed extending the term-definition.

The reason for this drawback of Beth tableaux is the following: an existential quantification at the right of the tableaux (a universal at the left) must be refuted (verified) by all instantiations using the terms that have occurred so far in the tableaux. If there are no functions, these terms are finite in number; but with functions the possible instantiations are denumerable. And this case cannot be handled because of the finitary character of the Beth tableaux rules.

If we drop this finitary restriction, we can analyse formulas with functions in a very elegant and perspicuous way. We must change only two rules: existencial quantification to the right, and universal quantification to the left. In both cases we prescribe that the next line of the tableau must contain all instantiations of the quantification using the terms — functions and variables — already employed in the tableau. In this way we get eventually a new line formed by a denumerable list of formulas having a quantifier less. The list must be handled as a single node of the tableau.

A semi-formal Beth tableau — either a closed or an open one — is no longer always a finite object. But it is still a constructively generated object, and this is the only important property in proving the four fundamental metatheorems of quantification theory; namely, completeness, cut-elimination, interpolation, and Herbrand. By the way, syntactic compactness is also valid because of the definition of a closed tableau as a finite development — denumerable nodes count as a single entity — with the same prime formulas at the left and right in every branch. In a closed tableau we delete all formulas which are not used and we obtain in this way a strictly finitary proof.

We show next how to prove those four metatheorems. The proofs are not more complicate than the usual ones and in one case, Herbrand, definitely shorter and easier.

Completeness: we follow the usual pattern. To be sure that, if necessary, every formula in a tableau is going to be analysed, we assign to each formula a left index expressing its depth in the tableau, and a right index expressing the place of the formula in its node from left to right. We add the rule that formulas should be handled following the order of the sum of their two indices.

Cut-elimination: we follow the usual pattern. As Lorenzen and Schütte have shown, the treatment of the two infinite rules present no difficulties.

Interpolation: we follow the usual pattern. As it is well known this proof is very cumbersone when functions are not eliminated introducing predicates.

Herbrand: in the semi-formal Beth tableaux we do not need to use the extremely complicated proof via Gentzen's extended Haupstsatz; we can use Hilbert and Bernay's elegant proof, but with the following essential simplification: instead of using their ε-theorem we argue as follows. Suppose that a formula in prenex normal form is deducible; using cut-elimination we prove that the formula obtained by deleting in the prefix all universal quantifiers and substituting in the nucleus all variables bound by those universal quantifiers by new functions whose arguments are the free variables of the original formula and the variables bound by an existential quantifier preceding in the prefix the corresponding quantifier, is also deducible. For example:

$$\vdash \bigvee y \bigwedge z \bigvee m \bigwedge \ell\ \mathfrak{A}(x,y,z,m,\ell),$$

$$\vdash \bigvee y \bigvee m\ \mathfrak{A}(x,y,f(y),m,g(y,m)).$$

But then the nth line of the closed tableau starting with this formula (n = number of existential quantifiers in the prefix) is a denumerable list of quantifier free formulas obtained from the new nucleus by substitution of terms built up from function signs and variables. Because of the finite character of this closed tableau only a finite number of formulas in the nth line have been used. We build the disjuntion thereof; this disjunction is also deducible. We continue as in Hilbert and Bernays.

Semi-formal Beth tableaux seem to be the most simple and elegant systematization of quantification theory.

REFERENCES.

Hilbert, D. and P. Bernays

1939, Grundlagen der Mathematik, Bd. II, Spriger-Verlag, Berlin.

Lorenzen, P.

1951, *Algebraische und logistische untersuchungen über freie verbände*, The Journal of Symbolic Logic, vol. 16, 81-106.

Raggio, A. R.

1974, A *simple proof of Herbrnad's theorem*, Notre Dame Journal of Formal Logic, vol. XV, nº 3 , 487-488.

Schütte, K.

1951, *Beweistheoretische erfassung der unendlichen induktion in der zahlentheorie*, Mathematische Annalen, Bd. 122, 369-389.

Departamento de Matemática
and
Centro de Lógica e Epistemologia
Universidade Estadual de Campinas
Campinas, São Paulo, Brazil.

Non-Classical Logics, Model Theory and Computability,
A.I. Arruda, N.C.A. da Costa and R. Chuaqui (eds.)

QUANTIFIER ELIMINATION IN FIELDS

by *J. R. SHOENFIELD*

Among the most important contributions of Model Theory to Algebra have been the results of Tarski on algebraically closed fields and real closed fields, the results of Robinson on differentially closed fields, and the results of Ax and Kochen on certain valued fields.

Even a preliminary examination of these results show that they have much in common. The (by no means original) thesis of this article is that the basic unifying concept is that of quantifier elimination. We will show that the central theorem of each set of results says that a certain theory admits quantifier elimination, and that all other results follow easily. Because of the limited space, we only treat algebraically closed fields and real closed fields in detail.

1. QUANTIFIER ELIMINATION.

We shall suppose that all (first-order) languages contain only countably many non-logical symbols and contain at least one constant. These assumptions are not essential; but they simplify the statements and proofs of the results, and are satisfied in all applications.

A theory T *admits quantifier elimination* (abbreviated QE) if for every formula ϕ of T there is a quantifier free formula ψ of T such that $\phi \leftrightarrow \psi$ is a theorem of T.

There are a number of methods for proving that specific theories admit QE. We give one which is particularly suited to the cases which we consider. We first need some definitions.

A *type* is a set Φ of formulas, each of which contains exactly one free variable. These formulas may contain, in addition to the symbols of the language being considered, names of individuals of a structure A. We then say Φ is *realized* in A if there is an individual a in A such that $\phi(a)$ is true for every formula $\phi(x)$ in Φ. We say Φ is *finitely realized* in A if every finite subset of Φ is realized in A.

A structure A is *countably saturated* if every type Φ containing names of only countably many individuals in A which is finitely realized in A is realized in A.

An isomorphism *between* A *and* B is an isomorphism of a substructure of A and a substructure of B. Such an isomorphism is *maximal* if it has no proper extension which is an isomorphism between A and B.

The following result is proved in Shoenfield 1970.

QUANTIFIER ELIMINATION THEOREM. *A theory* T *admits* QE *iff whenever* A *is a countable model of* T, B *is a countably saturated model of* T *and* F *is a maximal isomorphism between* A *and* B, *then the domain of* F *is* A.

The principal uses of QE are in establishing completeness and model completeness. Recall that T is *complete* if every sentence of T is decidable in T, where ϕ is *decidable* in T if either ϕ or $\neg\phi$ is a theorem of T. The theory T is *model complete* if whenever A and B are models of T such that A is a substructure of B, then A is an elementary substructure of B.

PROPOSITION 1. *Let* T *admit* QE. *Then* T *is model complete. If also every quantifier-free sentence of* T *is decidable in* T, *then* T *is complete.*

PROOF: Let A and B be models of T such that A is a substructure of B, and let $\phi(a_1, \ldots, a_n)$ be a sentence containing names $a_1, \ldots, a_n$ of individuals of A. We must show that $\phi(a_1, \ldots, a_n)$ is true in A iff it is true in B. Because T admits QE, we may suppose that ϕ is quantifier-free. But in this case the result is clear.

Now assume the additional hypothesis and let ϕ be a sentence of T.

Choose a quantifier-free ψ such that $\phi \Longleftrightarrow \psi$ is a theorem of T. Replacing each free variable of ψ by some constant, we may suppose that ψ is a sentence. Then ψ is decidable in T; so ϕ is also. Q.E.D.

We shall now give some applications of completeness.

An *axiomatization* of a structure A is a theory T such a sentence of T is a theorem of T iff it is valid in A. We can always obtain an axiomatization of A by taking as axioms of T all sentences true in A. However, we are generally interested in obtaining a simpler axiomatization. For this purpose, we select some sentences we know are true in A and adopt them as axioms. This gives a theory T having A as a model. To test whether T is an axiomatization, we use the following trivial result.

PROPOSITION 2. *If* A *is a model of* T, *then* T *is an axiomatization of* A *iff* T *is complete.*

The second application of complete theories is given by an equally trivial result.

PROPOSITION 3. *If* T *is complete, then every sentence which is true in one model of* T *is true in every model of* T.

For our third application, we recall two definitions. A theory T is *axiomatized* (*decidable*) if there is an algorithm for deciding whether or not a given formula is an axiom (theorem) of T. To make the notion of algorithm precise here, we need the theory of recursive functions. A basic result in that theory then leads to the following.

PROPOSITION 4. *If* T *is axiomatized and complete, then* T *is decidable.*

We conclude this section with a method for establishing the additional hypothesis in Proposition 1.

We say A is a *prime* structure for T if every model of T has a substructure isomorphic to A. (Note that A need not be a model of T.)

PROPOSITION 5. *If* T *has a prime structure, then every quantifier-free sentence of* T *is decidable in* T.

PROOF: Let A be the prime structure and let ϕ be a quantifier-free sentence. If ϕ is true in A, then it is true in every model of T and hence is a theorem of T. If ϕ is false in A, the same argument shows that $\neg\phi$ is a theorem of T. Q.E.D.

2. ALGEBRAICALLY CLOSED FIELDS.

We begin with the following problem: find a simple axiomatization of the complex field C.

First we must specify the language of C. (Warning: Theorems on QE are very sensitive to small changes in the language.) As non-logical symbols of the language of fields, we take the binary operation symbols + and $\cdot$ and the constants 0, 1, and -1. In this language we can easily write down the field axioms. For each positive integer n we can write down a sentence which says that every polynomial of degree n has a root. We then have the axioms of the theory ACF of algebraically closed fields.

THEOREM 1 (Tarski). ACF *admits* QE.

PROOF: Let A be a countable model of ACF; B a countably saturated model of ACF; F a maximal isomorphism between A and B. We must show the domain D of F is A.

Since D is a substructure of A, it is a subring containing 1. We know every isomorphism defined on such a subring can be extended to the smallest subfield including it. Hence D is a subfield of A.

By replacing B by an isomorphic field, we may suppose that F is the identity on D. It is then clearly enough to show that for every $a \in A$, there is a $b \in B$ and an isomorphism of the fields $D(a)$ and $D(b)$ which is the identity on D and takes a into b. The theory of fields tells us what conditions b must satisfy: if a is algebraic over D, b must be algebraic with the same minimal polynomial; if a is transcendental over D, b must be transcendental. In the first case, such a b exists because B is algebraically closed. Thus it suffices to show that some member of B is transcendental over D.

If $f(x)$ is a non-constant polynomial with coeficients in D,

$f(x) \neq 0$ may be considered as a formula with one free variable containing names of individuals in D. We want to show that the set Φ of these formulas is realized in B. Since D is countable (as a subset of A) B is countably saturated, we need only show that Φ is finitely realized in B. Thus we must show that, given finitely many non-constant polynomials, there is an element of B which is not a root of any of them. For this, it suffices to see that B is infinite. But if $b_1, \ldots, b_n$ were all the members of B, then

$$(x - b_1) \cdot \ldots \cdot (x - b_n) + 1$$

would have no root in B. Q.E.D.

We can now see why the Quantifier Elimination Theorem is so suitable for our purposes. It deals with extensions of isomorphisms, and this is a question about which field theory tells us a great deal. In more complicated cases, we shall be dealing with fields with additional structure. In such cases, the crucial new point is to find conditions under which the extended isomorphisms are isomorphisms for the additional structure.

COROLLARY 1. ACF *is model complete.*

PROOF: By Theorem 1 and Proposition 2. Q.E.D.

We give an application of Corollary 1. The following result is very important in algebraic geometry. An algebraic proof of it, while not difficult, is not entirely trivial.

PRINCIPLE OF ALGEBRAIC ZEROES. *Let* $f_1(x_1, \ldots, x_n), \ldots, f_r(x_1, \ldots, x_n)$ *be polynomials with coefficients in an algebraically closed field* F. *If they have a common zero in some extension* G *of* F, *then they have a common zero in* F.

PROOF: Replacing G by a larger field, we may suppose that G is algebraically closed. There is a sentence ϕ, using names of members of F, which says that $f_1, \ldots, f_r$ have a common zero. Then ϕ is true in G. By Corollary 1, F is an elementary substructure of G. Thus ϕ is true in

F. Q.E.D.

We still do not have our axiomatization of C, since ACF is not complete. We therefore add axioms saying that the characteristic is 0, thus obtaining the theory ACF(0).

COROLLARY 2. ACF(0) *is complete.*

PROOF: By the Theorem, ACF(0) admits QE. Moreover the rational field is a prime structure for ACF(0). Now apply Proposition 5 and 1. Q.E.D.

COROLLARY 3. ACF(0) *is an axiomatization of* C.

PROOF: By Corollary 2 and Proposition 2. Q.E.D.

COROLLARY 4. *If a sentence* ϕ *(of the language of fields) is true in* C, *then it is true in every algebraically closed field of characteristic* 0.

PROOF: By Corollary 2 and Proposition 3. Q.E.D.

Corollary 4 is a special case of a vague empirical principle called Lefschetz's Priciple: *any algebraic fact true in* C *is true in every algebraically closed field of characteristic* 0.

COROLLARY 5. ACF(0) *is decidable.*

PROOF: By Corollary 2 and Proposition 3. Q.E.D.

3. REAL CLOSED FIELDS.

Now let us attempt to find an axiomatization of the real field R. To apply our method, we must find a theory which admits QE and has R as a model. However, there is no such theory in the language of fields. For if $\phi(x)$ is a quantifier-free formula, then the set x in R such that $\phi(x)$ is true is either finite or cofinite. (This is easily verified by induction on the length of $\phi(x)$.) On the other hand, the set of x in R such that $\exists y(x = y \cdot y)$ is true is neither finite nor cofinite.

What we must do is introduce a new symbol so that $\exists y(x = y \cdot y)$ will be equivalent to a quantifier-free formula. We introduce the binary relation symbol $\leq$; then $\exists y(x = y \cdot y)$ is equivalent to $0 \leq x$.

We now have the language of ordered fields. In this language, we can write down the axioms for an ordered field. Now R, in addition to being an ordered field, is a real closed field; that is, every polynomial which has both positive and negative values has a root. It is easy to write down an infinite set of axioms which express this; we need one axiom for each degree of a polynomial. We thus obtain the theory RCF of real closed fields.

THEOREM 2 (Tarski). RCF *admits* QE.

PROOF: Let A be a countable model of RCF; B a countable saturated model of RCF; F a maximal isomorphism between A and B. We must show that the domain D of F is A.

As in the proof of Theorem 1, D is a subfield of A. (We need here the fact that the extension of an isomorphism from a ring to the smallest field including it is valid for *ordered* fields.) Also as in that proof, we may suppose that F is the identity on D.

By the properties of real closed fields, F can be extended to an isomorphism of the algebraic closure of D in A and the algebraic closure of D in B. Hence by the maximality of F, D is algebraically closed in both A and B.

Now let $a \in$ A. Again we must find $b \in$ B so that there is an isomorphism of the ordered field D(a) and D(b) which is the identity on D and takes a into b. We may suppose $a \notin$ D.

Let Φ consist of all formulas $x < d$, where $d \in$ D and $a < d$, and all formulas $d < x$, where $d \in$ D and $d < a$. We show that Φ is realized in B. Because B is countably saturated and D is countable, it is enough to realize a finite subset of Φ. For this, it is enough to see that if $d_1 < a < d_2$ with $d_1, d_2 \in$ D, then there is a $b \in$ B such that $d_1 < b < d_2$. But we can simply take $b = \frac{1}{2}(d_1 + d_2)$.

Let $b \in$ B realize Φ. Then for all $d \in$ D,

(1) $a < d \Rightarrow b < d$

and

(2) $d < a \Rightarrow d < b$.

Since $a \notin$ D, a is transcendental over D. From (1) and (2), $b \notin$ D; so b is transcendental over D. Hence we have a unique field isomorphism of D(a) and D(b) which is the identity on D and takes a into b. We must show that this isomorphism preserves order.

An element of D(a) is of the form $f(a)/g(a)$, where f and g are polynomials with coefficients in D; and the isomorphism carries this into $f(b)/g(b)$. It will therefore be sufficient to prove

(3) $0 \leq f(a) \Longleftrightarrow 0 \leq f(b)$

for every polynomial f with coefficients in D.

If (1) holds for f, it holds for all $d.f$ with $d \in$ D. Hence we may suppose that the leading coefficient of f is 1. If (1) holds for f_1 and f_2, it holds for $f_1 f_2$. Hence we may suppose that f is irreducible.

If f is constant, then (3) is clear. If f is of degree 1, then $f(x) = x - d$ for some $d \in$ D; so (3) follows from (1) and (2). Now suppose f has degree ≥ 2. Since f is irreducible, it has no roots in D. Then it has no roots in A or B; for these roots would be algebraic over D and hence in D. Since the leading coefficient of f is 1, $f(x) > 0$ for large values of x in D. Combining these facts with the fact that A and B are real closed, we see that f assumes only positive values for arguments in A or B. Thus (3) holds. Q.E.D.

The following corollaries of Theorem 2 are proved like the corresponding corollaries of Theorem 1.

COROLLARY 1. RCF *is model complete.*

COROLLARY 2. RCF *is complete.*

COROLLARY 3. RCF *is an axiomatization of* R.

COROLLARY 4. *If a sentence* ϕ *(of the language of ordered fields) is true in* R, *then it is true in every real closed field.*

COROLLARY 5. RCF *is decidable.*

We give an application of Corollary 1. In order to solve Hilbert's 17th problem, Artin proved the following result. Let $f(x_1, \ldots, x_n)$ be a polynomial with rational coefficients which assume only non-negative values when its arguments are real. Then f is a sum of squares of rational functions with rational coefficients.

To prove this, Artin considered the field G of rational functions of $x_1, \ldots, x_n$ with rational coefficients. If the conclusion is false, f is not a sum of squares in G. A theorem of Artin and Schreier then tells us that the field G may be ordered so that $f < 0$. Artin then used this to show that f must assume a negative value for some real arguments. This part of the proof follows readily from Corollary 1. First, the theory of real closed fields shows that G can be extended to a real closed field H; and R is a subfield of H. There is a sentence ϕ which says that f assumes a negative value; and this sentence is true in H, since $f < 0$. It follows that ϕ is true in R.

Finally, we can obtain an axiomatization for R in the language of fields. We take the axiomatization in the language of ordered fields, and replace each formula $s \leq t$ by $\exists x(s + x^2 = t)$.

4. OTHER CASES.

We shall not treat the other two cases in detail, since the additional material is mostly algebraic. This material is discussed in Robinson 1956 for differentially closed fields and in Kochen 1974 for valued fields. We only make a few remarks which will enable the reader to connect the material in Kochen 1974 with our methods.

The language of valued fields can be formulated in several ways. However it is done, it is possible to make assertions in this language about the residue class field and the value group. For example, let $\phi(x_1, \ldots, x_n)$ be a formula of the field language. Then there is a formula $\phi^*(x_1, \ldots, x_n)$ of the valued field language with the following property. Let F be a valued field; $a_1, \ldots, a_n$ integers of F; $\bar{F}$ the residue class field; $\bar{a}_1, \ldots, \bar{a}_n$ the cosets of $a_1, \ldots, a_n$ in $\bar{F}$. Then

$\phi(\bar{a}_1,\ldots,\bar{a}_n)$ is true in $\overline{F}$ iff $\phi^*(a_1,\ldots,a_n)$ is true in F.

In the valued field language we can form the theory T of valued fields which satisfy Hensel's Lemma and have residue class fields of characteristic 0. To make this into a theory admitting QE, we must add new symbols for each formula about the residue class field or the value group. Thus if $\phi(x_1,\ldots,x_n)$ is as above, we add a new relation symbol P and a new axiom

$$P(x_1,\ldots,x_n) \Longleftrightarrow \phi^*(x_1,\ldots,x_n).$$

We can now prove that the resulting theory T* admits QE, and use this to obtain, for example, the Ax-Kochen results on Artin's conjecture. Making more use of the results in Kochen 1974, we can get an axiomatization of the field of p-adic numbers. However, the problem of finding an axiomatization of the field of meromorphic series with coefficients in a finite field is still open.

REFERENCES.

Kochen, S.

1974, *The model theory of local fields*, Logic Conference Kiel 1974, Lecture Notes in Mathematics, Springer-Verlag, 384-425.

Robinson, A.

1956, Complete Theories, North-Holland Publishing Co., Amsterdam.

Shoenfield, J. R.

1970, A *theorem on quantifier elimination*, Symposia Mathematica, vol. V, 173-176.

Department of Mathematics
Duke University
Durham, North Carolina, U.S.A.

PART III

COMPUTABILITY

Non-Classical Logics, Model Theory and Computability,
A.I. Arruda, N.C.A. da Costa and R. Chuaqui (eds.)

On the Decision Problem of the Congruence Lattices of Pseudocomplemented Semilattices(*)

by *H. P. SANKAPPANAVAR*

1. Introduction.

The study of the decision problems for various classes of structures (both algebras and relational structures) - which began in the 1930's when Church gave the first undecidability result (see Church 1936) - has for its credit a vast literature, and an excellent survey of the work done until 1965 has appeared in Ershov et al 1965. However, the decision problems for classes of structures which are, in a natural way, associated with other classes of structures - for example, congruence lattices, subalgebra lattices, lattices of subspaces of geometries, lattices of varieties of algebras, automorphism groups, endomorphism groups, etc. - were considered only much later, in the early 1960's, except for results of Tarski and of Grzegorczyk which were published in 1949 and in 1951 respectively (see Tarski 1949, and Grzegorczyk 1951).

Tarski proved that the (first-order) theory of the lattices of subspaces of two-dimensional projective geometries (with points having homogeneous rational coordinates) is undecidable (Tarski 1949). The problem of whether this theory is recursively inseparable from the set of its finitely refutable sentences appears to be still open. Grzegorczyk (in the above

---

(*) This work was essentially done in 1973 at the University of Waterloo, Waterloo, Ontario, Canada and was supported by a Province of Ontario Graduate Fellowship.

mentioned paper) considered the decision problems for the theories of certain classes of algebras which are associated with topological spaces: he showed the essential undecidability of the theories of closure algebras, of Brouwerian algebras, of the algebras of bodies, of the algebras of convexity and of the semi-projective algebra. He also deduced Tarski's result mentioned above from one of his results. In 1962 Kargapolov initiated the study of decision problems for lattices of subgroups by showing that the theory of subgroup lattices, and hence also of the congruence lattices, of Abelian torsion-free reduced groups is undecidable (cf. Kargapolov 1962). As corollaries he mentions the undecidability of the theory of subgroup lattices of groups and of Abelian groups. In 1970 Kozlov proved that the theory of lattices of subgroups of finite Abelian p-groups is recursively inseparable (cf. Kozlov 1970). As consequences Kozlov lists also the undecidability of the theories of lattices of subgroups of Abelian p-groups, of finite Abelian groups, of torsion-free Abelian groups and of free Abelian groups. He asks whether the universal theory of lattices of subgroups of Abelian groups is decidable. Also in Taitslin 1970 is established the hereditary undecidability of the theory of the lattice of subgroups of the direct sum of two infinite cyclic groups and hence that of the theory of subgroup lattices of any class of groups which contains such a direct sum; he also gives some positive results, such as the decidability of the theories of subgroup lattices of finite Abelian groups with n generators, of finite Abelian p-groups with n generators and of Abelian groups which are direct sums of n isomorphic finite cyclic groups. Moving on to rings, Taitslin proved in 1968 (cf. Taitslin 1968a) that the theory of the lattice of ideals of a polynomial ring over a field with at least two unknowns is hereditarily undecidable, while that of a polynomial ring in one unknown is decidable; in fact, he observed the decidability of the ideal lattice of a Dedekind domain. In the same year Taitslin published another paper (Taitslin 1968b) in which be proved that the theory of the partially ordered set of simple ideals of a polynomial ring in at least three unknowns over a field is hereditarily undecidable and he mentions the case of two unknowns as an open problem. It is a consequence of the results proved in Rabin 1964 that the theory of congruence lattices of countable Boolean algebras is decidable. More recently, Burris and Sankappanavar have examined the decision problems of the theories of lattices of subrings of rings with unity (the case of Boolean algebras is a particular case), of congruence lattices of semilattices, semigroups and unary algebra, and of lattices of varieties

(cf. Burris and Sankappanavar 1975).

Boolean algebras are the only non-trivial proper subvariety (see e.g. Sankappanavar 1974) of the variety of pseudocomplemented semilattices, and the class of Boolean algebras is a subclass of the congruence-distributive pseudocomplemented semilattices. Since the theory of congruence lattices of countable Boolean algebras is decidable as mentioned above, a question naturally arose whether the theory of (a slightly larger class of) congruence lattices of congruence-distributive pseudocomplemented semilattices is also decidable. In this note we show that this theory is recursively inseparable (and hence undecidable). From this we deduce an undecidability result of Ershov and Taitslin 1963 which in turn is an improvement on an undecidability result due to Grzegorczyk (c.f. Grzegorczyk 1951). It is also observed that the theory of Heyting lattices and that of filter lattices of pseudo-complemented semilattices are recursively inseparable.

2. PSEUDOCOMPLEMENTED SEMILATTICES AND THEIR CONGRUENCE LATTICES.

An algebra $\mathcal{L} = \langle L, \wedge, 0 \rangle$ is a $\wedge$-*semilattice with zero* iff $\wedge$ is a binary operation on L, and 0 is a distinguished element of L, satisfying the identitities $x \wedge y = y \wedge x$, $x \wedge (y \wedge z) = (x \wedge y) \wedge z$, $x \wedge x = x$, and $x \wedge 0 = 0$. Let $\mathcal{L}$ be a $\wedge$-semilattice with zero and let $\leq$ be the partial ordering of $\mathcal{L}$. If $a, b \in L$, we define $[a, b] = \{x \in L : a \leq x \leq b\}$. An element x in L is a *pseudocomplement* of an element a in L iff x is the greatest element such that $x \wedge a = 0$ and is denoted by a^*. An algebra $\mathcal{S} = \langle S, \wedge, *, 0 \rangle$ is a *pseudocomplemented semilattice* iff $\langle S, \wedge, 0 \rangle$ is a $\mathcal{S}$-semilattice with zero and $*$ is a unary operation on S such that a^* is the pseudocomplement of a in S. The class of all pseudocomplemented semilattices is a variety and is denoted by PCS. If $S \in$ PCS, we write S is a PCS, and its congruence lattice is denoted by *Con S*. For a systematic study of the congruence lattices of PCS's one should refer to Sankappanavar 1974. We mention here only those concepts and results needed for our purpose.

In the sequel S denotes an arbitrary PCS. An element a in S is *closed* iff $a^{**} = a$, and the set of such elements is denoted by $B(S)$, while its complement in S by $N(S)$. The elements of $N(S)$ are said to be *non-closed*, or equivalently, an element a is non-closed iff $a < a^{**}$. $B(S)$ is a subalgebra of S and also forms a Boolean algebra in the usual sense. It is well-known that the class of Boolean algebras can be defined by a set of identities involving only $\wedge$, $*$ and 0, hence as a subvariety of the variety PCS. For $c \in B(S)$ define $D_c(S) = \{x \in S : x^{**} = c\}$. It is known that $D_c(S)$ is a subalgebra of S.

For θ in $Con\ S$, $(\theta)_B$ denotes the restriction of θ to $B(S)$. It should be noted that $(\theta)_B$ can be regarded either as a PCS - congruence or as a BA - congruence on $B(S)$ considering $B(S)$ as a PCS or as a Boolean algebra (in the usual sense) respectively. The mapping $** : S \to S$, $s \mapsto s^{**}$, is a PCS - homomorphism whose kernel is denoted $\Phi(S)$ or simply Φ, i.e. $\Phi(S) = \{\langle x,y\rangle : x, y \in S \text{ and } x^{**} = y^{**}\}$. For $a \in S$ we define the congruence $\hat{a}$ (or $(a)\hat{}$) on S as follows:

$$\langle x,y\rangle \in \hat{a} \text{ iff } x \wedge a = y \wedge a, \quad x,y \in S.$$

LEMMA 1. *For* $a \in S$, $(\hat{a})_B = ((a^{**})\hat{}\ )_B$

LEMMA 2. *If* $\alpha, \beta \in Con\ S$ *are such that* $\alpha \geq \Phi$, $\beta \geq \Phi$ *and* $(\alpha)_B = (\beta)_B$ *then* $\alpha = \beta$.

The proofs of the above lemmas are not hard and hence omitted. We also need the following proposition which is proved in Sankappanavar 1974.

PROPOSITION *Con S is distributive iff S satisfies the condition* (D) $\forall x \forall y\ (x < y^{**} \to x \leq y$ *or* $y \leq x)$.

We say that S is *congruence - distributive* iff $Con\ S$ is distributive. Let κ be a cardinal. A PCS S is a κ - *Boolean algebra* iff S is congruence - distributive and $|D_c(S)| \leq \kappa$ for every $c \in B(S)$. It follows that a PCS is a Boolean algebra iff it is a 1 - Boolean algebra. Δ_S and ∇ denote the equality relation on S and $S \times S$ respectively. $\theta(a,b)$ denotes the congruence generated by $\langle a, b\rangle$

3. Logical terminology and the basic theorem.

By a *language* we mean a first - order language with equality (for these and other related notions see Shoenfield 1967) which has only a finite number of non - logical symbols. If L is a language, a *theory T in L* is a set of sentences of L which is closed under logical deduction. A sentence σ in L is *finitely refutable* in a theory T iff there exists a finite model of T in which $\neg\sigma$ is true. We denote by T_{fin} the set of all sentences in L which are true of all the finite models of T, and by T_f the set of all finitely refutable sentences in L. We say that T is *recursively inseparable* iff there exists no recursive set of sentences A in L such that $T \subseteq A$ and A is disjoint from T_f. It is clear that if T is recursively inseparable then T and T_{fin} are undecidable.

Let L be a language with one binary predicate symbol ρ, L_1 another language (not necessarily different from L). Let $\delta(x)$ and $\rho(x, y)$ be formulas of L_1 with one and two free variables respectively. For every structure M_1 of L_1 with universe M_1, we define a structure of L *induced by* δ and ρ - which is denoted by $M_1(\delta, \rho)$ - as follows:

$$M_1(\delta, \rho) = \langle D; R\rangle$$

where

$$D = \{s \in M_1 : M_1 \models \delta(s)\}$$

and

$$R = \{\langle s, t\rangle \in M_1^2 : s, t \in D \text{ and } M_1 \models \rho(s, t)\}$$

The following theorem - which is taken from Burris and Sankappanavar 1975- is the basic tool in the next section.

THEOREM 3.1. *Let T be a theory in a language L with the property that T is recursively inseparable. Let T_1 be a theory in L_1. Assume that $\delta(x)$ and $\rho(x, y)$ are two formulas in L_1 such that*

(1) *for every finite model N of T there exists a finite model M_1 of T_1 such that the induced structure $M_1(\delta, \rho) \cong N$, and*

(2) *for every model M_1 of T_1 the induced structure $M_1(\delta, \rho)$ is a model of T.*

Then T_1 is recursively inseparable.

4. Elementary theory of congruence lattices.

Let L_1 denote the language of lattices, i.e. L_1 has two non-logical symbols $\wedge$ and $\vee$. One can write in L_1 formulas $Coatom(x)$, $Irr_\wedge(x)$ which say respectively that "x *is a coatom*", "x *is* $\wedge$-*irreducible*".

We denote by DCON the class of all (distributive) congruence lattices of congruence-distributive PCS's and Th(DCON) denotes the theory of DCON in L_1, i.e. the set of all sentences in L_1 that are true of DCON.

THEOREM 4.1 *Th*(DCON) *is recursively inseparable.*

PROOF: Let T' denote the theory of an irreflexive, symmetric binary relation R. It is shown in Ershov 1965 that T' and T'_f are recursively inseparable. Let ξ be the sentence

$$\exists x \exists y \exists z (x \neq y \;\&\; x \neq z \;\&\; y \neq z)$$

and let T be the theory axiomatized by $T' \cup \{\xi\}$. Then since T is a finite extension of T', T and T_f are recursively inseparable and so T qualifies to be the theory T of Theorem 3.1.

Let $M = \langle A, R \rangle$ be a finite model of T, so $|A| \geq 3$. For every pair $a, b \in A$ such that $\langle a, b \rangle \in R$ (and hence $a \neq b$ necessarily), choose a new symbol t_{ab} and require that $t_{ab} = t_{ba}$. Let $A_1 = \{t_{ab} : \langle a,b \rangle \in R\}$ and let 2^A denote the power set of A and 2^A denotes: the power set algebra $\langle 2^A, \cap, ', 0 \rangle$ regarded as a PCS (where 0 stands for the empty set and we use 1 for $0'$ which is A). Letting $S = 2^A \cup A_1$ we will make S into the universe of a PCS. Define an operation $\wedge : S \times S \to S$ as follows:

(i) if $s, t \in 2^A$, $s \wedge t = s \cap t$;

(ii) if $a, b \in A$ with $\langle a, b \rangle \in R$, $t_{ab} \wedge t_{ab} = t_{ab}$;

(iii) if $a, b, c \in A$ with $\langle a, b \rangle \in R$ and $\langle a, c \rangle \in R$ then $t_{ab} \wedge t_{ac} = t_{ac} \wedge t_{ab} = \{a\}$;

(iv) if $a, b, c, d \in A$ such that $\{a,b\} \cap \{c, d\} = \emptyset$, $\langle a, b \rangle \in R$ and $\langle c,d \rangle \in R$, then $t_{ab} \wedge t_{cd} = 0$;

and

(v) if $a, b \in A$ with $\langle a, b \rangle \in R$ and $x \in 2^A$ then

$$t_{ab} \wedge x = x \wedge t_{ab} = \begin{cases} \{a\} & \text{if } x = \{a\}, \\ \{b\} & \text{if } x = \{b\}, \\ t_{ab} & \text{if } x = \{a, b\}, \\ x \wedge \{a,b\} & \text{if } x \neq \{a\},\{b\},\{a, b\}. \end{cases}$$

We take the $0 \in 2^A$ as a distinguished element in S and define $*: S \to S$ as follows:

(i) if $s \in 2^A$, $s^* = s'$;

(ii) if $s \in A_1$, $s^* = \{a, b\}^*$ where $s = t_{ab}$.

Then it is easy to verify that $\mathcal{S} = \langle S, \wedge, *, 0 \rangle$ is indeed a PCS with

$B(\mathcal{S}) = 2^A$ and $N(S) = \{t_{ab} : \langle a,b\rangle \in R\}$. We also note that $\Phi = \{\langle t_{ab}, \{a,b\}\rangle : \langle a,b\rangle \in R\}$. It is clear that $\mathcal{S}$ satisfies condition (D) and hence is congruence-distributive. Since A is finite, S is also finite. The above construction is illustrated in Figure 1, where $A = \{a, b, c, d\}$ and $R = \{\langle a,b\rangle, \langle a,c\rangle, \langle b,d\rangle, \langle c,d\rangle, \langle b,a\rangle, \langle c,a\rangle, \langle d,b\rangle, \langle d,c\rangle\}$.

Figure 1

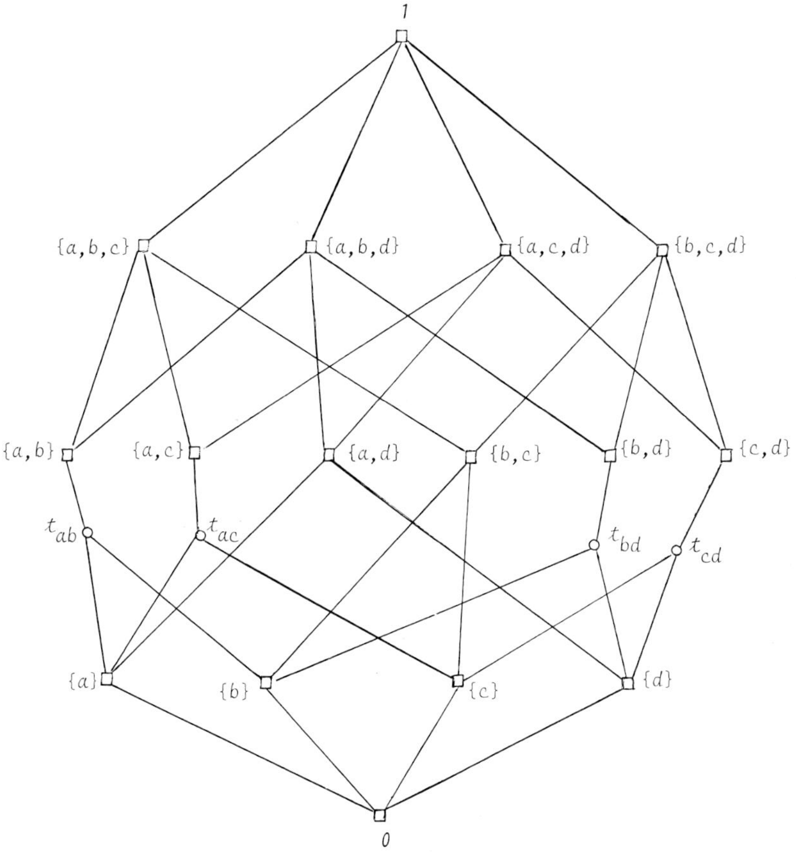

Let us choose $Con\ \mathcal{S}$ for the M_1 of Theorem 3.1 and consider the following formulas in L_1, where $x \prec y$ is an abbreviation for the formula $x = x \wedge y \ \&\ x \neq y \ \&\ \forall z((x = x \wedge z \ \&\ z = z \wedge y) \to (z = x \ \ or \ \ z = y))$.

$$\delta(x) \overset{def}{\longleftrightarrow} Coatom\ (x)$$

and

$$\rho(x, y) \overset{def}{\longleftrightarrow} \delta(x) \ \&\ \delta(y) \ \&\ \exists z(Irr_\wedge(z) \ \&\ z \prec x \wedge y).$$

For $a \in A$ it is clear that $\{a\}^\wedge$ is a coatom in $Con\ \mathcal{S}$ since it has two congruence classes, namely $[\{a\}, 1]$ and $[0, \{a\}^*]$. We claim that $\delta(x)$ picks out precisely the congruences of the form $\{a\}^\wedge$ with $a \in A$.

CLAIM 1. $Con\ \mathcal{S} \models \delta(\Psi)$ iff $\Psi = \{a\}^\wedge$ for some $a \in A$. To prove this claim, if $a \in A$ then it is already noted that $\{a\}^\wedge$ is a coatom in $Con\ \mathcal{S}$ and so $Con\ \mathcal{S} \models \delta(\{a\}^\wedge)$. Conversely, suppose $Con\ \mathcal{S} \models \delta(\Psi)$; then Ψ is a coatom in $Con\ \mathcal{S}$. Since 2 is the only simple PCS, Ψ has exactly two congruence classes, namely $[1]$ and $[0]$ (2 is the PCS whose universe is $\{0, 1\}$). It follows that $[1] = [t, 1]$ for some $t \in S$ since S is finite and so $\Psi = \hat{t}$. If $r \in S$ and $r < t$ then $r = 0$, for, if $r \neq 0$ then $\hat{t} < \hat{r} < \nabla$ since $\langle r, t\rangle \in \hat{r}$ but $\langle r, t\rangle \notin \hat{t}$ and $\langle 0, r\rangle \in \nabla$; but $\langle 0, r\rangle \notin \hat{r}$, this is impossible since $\hat{t}$ is coatom. Thus it follows that t is an *atom* in S. Since, by the construction of S, atoms in S are precisely of the form $\{a\}$ with $a \in A$ we have $t = \{a\}$ for some $a \in A$, which proves Claim 1.

CLAIM 2. For $a, b \in A$, $\langle a, b\rangle \in R$ iff $Con\ \mathcal{S} \models \rho(\{a\}^\wedge, \{b\}^\wedge)$. To prove this, first suppose $\langle a, b\rangle \in R$. It is clear that $(t_{ab})^\wedge \geq \Phi$ and $\{a\}^\wedge \wedge \{b\}^\wedge \geq \Phi$; since $((t_{ab})^\wedge)_B = \{a, b\}^\wedge$ by Lemma 1 and $(\{a\}^\wedge \wedge \{b\}^\wedge)_B = (\{a\}^\wedge)_B \wedge (\{b\}^\wedge)_B = \{a, b\}^\wedge$, we have $((t_{ab})^\wedge)_B = (\{a\}^\wedge \wedge \{b\}^\wedge)_B$. Therefore by Lemma 2 we get $(t_{ab})^\wedge = \{a\}^\wedge \wedge \{b\}^\wedge$. Now consider the congruence $\{a, b\}^\wedge$ in $Con\ \mathcal{S}$. Since $t_{ab} \leq \{a, b\}$, we have $\{a, b\}^\wedge \leq (t_{ab})^\wedge$; since $\langle t_{ab}, \{a, b\}\rangle \in (t_{ab})^\wedge$ and $\langle t_{ab}, \{a, b\}\rangle \notin \{a, b\}^\wedge$ we indeed have $\{a, b\}^\wedge < (t_{ab})^\wedge$. In fact, since $(t_{ab})^\wedge = \{a, b\}^\wedge \vee (\{\langle t_{ab}, \{a,b\}\rangle\} \cup \Delta_s)$ and $(\{\langle t_{ab}, \{a, b\}\rangle\} \cup \Delta_s)$ is an atom in $Con\ \mathcal{S}$, we conclude that $\{a, b\}^\wedge \prec (t_{ab})^\wedge$. Next we claim that $\{a, b\}^\wedge$ is $\wedge$-irreducible; for, since the only congruences greater than $\{a, b\}^\wedge$ are $(t_{ab})^\wedge$, $\hat{a}$, $\hat{b}$ and ∇, it is clear that $\Psi < \{a, b\}^\wedge$ implies $\Psi \geq (t_{ab})^\wedge$, proving the claim. Thus $Con\ \$ \models \rho(\{a\}^\wedge, \{b\}^\wedge)$. To prove the converse, suppose $\langle a,b\rangle \notin R$. It is clear that the congruence classes of $\{a\}^\wedge \wedge \{b\}^\wedge$ are precisely $[\{a\}, 1] \cap [\{b\}, 1]$, $[\{a\}, 1] \cap (S - [\{b\}, 1])$, $(S - [\{a\}, 1]) \cap [\{b\}, 1]$ and $(S - [\{a\}, 1]) \cap (S - [\{b\}, 1])$. Since $\langle a,b\rangle \notin R$, we see that $[\{a\}, 1] \cap [\{b\}, 1] = [\{a, b\}, 1]$. Also we have $[\{a\}, 1] \cap (S - [\{b\}, 1]) = [\{a\}, \{b\}^*]$, $(S - [\{a\}, 1]) \cap [\{b\}, 1] = [\{b\}, \{a\}^*]$, and $(S - [\{a\}, 1]) \cap (S - [\{b\}, 1]) = [0, \{a, b\}^*]$. Then it follows that $\{a\}^\wedge \wedge \{b\}^\wedge = \{a,b\}^\wedge$. Let $\Psi \prec \{a\}^\wedge \wedge \{b\}^\wedge$ and we wish to show that Ψ is not $\wedge$-irreducible. We have $\Psi \prec \{a, b\}^\wedge$ and since $|A| \geq 3$, $\{a, b\} \neq 1$. If $\Psi_B = (\{a, b\}^\wedge)_B$ then $\langle\{a, b\}, 1\rangle \in \Psi_B \leq \Psi$ and so $\theta(\{a, b\}, 1) \leq \Psi$ whence $\{a, b\}^\wedge = \Psi$ which is impossible; hence $\Psi_B < (\{a, b\})_B$ and consequently there exist $u, v \in B(\$)$, $u \neq v$, such that $\langle u, v\rangle \in (\{a, b\}^\wedge)_B$ and $\langle u,v\rangle \notin \Psi_B$. Now we claim that $\Psi \geq \Phi$. For if $\Psi \ngeq \Phi$ then for some non-closed n, $\langle n,n^{**}\rangle \notin \Psi$. Let $\beta = \{\langle n, n^{**}\rangle\} \cup \Delta_S$ which is clearly a congruence on S. Since $\Phi \leq \{a\}^\wedge$ and $\Phi \leq \{b\}^\wedge$, it follows that $\Phi \leq \{a, b\}^\wedge$ which implies that $\beta \leq \{a,b\}^\wedge$. Thus we get $\Psi \vee \beta \leq \{a, b\}^\wedge$; but $\Psi \vee \beta \neq \{a, b\}^\wedge$ because $(\Psi \vee \beta)_B = \Psi_B$ and so $\langle u,v\rangle \notin \Psi \vee \beta$. We have thus shown that $\Psi < \Psi \vee \beta < \{a, b\}^\wedge$, contradicting the fact that $\Psi \prec \{a, b\}^\wedge$ - hence proving that $\Psi \geq \Phi$. From this it follows that Ψ is the meet of maximal elements since $[\Phi, \nabla] \cong Con\ (B(\$))$, implying that Ψ is not $\wedge$-irreducible; hence $Con\ \$ \not\models \rho(\{a\}^\wedge, \{b\}^\wedge)$. This proves Claim 2.

From claims 1 and 2 it follows that if $D = \{\{a\}^\wedge : a \in A\}$ and $P = \{\langle\{a\}^\wedge, \{b\}^\wedge\rangle : Con\ \$ \models \rho(\{a\}^\wedge, \{b\}^\wedge)\}$ then $\langle D,P\rangle \cong \langle A,R\rangle$, hence (1) of Theorem 3.1 is satisfied; while (2) of that theorem is easily verified. Therefore Theorem 4.1 is proved.

In fact, we have proved the following stronger result once it is noted that the PCS S constructed in the above proof is a 2-Boolean algebra.

THEOREM 4.2. *The theory of the class of congruence lattices of 2-Boolean algebras is recursively inseparable.*

COROLLARY 4.3. *Let K be any class of lattices which contains lattices L with any (or all) of the following properties:*

(1) *L is algebraic, distributive and atomic and has 0 and 1,*

(2) *every interval in L is pseudocomplemented, and*

(3) *L has an element f such that [0, f] is a complete atomic Boolean algebra, f is the meet of all coatoms in L and [f, 1] is an algebraic sublattice whose compact elements form a complemented sublattice.*

Then Th(K) is recursively inseparable.

The above corollary is immediate from the fact that the congruence lattices of 2 - Boolean algebras have these properties (see Sankappanavar 1974). From this corollary it follows immediately that the theory of Heyting lattices is recursively inseparable, which is an improvement on a result of Ershov and Taitslin 1963 that the theory of distributive lattices is recursively inseparable.

We also obtain the hereditary undecidability of the theory of a very restricted class of finite distributive lattices as given in the following corollary.

COROLLARY 4.4 *Let F be the class of finite distributive lattices L with 0 and 1 such that*

(i) *L contains an element f such that [0, f] and [f, 1] are Boolean sublattices, and*

(ii) *the number of* $\vee$ *- irreducible elements* $\nleq f$ *in L is equal to the number of coatoms in L.*

Then Th(F) is hereditarily undecidable.

Using the same construction and the same $\delta(x)$ and $\rho(x, y)$ as in the proof of Theorem 4.1, Claims 1 and 2 can also be proved with $F(S)$, the lattice of filters of S, in place of $Con\ S$ and thus we obtain the following.

THEOREM 4.5 *The theory of filter lattices of 2 - Boolean algebras (and hence of all PCS's) is recursively inseparable.*

5. CONCLUDING REMARKS.

We observe that the PCS's S constructed in the proof of Theorem 4.1 can have distinct non - closed elements which are not disjoint. Call a 2-Boolean algebra as a *neo - Boolean algebra* iff any two distinct non – closed elements in it are disjoint. The above observation leads to the following problem which we suspect has an affirmative solution.

PROBLEM. Is the theory of the class of congruence lattices of neo - Boolean algebras decidable ?

It is also of interest to find a lattice-theoretic characterization of the congruence lattices of 2 - Boolean algebras.

The author would like to express his gratitude to S. Burris for the encouragement.

REFERENCES.

Burris, S. and H. P. Sankappanavar

1975, *Lattice - theoretic decision problems in Universal Algebras*, Algebra Univ. 5, 163 - 177.

Church, A.

1936, *A note on the entscheidungsproblem*, The Journal of Symbolic Logic 1, 40 - 41.

Ershov, Y. L., J. A. Lavrov, A. D. Taimanov, and M. A. Taitslin

1965, *Elementary Theories*, Russian Math. Surveys, 20, 35 - 105.

Ershov, Y. L., and M. A. Taitslin

1963, *Some unsolvable theories*. (in Russian), Algebra i Logika, 2, 37 - 41.

Gratzer, G.

1971, Lattice Theory, W. H. Freeman and Company, San Francisco.

Grzegorczyk, G.

1951, *Undecidability of some topological theories*, Fund. Math.,38,137-152.

Kargapolov, M. I.

1962, *On the elementary theory of lattices of subgroups*, Algebra i Logika, 1, 46 - 53.

Kozlov, G. T.

1970, *The undecidability of the theory of lattices of subgroups of finite Abelian p - groups*, Algebra i Logika, 9, 167 - 171.

Rabin, M. O.

1964, *Decidability of second - order theories and automata on infinite trees*, Trans. Amer. Math. Soc., 141, 1 - 34.

1965, *A simple method for undecidability proofs and some applications*, Logic, Methodology and Philosophy of Science, Proceedings of the 1964 International Congress,Bar Hillel ed., Amsterdam (1965),58-68.

Sankappanavar, H. P.
1974, A study of congruence lattices of pseudocomplemented semi-lattices, Ph. D. Thesis, University of Waterloo, Ontario, Canada.

Shoenfield, J. R.
1967, Mathematical Logic, Addison Wesley, Reading.

Taitslin, M. A.
1968a, *Elementary lattice theories for ideals in polynomial rings*, Algebra and Logic, 7, 127 - 129.
1968b, *On simple ideals in polynomial rings*, Algebra and Logic, 7, 394-395.
1970, *On elementary theories of lattices of subgroups*, Algebra and Logic, 9, 285 - 290.

Tarski, A.
1949, *Undecidability of the theories of lattices and projective geometries*, The Journal of Symbolic Logic, 14, 77 - 78.

Tarski, A., A. Mostowski and R. M. Robinson
1953, Undecidable Theories, North - Holland, Amsterdam.

Instituto de Matemática
Universidade Federal da Bahía
Salvador, Bahía, Brazil.

Non-Classical Logics, Model Theory and Computability,
A.I. Arruda, N.C.A. da Costa and R. Chuaqui (eds.)

Polynomially Bounded Quantification over Higher Types and a New Hierarchy of the Elementary Sets (*)

by *JANOS SIMON*

It is well known that nondeterminism and existential quantification are related. In particular NP can be obtained by polynomially bounded quantification over predicates on strings, were the predicates are in P. Meyer and Stockmeyer suggested considering the analogue of the arithmetic hierarchy, where the alternating quantifiers are all polynomially bounded. It is not known whether the the resulting hierarchy is proper. In this paper we consider polynomially bounded quantification over sets and higher types, and show that one obtains a proper hierarchy of the elementary recursive languages. In particular, with a single existential set quantifier (and predicates in P) one obtains exactly the nondeterministic exponential time recognizable languages. Existential quantification over type i corresponds to i levels os exponentiation of the time required to accept the set by a nondeterministic Turing machine (Tm). The results may be considered as a characterization of the computational power of the predicate '$\in$'.

1. Introduction.

The connection between nondeterministic computations and existential

(*) This research was supported in part by grant 70/755 from Fundação de Amparo à Pesquisa do Estado de São Paulo (FAPESP),Brazil.

quantification is well known (Scott 1968). Given a nondeterministic TmT' and an input x we add, as a separate input, the string of choices y that T' makes in an accepting computation. We may then easily build a TmT which operates deterministically and accepts exactly the same set as T', in the same number of moves, whenever y is (an encoding of) the series of choices that T' uses. T rejects if y does not represent an accepting sequence of choices of moves. Thus the set accepted by the nondeterministic machine is

$$L(T') = \{x \mid (\exists y)\ T'(x,y) \text{ halts and accepts}\}$$

This connection was exploited in Meyer and Stockmeyer 1972 and 1973 to exhibit an analogue of the Kleene hierarchy for P, the set of polynomial time recognizable languages, by defining

$$\Delta_0^P = \Sigma_0^P = \Pi_0^P = P$$

$$\Sigma_{i+1}^P = \text{class of languages definable as } \{y \mid \exists^P x\ R(x,y),\ R \text{ in } \Pi_i^P\}$$

$$\Pi_{i+1}^P = \text{class of languages definable as } \{y \mid \forall^P x\ S(x,y),\ S \text{ in } \Sigma_i^P\}$$

where $\exists^P x (\forall^P x)$ means that there is a polynomial $p(\ )$ such that the quantification ranges only over strings x with $|x| \leq p(|y|)$ ($|x|$ denotes the length of x). The polynomial is fixed for a given set.

$$\Delta_{i+1}^P = \text{class of sets definable by } \{y \mid y \text{ is accepted by polynomially bounded Tm with a } \Sigma_i^P \text{ oracle}\}$$

Then Σ_0^P sets are languages in P, Σ_1^P are languages in NP, and Π_1^P sets are languages whose complement is in NP.

Many of the properties of the Kleene hierarchy hold for these classes. It is not known, however, whether the hierarchy is proper, since it is not known whether P = NP ; see Stockmeyer 1975 for further details.

We present in this paper a proper hierarchy of the elementary recursive languages, by an extension of these definitions. This will be done by a generalization of the polynomial hierarchy by using quantifiers over polynomially bounded objects of higher type, in a manner analogous to the defition of the analytical hierarchy generalizing the arithmetical hierarchy.

We will quantify over sets, classes of sets, etc., where the elements of the sets are polynomially bounded. Existential quantification over type i will correspond to i levels of nondeterministic time.

We develop now some notation in order to present these results.

DEFINITION. *A collection of sets of strings,* $A(y)$, *is polynomially bounded iff* $\exists p(\ )$, $p(\ )$ *a polynomial,* $x \in A(y)$ $|x| < p(|y|)$ ($|x|$ *denotes the length of* x).

A string is an object of type 0. An object of type $i+1$ is a collection of objects of type i. For $i > 1$ we say that an object of type i is *polynomially bounded* by its parameter y iff the objects that it is a collection of are polynomially bounded.

The functions $t_i(n)$ are defined by

$$t_0(n) = n, \quad t_{i+1}(n) = 2^{t_i(n)} .$$

Thus $t_i(n)$ stands for i levels of exponentiation.

NTIME $[f(x)]$ denotes the language accepted by nondeterministic Turing machines within time $f(x)$ and DTIME $[f(x)]$ the languages accepted within time $f(x)$ by deterministic Turing machines. Let

$$\text{NEXPTIME} = \bigcup_{p \text{ a polynomial}} \text{NTIME}\left[2^{p(n)}\right];$$

and, in general

$$S_i = \bigcup_{p \text{ a polynomial}} \text{NTIME}\left[t_i(p(n))\right].$$

It is well known that the elementary languages are exactly the languages in

$$\bigcup_{i=1}^{\infty} \text{DTIME}\left[t_i(n)\right]$$

and it is easy to show that

$$\bigcup_{i=1}^{\infty} \text{NTIME}\left[t_i(n)\right] = \bigcup_{i=1}^{\infty} \text{DTIME}\left[t_i(n)\right]$$

since

$$\text{NTIME}\,[t_i(n)] \subseteqq \text{DTIME}\,[t_{i+1}(n)].$$

Also for all polynomials $p(\ )$, for sufficiently large n,

$$t_i(p(n)) < t_{i+1}(n)$$

We saw that NP (i.e.

$$\bigcup_{p \text{ a polynomial}} \text{NTIME}\,[t_0(p(n))])$$

can be characterized as the class of languages L which could be expressed as

$$L = \{y \mid \quad x: \exists |x| < p(|y|) \text{ and } R_L(x,y)\}$$

for some predicate $R_L \in P$.

Our result is an extension of this characterization to the classes

$$S_i = \bigcup_{p \text{ a polynomial}} \text{NTIME}\,[t_i(p(n))]$$

namely, that any language $L \in S_i$ may be obtained by a single polynomially bounded existential quantification of a simple predicate. In the case $i = 0$, the quantifier ranges over polynomially bounded objects of type 0 and predicate is computable in deterministic polynomial time, while in the general case, to describe a language $L \in S_i$, we quantify over polynomially bounded objects of type i, and allow a constant number of polynomially bounded quantifiers over objects of type at most $i-1$. In addition, the matrix (the quantifier-free portion of the formula) becomes a fixed Boolean combination of deterministic polynomial time computable predicates and membership predicates - or, in other words, if we make the convention that predicates of the form "$u \in V$" may be evaluated in unit time, the matrix is deterministic polynomial time computable in this new sense.

Let us use $\exists^P_i x$ to denote "there is a polynomially bounded object x of type i", and $\forall^P_i x$ to denote universal quantification. We shall write Q^P_i for an unspecified quantification (i. e. Q^P_i means $\forall^P_i$ or $\exists^P_i$).

Say that a language L is ${}_i\Sigma^P_1$ if it can be expressed as

$$L = \{x \mid \exists^P_i z\, Q^P_{j_1} y_1 Q^P_{j_2} y_2 \dots Q^P_{j_n} y_n \;\; R\;(x, z, y_1, \dots y_n)\}$$

where $j_k < i$, the quantifiers are polynomially bounded by $|x|$, n and the $Q^P_{j_i}$ are fixed, and R_L is a predicate in P, except for occurrences of '$\in$'. This is the obvious generalization of the analytical and higher type hierarchies for polynomial time bounded computations. We could similarly define ${}_i\Pi^P_j$ classes and ${}_i\Delta^P_j$ classes.

Using the notation, our main theorem is

THEOREM. $\bigcup_{p \text{ a polynomial}} \text{NTIME}[t_i(p(n))] = {}_i\Sigma^P_1$.

In other words: a language may be described by a formula of type ${}_i\Sigma^P_1$ (i. e. a single existential quantification over polynomially bounded objects of type i, followed by a fixed number of polynomially bounded quantifiers of lower types and a deterministic polynomial time computable matrix) if and only if it can be recognized by a nondeterministic Turing machine in time bounded by i levels of exponentiation of a polynomial.

In particular we have:

LEMMA 1. $\text{NEXPTIME} = {}_1\Sigma^P_1$.

In order to prove our result, we will proceed by induction. The case $i = 1$, the start of the induction, is the Lemma above, which we prove in the next section.

2. Exponential time is a single set quantifier.

PROOF OF LEMMA 1: Let $L \in \text{NTIME}[2^{p(n)}]$, i.e. $L \in S_1$. For clarity, let us assume that $p(n) = n$ – it will become clear that our argument does not use the fact that we use this particular polynomial, but the formulas are somewhat cleaner. For every $y \in L$, we shall describe an accepting computation of the TmM that recognizes y, i. e. our predicate will be

$L = \{y \mid \exists X$: X represents an accepting computation of M on input y taking exponential time at most$\}$.

X will be the set of ID's of M's computation (i. e. a description of M's tape — including the square scanned and the state of the finite control). The only problem is that the tape used may be of exponential length and our quantifiers (except for $\exists X$) range over strings of polynomially bounded length.

We overcome this difficulty by representing each ID as 2^n elements, where each element stands for a single square of the Tm tape. Each element will be a triple (i. e. a string with the components separated by markers, \$) of the form

$$\text{time \$ position \$ square.} = (t,p,s).$$

(t,p,s) will represent the contents of the p-th square of the Tm tape at time t. We assume the Turing machine to have a single infinite tape and a single read/write head (so that *position* is well defined) and if at time t the Tm is scanning the p-th square in state q, we shall represent the contents of that square as the pair (σ,q) where σ is the symbol that the square contains. It is easily seen that the length of t and p will be polynomially bounded.

In order to ensure that X is the set of ID's, as claimed, we shall write a predicate expressing the fact that for all strings $w \in X$,

a) w is of the form (t,p,s) as described above (ensured by the formula A defined below);

b) for all times we have a complete description of M's ID (and only one), denoted by B;

c) the set of such ID's is an accepting computation of M on input y, guaranteed by C.

Detailed descriptions of each of these are:

$$A = \forall^P w \in X\ \exists^P t\ \exists^P p\ \exists^P s:\ w = t\,\$\,p\,\$\,s\ \wedge\ s \in \Gamma \text{ where}$$

Γ = {M's tape alphabet} $\cup$ {M's tape alphabet} $\times$ {M's state set}.

Actually $\exists^P s$ is not necessary — one could simply write out all possibilities for s since Γ is finite — we used it only for conciseness of exposition. We shall use this as an abbreviation for the complete expression $\bigcup_{s \in \Gamma}$, which, when expanded, only multiplies the length of the formula by a constant.

$$B = \forall^{\mathrm{P}} t\, \forall^{\mathrm{P}} p (t \le 2^{cn} \wedge p \le 2^{cn}) \rightarrow \exists s\, \exists^{\mathrm{P}} w\, [w = t\$p\$s \wedge w \in X] \wedge$$
$$\forall^{\mathrm{P}} t\, \forall^{\mathrm{P}} p\, \forall^{\mathrm{P}} w\, \forall^{\mathrm{P}} w'\, \forall s\, \forall s'\, [w = t\$p\$s \wedge w' = t\$p\$s'] \rightarrow w = w'.$$

C) An accepting computation is characterized by:

c1) at time $t = 0$ we have the initial ID of M with input y;

c2) at time $t = 2^{cn}$ M is in an accepting state;

c3) if a configuration is assumed at a time $t > 0$, it is attained as the result of a sequence of legal moves from the initial configuration. Such descriptions of accepting computations were used in Meyer and Stockmeyer 1972, and 1973 for the study of lower bounds.

c1) is ensured by

$$c1 = 0\$1\$y_1 \times q_i \in X \wedge 0\$2\$y_2 \in X \wedge 0\$3\$y_3 \in X \wedge \ldots \wedge$$
$$0\$n\$y_n \in X \wedge \forall^{\mathrm{P}} p (p > n \wedge p \le 2^{cn}) \rightarrow 0\$p\$b \in X \text{ ,}$$

where the input is $y = y_1 y_2 \ldots y_n$, q_i is the initial state of M, and b denotes the symbol for blank in M's tape alphabet.

$$\text{c2): } c2 = \exists^{\mathrm{P}} w\, \exists^{\mathrm{P}} t\, \exists^{\mathrm{P}} p\, \exists s\, [w \in X \wedge w = t\$p\$s \wedge s = a \times q_f] \text{ ,}$$

where a stands for some symbol in M's tape alphabet and q_f denotes the unique final state of M, such that M upon entering q_f accepts and stops (it can be assumed wlog that M has such behavior).

c3): To make the expressions easier, let us assume that M may either rewrite a square or move its head but not both. This causes the running time to double at most. Now (t,p,s) is part of M's configuration at time t if

(1) s does contain a state component ($s = a \times q$), at the previous moment M was either scanning the p-th square in state q', read b on it, and one of the valid moves of M upon reading a b in state q' is to rewrite it as an a and go to state q; or M was scanning one adjacent square, and an allowable transition was to move the head to to the p-th square.

(2) s does not contain a state component ($s = a$) and at the previous move either M was scaning the p-th square, reading a, and its state caused the head to move; or M was scanning another square and did not move onto the p-th square, which at time $t - 1$ contained a.

We must, in addition, require that only one move was made (*c. g.* if $\delta(q,a) = \{(c,q',0),(a,q'',L),(a,q''',R)\}$ then it is not the case that $(t, p, q \times a) \in X$ and more than one of $(t+1, p, q' \times c)$, $(t+1, p-1, b \times q'')$, $(t+1, p+1, q''' \times d) \in X$. It can be seen easily that this requirement is satisfied (given our previous restrictions) by

(3) at all times exactly one square has a state specified in it.

The expressions that stand for the facts above are:

$C3(1) = \forall^P t \forall^P p \forall s \in \Sigma \times Q,\ 0 < t \le 2^{cn},\ t\$p\$s \in X \Rightarrow$

$[s = a \times q \Rightarrow ([(t-1\$p\$b \times q') \in X \wedge (a \times q, s) \in (b,q')]$

(state change on some square)

$\vee (\exists c,q''\ \ c \in \Sigma,\ q'' \in Q,\ (q,R) \in \delta(c,q'')$

$\wedge (t-1\$p-1\$c \times q'') \in X \wedge (t-1\$p\$a) \in X]$ (or come from the left)

$\vee [\exists d,q'''\ \ d \in \Sigma,\ q''' \in Q,\ (q,L) \in \delta(d,q''')$

$\wedge (t-1\$p+1\$d \times q''') \in X \wedge (t-1\$p\$a) \in X])]$ (or come from the right);

$C3(2) = \forall^P t \forall^P p \forall s \in \Sigma,\ 0 < t \le 2^{cn},\ t\$p\$s \in X \Rightarrow$

$[s = a \Rightarrow [[t-1\$p\$a \times q) \in X \wedge q \in Q \wedge [(a,q',R) \in \delta(a,q)$

$\wedge (t-1\$p+1\$c) \in X \wedge (t\$p+1\$c \times q') \in X]$

$\vee [(a,q'',L) \in \delta(a,q) \wedge (t-1\$p-1\$d) \in X$

$\wedge (t\$p-1\$d \times q'') \in X]]$ (head was here but moved)

$\vee [(t-1\$p\$a) \in X$ (actually this suffices, we show the possibilities)

$\wedge [((t-1\$p-1\$b) \in X \wedge (t-1\$p+1\$c) \in X,\ b,c \in \Sigma)$

$\vee ((t-1\$p-1\$b \times q') \in X$ (head was to the left)

$\wedge ((t\$p-1\$c \times q'') \in X \vee (t\$p-2\$c' \times q''') \in X))$ (did not move to p)

$\vee ((t-1\$p+1\$b \times q') \in X$ (head was to the right)

$$\wedge((t\$p+1\$c'' \times q'') \in X \vee (t\$p+2\$c''' \times q''') \in X))]\,]\,]$$

$$C3(3) = \forall^{P} t\, \exists^{P} p\, \forall^{P} p' ,\ \exists\, a \times q \in \Gamma$$

$$(t\$p\$a \times q) \in X \wedge ((t\$p'\$s) \in X \wedge p \neq p' \Rightarrow s \in \Sigma).$$

Now the language L may be expressed as

$$L = \{y \mid \exists^{P} X\ \ A \wedge B \wedge C1 \wedge C2 \wedge C3(1) \wedge C3(2) \wedge C3(3)\}.$$

It is clear that each of A, B, ..., C3(3) contains only predicates testable in polynomial time (remember that $w \in X$ is counted as a unit operation).

Also, only 4 quantifiers were used —if we count alternation of quantifiers, only 3 have been used. (A slightly more clever encoding would use only $\forall\exists$ prefixes. Remember that we do not count quantifiers over constant sized sets.) This proves that every language recognizable in $\mathrm{NTIME}[2^{p(n)}]$, $p(\ )$ a polynomial, has the claimed representation.

This proves half of our lemma. The other half follows from a simple counting argument: in a ${}_1\Sigma_1^P$ formula, we have a polynomially bounded set. If the bounding polynomial is $p(\ )$, the set has at most $2^{p(n)}$ elements. Given the input x, the nondeterministic Turing machine that will accept the language denoted by the formula writes, on one of its tapes a vector of length $2^{p(n)}$, with a 1 in position i iff i is in the set. This takes exponential time, and the rest of the simulation is straightforward: existential quantification over strings is treated by writing the string down, while universal quantification is taken care of, by trying all strings of bounded length. The evaluation of the predicate takes polynomial time, except for subexpressions of the type $x \in X$, which are decided by looking up the x-th position of the vector. All of this takes at most exponential time, so that our lemma is proved.

3. THE MAIN THEOREM.

First, we want to show that all the languages acceptable by a Tm in nondeterministic time

$$\left.2^{\cdot^{\cdot^{2^{p(n)}}}}\right]i$$

(for input of length n) can be expressed by a predicate.

$$_{i}\Sigma_1^P \quad , \text{ i. e.}$$

$$\exists_i^P z \; Q_{j_1}^P y_1 \ldots Q_{j_k}^P y_k \; R_L(y_1, \ldots, y_k, z, x)$$

that is, the predicate contains a single polynomially bounded existential quantifier over objects of type i, and a fixed finite number of polynomially bounded quantifiers over objects of type at most $i-1$, and the predicate preceded by these quantifiers is computable in deterministic polynomial time (assuming all membership predicates to take unit time).

We shall proceed by induction. The case $i=1$ has just been proved.

Our approach to the general case will be analogous to the case of exponential time: we shall have an object z of type i, representing an accepting computation of the Tm M (that recognizes the language within the given bound) on input x. Again, we shall want to describe an ID as a triple (time, position, square) and write down the predicates A-C3(3) that will ensure that X indeed represents such a sequence. The problem is that the first two components now have length

$$2^{2^{\cdot^{\cdot^{2^{p(n)}}}}}\Big]{i-1} = t_{i-1}(p(n))$$

and all strings must have polynomially bounded length.

The trick we will use is to use encodings for the numbers, just as we used encodings for the ID's. Intuitively, the numbers "stay one level below" so that the procedure may be carried out inductively.

More precisely, we shall prove that with quantifiers of type $i-1$ we may obtain objects of type $i-1$ that will represent numbers. Also, the operations of forming pairs and triples of such objects can be described within the same formalism, and we may express the relations

$$a = b$$

$$a = 0$$

$$a = b+1$$

a is the first (second) components of a pair $w = (a,b)$, a b, w representing numbers up to $t_{i+1}(n)$, using a finite number of quantifiers of type at most $i-1$, over polynomial time bounded predicates. Triples can be handled as the result of two pairing operations.

It is only a matter of going over our previous proof (for the case $i=1$) to check that only the predicates above were used to write down formulae A - C3(3), so that if we prove the claim above we shall have proven the theorem.

LEMMA 2. *If* A *and* B *are polynomially bounded objects* (pbos) *of type* i, *and* $x \in \Gamma$, *there are* pbos *of type* i *(perhaps with a higher bound)* C *and* D *such that*

1) C *represents* (A,B),

2) D *represents* (A,x).

Moreover, the predicate $C = (A,B)$, $(D = (A,x))$ $x =$ *first (second) component of* C (D) *are expressible using polynomially bounded quantifiers of types at most* $i-1$ *(except for* $i=0$ *when an existential quantifier is needed for the last one).*

PROOF: By induction on i.
For $i = 0$, $C = A \$ B$, $D = A \$ x$; $w = A \$ B$ is deterministic polynomial time computable, as is the predicate $(w,y)\ [w = x \ \$ \ y]$

For $i > 0$, $A = \{a_j\}$, $B = \{b_k\}$ where a_j, b_k are objects of type $i-1$. By induction $c_{jk} = (a_j,b_j)$ is defined, as well as $d_j = (a_j,x)$. Define $C = \{c_{jk}\}$, $D = \{d_j\}$. Now $C = (A,B)$ iff $\forall w \in C$, $w = (a_j,b_j)$ $\quad a_j \in A$, $b_k \in B$ and

$$A = \{a_j \mid \exists b_k \, [b_k \in B \quad (a_j,b_j) \in C\,]\}.$$

This lemma justifies the use of triples at all levels. Now let us define representation for numbers. For $i = 0$ we represent a number by writing it down. Since our numbers were bounded by 2^{cn} (remember the simplifying assumption that our polynomials were linear), this could be done in polynomial space.

For $i=1$, a number will be represented as a set {(position, digit)}, with the obvious meaning: *position* will be a string of polynomial length. This enable us to write $\sim 2^n$ positions (i. e. our number will have length $\sim 2^n$) and represent values up to 2^{2^n}.

In general, we will assume as inductive hypothesis that the representation of numbers up to $t_{i+1}(n)$ (i. e. of length $t_i(n)$) can be done using an object T_i of type i ($i \geq 1$) and the elements of T_i are of the form (p_{ij}, digit) where p_{ij} represents a number (the j-th position of the number represented by T_i) and *digit* is the corresponding digit. Moreover, we assume that the set T_i may be defined using a single existential quantifier of type i (and quantifiers of smaller types). (Since we shall use T_i in a formula which has a quantifier of type $i+1$, this will cause no problems.) Then we define numbers of length $t_{i+1}(n)$ as sets $\{(T_i, \text{digit})\}$ where T_i will denote a position. In order to get numbers we must

a) ensure that all digits are defined;

b) be able to deal with these objects as numbers $\leq t_{i+1}(n)$, i. e.

  i) test for $= 0$,

  ii) test for $= t_{i+2}(n) - 1$,

  iii) test for equality of two "numbers",

  iv) test for successor relationship,

  v) test for $<$.

Note that if we can prove that a) and b) can be done using quantifiers of type $i+1$ at most, we not only shall have proven that we can define numbers but will have proven our theorem, since the predicates b)i) - b)iv) are exactly those needed for the proof of the main theorem.

BASIS: ($i = 1$)

a) $(\forall^P z)\,[(z,0) \in T_\ell \vee (z,1) \in T_\ell \wedge \neg((z,0) \in T_\ell \wedge (z,1) \in T_\ell)]$
("T_ℓ is a number").

b) i) $T_i = 0 \stackrel{\text{def}}{=} (\forall^P z)\,[(z,0) \in T_i]$

ii) $T_i = t_i(n) - 1 \stackrel{\text{def}}{=} (\forall^P z)\,[((z,1) \in T_i)]$

iii) $T_i = T_j \stackrel{\text{def}}{=} (\forall^P z)\,[(z,0) \in T_i \Longleftrightarrow (z,0) \in T_j]$

iv) $T_i = T_j + 1 \stackrel{\text{def}}{=} (\exists^P z)\,[(z,1) \in T_i \wedge (z,0) \in T_j \wedge$
$(\forall^P z')\,[z' < z \Longrightarrow ((z',0) \in T_i \wedge$
$(z',1) \in T_j) \wedge (z' > z \Longrightarrow$

$$((z',0) \in T_i \Longleftrightarrow (z',0) \in T_i))]]$$

(i.e. there is a position z which becomes 1 — all digits higher than z remain the same, while the digits below change from 1 to 0. In T_j all digits up to the $(z-1)$st were 1's).

v) $T_i > T_j \; (\exists^P z) [(z,1) \in T_i \wedge (z,0) \in T_j \wedge (\forall^P z') [z' < z \rightarrow$
$$((z',0) \in T_i \rightarrow (z',0) \in T_j)]].$$

This concludes the proof for $i=1$. Let us suppose the theorem holds for i.

INDUCTION $(i+1)$

a) $(\forall^P_i z) [((z,0) \in T_\ell \vee (z,1) \in T_\ell) \wedge ((z,0) \in T_\ell \Longleftrightarrow (z,1) \notin T_\ell))$.

b) It is easy to see, in a similar way, that i) - v) still hold — of course whenever we use < for example, we shall have to substitute the appropriate formula, but by the induction hypothesis such a formula exists and has quantifiers of type i at most.

This completes the proof of the existence of number representations. It also shows that < , = , +1, etc may be defined for objects representing numbers smaller then $t_{i+2}(n)$, and such a formula may be written by using quantifiers of type i at most.

Now we may substitute these representations into formulae A - C3(3) of our proof of the representation theorem for $i=1$, and obtain half of the equivalence for general i, i. e. if a language L is accepted by a nondeterministic Tm within time $t_i(n)$, it may be written as

$$L = \{y \mid \exists^P_i Z \; Q^P_1 x_1 \ldots Q^P_k x_k \; R_L(x_1,\ldots,x_k, Z, y)\}$$

where R_L is deterministic polynomial time computable, and the Q^P_j's are polynomially bounded quantifiers over objects of type at most $i-1$.

We still have to show the reverse implication, i. e. that if a language L can be characterized by

$$L = \{ y \mid \exists^P_i Z \; Q^P_1 x_1 \ldots Q^P_k x_k \; R_L \; (x_1,\ldots,x_k, Z, y)\}$$

then $L \in \mathrm{NTIME}[t_i(p(n))]$ for some polynomial $p(\;)$.

The proof is quite straightforward, and we just sketch it. First we prove a simple lemma.

LEMMA 3. *The number of objects of type i polynomially bounded by p is:*

$$n_i = 2^{2^{\cdot^{\cdot^{2^{p(n)}}}}}\Big]i \quad (\textit{binary alphabet}).$$

PROOF: For sets $n_1 = 2^{p(|y|)}$. The number of objects of type $i+1$ is the cardinality of the power set of objects of type i, i.e. 2^{n_i}.

Thus the cardinality n_i of an object of type i is $\leq t_i(p(n))$. We now sketch a Turing machine M, that recognizes L. First it writes down (nondeterministically) the set Z, in tape and time n_i. For the quantifiers of lower types, it acts as follows: if the quantifier is existential, it writes down a choice for the objects; if it is universal it goes into a loop - starting with the empty set, it tests whether the formula is true, if so adds a new element to the object (while this is possible) and repeats the procedure. If the formula is false, M rejects. If all possible values for the set satisfy the formula, the universal quantifier was checked. Since the predicate itself is computable in polynomial time, the validity of $a \in X$ can be checked by scanning the tape, and the enumeration of an object of type $i-1$ takes at most $t_i(n)$ steps, the running time of the nondeterministic procedure is bounded by $t_i(p'(n))$ for some polynomial $p'(\ )$, as claimed. This concludes the proof.

4. Conclusions.

We obtained a new characterization of the languages accepted within i levels of exponentiation by nondeterministic Tms. Although the proofs are quite messy, the underlying ideas are clean and the result is intuitively satisfying. Our theorem may be thought of, as characterizing the power of the symbol 'belong to' in an environment where all other operations are polynomially bounded. It is also satisfying, that while one cannot carry the analogy between the arithmetical hierarchy and the polynomial hierarchy far enough to prove proper inclusions, it is at least possible to prove that going to the analog of the analytical sets, one does obtain a proper hierarchy:

$${}_1\Sigma_1^P \supsetneq {}_0\Sigma_1^P$$

since NEXPTIME $\supsetneq$ NP.

Finally, although many other hierarchies of the elementary sets are known, they all correspond to deterministic resource bounds, so that it is a very difficult open problem to compare the levels in our hierarchy to, say the levels of the Meyer-Ritchie hierarchy, since this would answer the question

$$\text{NTIME}\ t_i\,[p(n)]\,] = \text{DTIME}\ t_i\,[p(n)]\,].$$

References.

Meyer, A. R. and L. J. Stockmeyer

1972, *The equivalence problem for regular expressions with squaring requires exponential tape*, Proc. 13th IEEE SWAT Conf., 125-129.

1973, *Word problem requiring exponential tape*, Proc. 5th STOC., 1-9

Scott, D.

1968, *Some definitional suggestions for automata theory*, JCSS, 1,2, 187-192.

Simon, J.

1974, On some central problems in computational complexity, TR 75-224, Dept. of Comp. Sci. Cornell U.

Stockmeyer, L. J.

1975, The polynomial-time hierarchy, IBM Res. Rep. RC 5379.

Departemento de Ciências da Computação
Universidade Estadual de Campinas
Campinas, São Paulo, Brazil.

Non-Classical Logics, Model Theory and Computability,
A.I. Arruda, N.C.A. da Costa and R. Chuaqui (eds.)

ON RANDOM R. E. SETS

by *ROBERT M. SOLOVAY*

ABSTRACT: Chaitin has recently proposed two explications of the entropy or information content of a recursively enumerable set. The first, $I(A)$, is (roughly) the minimal length of a program for a universal computer that enumerates A. The second, $H(A)$, is the negative log of the probability that an appropriately universal computer connected to a random sequence of 0's and 1's (with the usual distribution) will enumerate A.

It is easy to see that $H(A) \leq I(A) + O(1)$. Our main theorem asserts that $I(A) \leq 3H(A) + O(\log H(A))$. The proof may be viewed as a constructivization of a theorem of de Leeuw, et. al. that states if a set is recursively enumerated with positive probability by a computer attached to a source of a random string of 0's and 1's, then it is, in fact, recursively enumerable.

This paper arose from a question of Chaitin concerning the relationship between two measures of the "*entropy*" or "*information content*" of a recursively enumerable (r.e.) set. Our proof can best be viewed as a "*constructive*" sharpening of an old theorem of de Leeuw, Moore, Shannon, and Shapiro 1956.

(The word *constructive* should be taken with a grain of salt. I have no idea how to formulate, much less prove, the main theorem in the framework of constructive mathematics, à la Bishop (cf. Bishop 1967)).

The theorem of de Leeuw, Moore, Shannon, and Shapiro 1956 concerns a Turing machine, M, equiped with a *random number generator* and a write-only output tape. The random number generator emits an infinite sequence of 0's and 1's with no correlations among the successive outputs, and with 0 and 1 equally probable as output at any stage.

From time to time, M writes a non-negative integer on its output tape, with the successive integers separated by commas. In this way, in the course of time M enumerates a set A.

Let $S \subseteq \omega$. Suppose that the probability that S is the set enumerated by M is greater than zero. Then, Theorem 2 of de Leeuw et. al. 1956 states that S is r. e..

Our improvement is as follows. We give a recursive procedure that constructs from a Gödel number, e, for M and a lower bound, $1/n$, for the probability that M enumerate S, a finite set D. (In fact, D, will have at most $1/2(n+1)^3$ elements.) D will contain some Gödel number for the r.e. set S.

The proof of de Leeuw et. al. 1956 will be recalled in Section 1. One can show that our more precise result can not be obtained by the methods of de Leeuw et. al. .

One of the key tools in our proof is a recent Lemma of D. A. Martin which we present in Section 2. Indeed this result was "*commissioned*" for use in this proof.

We close this introduction with a rough description of how our principal result bears on the problem raised by Chaitin. (Precise description of the relevant concepts will be recalled in Section 5.) Let M_o be a *universal* Turing machine equiped with a random oracle. Let $P(A)$ be the probability that M_0 enumerates A. Let

$$H(A) = -\log_2 P(A).$$

Let $I(A)$ be the minimal number of bits of information needed to specify a program for enumerating A. Then Chaitin has proved the following:

PROPOSITION. 1) $H(A) \leq I(A) + C_0$,

2) *For infinitely many* A, *we have*

$$H(A) \leq I(A) - \log_2 I(A) + C_1 \log_2 \log_2 I(A).$$

Here C_0 and C_1 are certain absolute constants. Of the two results, 1) is trivial, while 2) is quite nontrivial, and is one of the main results of Chaitin 1976.

Chaitin's question was as follows: Is there a recursive function h

such that

$$I(A) \le h(H(A))\ ?$$

A positive answer follows easily from our main theorem (cf. Section 5).

$$I(A) \le 3H(A) + O(\log H(A)).$$

The coefficient 3 comes from the exponent 3 in our main theorem. I do not know if it is optimal.

This paper would not exist without the question of Greg Chaitin and the mathematical assistance of Tony Martin. Their help is gratefully acknowledged.

1. PRELIMINARIES.

1.1. ω is the set of non-negative integers. Except in Section 5 we follow the usual conventions in which each integer n is the set of all smaller integers. If A is any set, ${}^A2$ is the collection of all functions A into 2. If $f \in {}^A2$ and $B \subseteq A$, then $f|B \in {}^B2$ is the restriction of f to B.

Let ν be the measure on 2 such that $\nu(0) = \nu(1) = 1/2$. Let μ be the product measure on ${}^\omega 2$ which is the product of countably many copies of ν.

A subset $A \subseteq {}^\omega 2$ is *finitely based* if for some integer n and some subset D of ${}^n2$,

$$(1)\quad A = \{f \in {}^\omega 2 : f|n \in D\}.$$

(An alternate description, which we shall not use, is that A is finitely based iff it is both open and closed in the usual product topology on ${}^\omega 2$.)

Let us call the pair $\langle n, D\rangle$ a code for A. Then we can effectively tell whether two codes are codes for the same set, compute the measure of a set from its code, and compute a code for a Boolean combination of A and B from codes for A and B. Via a suitable Gödel numbering, we may as well assume that our codes for finitely based sets are integers.

1.2. Here is a precise statement of our main theorem. If $f \in {}^\omega 2$, let $W_e(f)$ be the set recursively enumerable in f with Gödel number e.

THEOREM. *There is a recursive function, $h(e,n,k,\ell)$ such that:*

if $A \subseteq \omega$, *and*

$$\mu(\{f : W_e(f) = A\}) \geq 1/n,$$

then for some $k < n+1$ *and some* $\ell < (n^2+n)/2$, *we have* $A = W_m$ (*for* $m = h(e,n,k,\ell)$.)

The finite set mentioned in the introduction is then

$$D_{e,n} = \{h(e,n,k,\ell) : k < n+1 \ ,\ \ell < (n^2+n)/2\}.$$

The function h will be obtained as follows. Uniformly in $e,n,k,$ we will construct an array of r.e. sets

$$A_0, \ldots, \quad A_{N-1}$$

where $N = (n^2+n)/2$. (This will be carried out in Section 3.) The number k encodes a certain finite amount of information needed to make the construction proceed correctly. Since our construction is uniform in $k,n,e,$ there will be a recursive function $h(e,n,k,\ell)$ which gives a Gödel number for A_ℓ qua r.e. set. We show in Section 4 that if k has its correct value, k_0, (to be defined below) then if A is as in the statement of the theorem, $A = A_\ell$ for some ℓ with $\ell < N$.

The estimate we actually get on the size of $D_{e,n}$ is $n(n+1)^2/2$, a bit better than what we claimed in the introduction. Obviously we can not do better than n. Can we get by with a set of size $O(n^2)$?

1.3. From now on, we suppress e from our notation, and write $W(f)$ rather than $W_e(f)$. (Everything we do will be done uniformly in the suppressed parameter e.) Also, we shall take $W(f)$ to be the characteristic function of the eth r.e. set, relative to f. We let $W(f,t)$ be the characteristic function of the portion of $W(f)$ ennumerated before stage t. For our purposes, the crucial properties are:

1) $t_1 \leq t_2 \Rightarrow W(f,t_1)(n) \leq W(f,t_2)(n)$.

2) $W(f)(n) = \lim_{t \to \infty} W(f,t)(n)$.

3) $W(f,t)(n) = 1 \Rightarrow n < t$.

4) $W(f,t)$ can be effectively computed from t and $f|t$, uni-

formly in e.

1.4. If $A \subseteq {}^{\omega}2$, let χ_A denote the characteristic function of A. Thus $\chi_A(f) = 1$ if $f \in A$ and $\chi_A(f) = 0$ if $f \notin A$.

We say that a sequence of sets $\langle A_n : n \in \omega \rangle$ converges to a set A if the corresponding sequence of characteristic functions converges pointwise. I.e., $\lim_{n\to\infty} \chi_{A_n}(f) = \chi_A(f)$ for all $f \in {}^{\omega}2$. In that case, $\mu(A_n) \to \mu(A)$. (For example, this follows from the bounded convergence theorem (cf. Halmos 1964, Theorem 26D) applied to the sequence of characteristic functions.)

1.5. For $g \in {}^{\omega}2$, let

$$\rho(g) = \mu(\{f : W(f) = g\}).$$

For $s \in {}^{n}2$, let

$$\rho(s) = \mu(\{f : W(f)|n = s\}).$$

For $s \in {}^{n}2$, $t \in \omega$, let

$$\rho(s,t) = \mu(\{f : W(f,t)|n = s\}).$$

The following lemma follows easily from the facts recalled in 1.4.

LEMMA. 1) $\lim_{m\to\infty} \rho(g|n) = \rho(g)$, $g \in {}^{\omega}2$;

2) $\lim_{t\to\infty} \rho(s,t) = \rho(s)$, $s \in {}^{m}2$, $m \in \omega$.

1.6. The following theorem is a special case of Theorem 49B of Halmos 1964. It is used in the original proof of de Leeuw, et.al of their theorem, which we shall recall in a moment. We will not need it for the proof of our own theorem (which, of course, implies the de Leeuw theorem).

First some notation. If $s \in {}^{m}2$ and $f \in {}^{\omega}2$, we use the notation $s^\frown f$ for the concatenation of s and f. I. e., $s^\frown f$ is that $h \in {}^{\omega}2$ such that $h(i) = s(i)$ for $i < m$, and $h(m+i) = f(i)$. We write

$$\rho(g;s) = \mu(\{f : W(s^\frown f) = g\})$$

($\rho(g;s)$ is the probability that our machine will output g given that the initial portion of our random input is s.)

THEOREM. *For almost all* f,

$$\lim_{n\to\infty} \rho(g;f|n) = \chi_B(f)$$

where $B = \{f\colon W(f) = g\}$.

1.7. We can now present the proof of de Leeuw et. al.

THEOREM. (de Leeuw et.al.) *Let* $g \in {}^{\omega}2$. *If* $\rho(g) > 0$, *then* g *is the characteristic function of some* r.e. *set.*

PROOF: First, consider the case when $\rho(g) > 1/2$. Then if $A = \{n \in \omega\colon g(n) = 1\}$, we can enumerate A by the following rule: Put $n \in A$ if the probability that n is written on the output tape is $> 1/2$. More formally

$$A = \{n\colon (\exists t)\ \mu(\{f\colon n \in W(f,t)\}) > 1/2\} .$$

This expression shows that A is r.e. (cf. 1.3.4.).

In the general case, we apply Theorem 1.6, and the fact that $\rho(g) > 0$, to get an s so that $\rho(g;s) > 1/2$. If we now put $W'(f) = W(s\hat{\ }f)$, then the arguments of the preceding paragraph apply to g, W' so again A is r.e..

1.8. The argument of de Leeuw et. al is too non-constructive to yield our theorem. For example, the following constructivization of the proof is *false*. (It is easily refuted using the recursion theorem.)

There is a recursive function $h(e,n)$ such that $h(e,n)$ encodes a finite set of finite sequences, $D_{e,n}$, and if $\rho_e(g) > 1/n$, then for some $s \in D_{e,n}$, $\rho_e(g;s) > 1/2$.

1.9. The following trivial lemma will be needed in a moment.

LEMMA. *We can effectively find a sequence of* $n+1$ *open intervals of width less than* $1/n$ *whose union contains* $[0,1]$.

PROOF: Let $a_i = i/n - (2i+1)\varepsilon$, $b_i = \frac{i+1}{n} - (2i+2)\varepsilon$.

Let $I_i = (a_i, b_i)$. Finally let $\varepsilon = \frac{1}{5n^2}$. Then $I_0, \ldots, I_n$ have the desired properties.

1.10. We can now define the integer k_0 mentioned earlier. Let g_1, $\ldots$, g_r be distinct functions in ${}^{\omega}2$ with $\rho(g_i) \geq 1/n$. We put

$$\lambda = \sum_{i=1}^{r} \rho(g_i).$$

Roughly speaking, λ tells us how much of the measure space we must take care of in our construction. Then for some k_0, $\lambda \in I_{k_0}$. (So roughly speaking, k_0 encodes the $\log_2 n$ high order digits of the binary expansion of λ .)

During the course of our construction k allows us to wait till about λ of the measure space is *active*. Too large a k, and we will wait forever. Too small a k, and we will not take care of all the measure space that we should. With the correct k_0, the construction will work correctly.

2. How to avoid putting all your eggs in one basket.

2.1. This section gives an exposition of a lemma of Martin used in the proof of the main theorem. We begin by indicating the problem the lemma is designed to solve.

Suppose we are at some stage s in the construction. We have constructed so far, $A_0(s), \ldots, A_{N-1}(s)$. Put

$$C_i(s) = \{f \in {}^{\omega}2 : A_i(s) \subseteq W(f,s)\}.$$

The significance of $C_i(s)$ is that we can only hope to make A_i equal to $W(f)$ by adding elements to $A_i(s)$ if $f \in C_i(s)$.

If, for example, there is a set Y of measure $\geq 1/n$ such that $Y \cap C_i(s) = \emptyset$, all $i < N$, we are in a potentially disastrous situation where if $W(f) = A$ for all $f \in Y$, we may not be able to make one of the A_i's equal to A.

On the other hand, it may happen that for some set Y of measure $\geq 1/n$, we have $W(f,s) = A_0$ for all $s \geq s_0$, $f \in Y$. If this happens we must, eventually, make one of the $A_i(s)$'s with $Y \cap C_i(s) \neq \emptyset$, equal to A_0. But then C_i will contract, possibly to Y, as we add elements to A_i.

2.2. To study this situation, we formulate an infinite two person game. A configuration in this game is N-tuple of finitely based sets: $< C_0, \ldots, C_{N-1} >$. The initial configuration is $< {}^{\omega}2, \ldots, {}^{\omega}2 >$.

Player I on his turn plays a code for a finitely based set Y of measure $> \frac{1}{n+1}$. If $Y \cap C_i = \emptyset$ for all $i < N$, I wins. If not, II replaces C_i by Y, for some C_i with $Y \cap C_i \neq \emptyset$, and the game continues. II wins if he can prevent I from winning as described above for the entire, infinitely long game.

The game differs slightly from the intuitive motivation of 2.1 in that we have replaced $\geq 1/n$ by $> \frac{1}{n+1}$. The reason is that a set which has measure $1/n$ at $s = \infty$, might have measure $< 1/n$ for finite s (but will eventually have measure $> \frac{1}{n+1}$).

THEOREM. (Martin) *Let* $N = \frac{1}{2}n.(n+1)$. *Then player* II *has a recursive winning strategy for the game just described.*

2.3. The proof is based on a property of configurations, which we dub property M. It is effectively checkable whether a configuration has property M. Moreover, the initial configuration has property M, and if a configuration has property M, then for any move of I, II has at least one reply that results in a configuration with property M. A recursive strategy for II is to pick C_i minimal so that $C_i \cap Y \neq \emptyset$ and the resulting configuration has property M. (Another recursive winning strategy is easily extracted from our proof that property M can be preserved.)

2.4. We let $T = \{ < i,j > : 1 \leq j \leq i \leq n \}$. Let $< Y_0, \ldots, Y_{N-1} >$ be a configuration. A bijection $h: T \to N$ determines an arrangement of the Y's as a triangular array: $Z_{i,j} = Y_{h(i,j)}$. (Cf. Figure 1.)

$$
\begin{array}{lllll}
Z_{1,1} & & & & \\
Z_{2,1} & Z_{2,2} & & & \\
\dots & \dots & \dots & \dots & \dots \\
Z_{n,1}, & Z_{n,2}, & \dots\ \dots & Z_{n,n-1}, & Z_{n,n},
\end{array}
$$

Figure 1.

DEFINITION. *The configuration* $\langle Y_1,\dots,Y_N\rangle$ *has property* M *if for some bijection* $h\colon T \to N$, *the resulting triangular array has the following property:*

Let $1 \le i_1 \le i_2 \le \dots \le i_k \le j$; *then*

$$\mu(Z_{j,i_1} \cup \dots \cup Z_{j,i_k}) > \frac{k}{n+1}\,.$$

2.5. The following lemma is now evident.

LEMMA. 1) *The initial configuration* $\langle {}^{\omega}2,\dots, {}^{\omega}2\rangle$ *has property* M.

2) *Let* $\langle Y_1,\dots,Y_N\rangle$ *have property* M. *Suppose* $Y_i \subseteq Y_i' \subseteq {}^{\omega}2$. *Then* $\langle Y_1',\dots, Y_N'\rangle$ *has property* M.

3) *We can tell effectively whether a sequence of integers* $\langle x_1,\dots,x_N\rangle$ *is a code for a configuration with property* M.

2.6. The following lemma will complete the proof of Theorem 2.2.

LEMMA. *Let* $\langle Y_1,\dots,Y_N\rangle$ *be a configuration with property* M. *Let* Y *be a finitely based subset of* ${}^{\omega}2$, *with* $\mu(Y) > \frac{1}{n+1}$. *Then there is an* $i \le N$ *such that* $Y_i \cap Y \ne \emptyset$, *and the configuration obtained by replacing* Y_i *by* Y *still has property* M.

PROOF: We arrange the Y_i's in a triangular array, $Z_{i,j}$, with the property guaranteed by the Y's having property M. Then the union of the Z's in the botton row has measure $> \frac{n}{n+1}$. Whence, some $Z_{n,j}$ has nonempty intersection with Y.

Let j be minimal such that some $Z_{j,i}$ has non-empty intersection with Y. Permuting the jth row, if necessary, we may assume $Y \cap Z_{j,j} \neq \emptyset$. We show that if we replace $Y_{h(j,j)}$ by Y the resulting configuration has property M.

If $j = 1$, this is evident. If $j > 1$, we put $Z'_{k,\ell} = Z_{k,\ell}$, unless $k = j$ or $j-1$. Put $Z'_{j-1,\ell} = Z_{j,\ell}$, for $\ell \leq j-1$, and $Z'_{j,\ell} = Z_{j-1,\ell}$ for $\ell \leq j-1$. Finally, put $Z'_{j,j} = Y$. We show that the triangular array Z' establishes that the new configuration has property M.

What rows could cause trouble? The only rows that do not appear in the array are the $j-1^{st}$ and j^{th}. The $j-1^{st}$ row is o.k., because the old j^{th} is o.k. The only way the jth row could cause trouble is if for some $1 \leq i_1 < \ldots < i_k \leq j-1$.

(1) $\mu(Z_{j-1,i_1} \cup \ldots \cup Z_{j-1,i_k} \cup Y)$ is $\leq \frac{k+1}{n+1}$.

But by our choice of j, $Y \cap Z_{j-1,i} = \emptyset$, all $i \leq j-1$. Thus (1) is equal to

$$\mu(Z_{j-1,i_1} \cup \ldots \cup Z_{j-1,i_k}) + \mu(Y).$$

Since the $j-1^{st}$ row of the Z's is o.k., and $\mu(Y) > \frac{1}{n+1}$, this is $> \frac{k}{n+1} + \frac{1}{n+1}$, as desired.

2.7. We shall need the following extension of lemma 2.6.

LEMMA. *Let* $\langle Y_1,\ldots,Y_N\rangle$ *be a configuration with the property* M. *Let* $Z_1, \ldots, Z_k$ *be pairwise disjoint finitely based sets, each of measure greater than* $\frac{1}{n+1}$. *Then we can find distinct integers* $i_1,\ldots,i_k$ *such that*

1) $Y_{i_j} \cap Z_j \neq \emptyset$, $1 \leq j \leq k$.

2) *If we put* $Y'_{i_j} = Z_j$, $Y'_r = Y_r$ *for* r *not an* i_j , *then*

$\langle Y'_1, \ldots, Y'_N \rangle$ *again has property* M

(again, given codes for everything in sight, we can effectively find $i_1, \ldots, i_k$*).*

PROOF: It will help if we think of N cubbyholes, which at the start contain the configuration $Y_1, \ldots, Y_N$. We apply Lemma 2.6 successively to each of $Z_1, \ldots, Z_k$ in turn. At the jth turn, Lemma 2.6 singles out a cubbyhole whose contents have non-empty intersection with Z_j and so that after replacing these contents by Z_j we still have property M. We then perform this replacement.

For our current lemma we must show no cubbyhole has contents replaced more than once . But this is evident since the Z_i's are pairwise disjoint. After a cubbyhole is first used, at time i, it will contain Z_i. If it were used again at time $j > i$, we would have to have $Z_i \cap Z_j \neq \emptyset$, contrary to our hypothesis.

3. THE CONSTRUCTION.

3.1. We shall be simultaneously enumerating an array of r.e. sets, $\langle A_0, \ldots, A_{N-1} \rangle$. We let $A_i(s)$ be those elements placed into A_i before stage s. (We will arrange matters so that $A_i(s) \subseteq s$.)

Let $C_i(s)$ =

(1) $\{f: \forall n(n \in A_i(s) \rightarrow W_i(f,s)(n) = 1\}$.

We will arrange our construction so that, for all s, the array $\langle C_0(s), \ldots, C_{N-1}(s) \rangle$ has property M. (This is true at $s = 0$, by Lemma 2.5 1). If it holds at s, and we add no new elements to any A_i at stage s, it will hold at $s+1$ by Lemma 2.5 2).) At each step s, we first check whether or not s is opportune for adding new elements to the A_i's. If s is opportune, we will add elements to the A_i's in an attempt to pursue those f such that $\rho(f) \geq 1/n$. We use Lemma 2.7 in doing this so that property M is preserved.

3.2. We now give the precise construction. Recall that our construction depends on the parameters e, n, and k. Here $k < n+1$. Let $I_k = (a_k, b_k)$ be as in Section 1.9. Let $N = \frac{1}{2}n(n+1)$. To start things off, let's agree that -1 is opportune and that $A_i(0) = \emptyset$, $i < n$.

We now define what it means for a stage, s, to be *opportune*. Let t be the largest number $< s$ that is opportune. Let $m = (s)_0$, $\ell = (s)_1$ (where, as usual, $(x)_i$ is the exponent of p_i in the prime power factorization of x, and $(0)_i = 0$ by convention). Then s is opportune if

1) $t + 2 \leq m \leq \ell$; $n^2 + n < m$.

2) Let $h_1, \ldots, h_r$ be those binary sequences of length ℓ such that $\rho(h_i, s) \geq \frac{1}{n} - \frac{1}{m}$. Then

$$a_k < \sum_{i=1}^{r} \rho(h_i, s) < b_k.$$

Note that we can effectively determine if s is opportune.

3.3. If s is not opportune, we put $A_i(s+1) = A_i(s)$, and proceed to stage $s+1$.

Now suppose that s is opportune. Since we only add elements at opportune stages, we have $A_i(s) \subseteq t+1 < \ell$.

Let $Y_i = \{f\colon W(f,s)|\ell = h_i\}$. By 1) of the definition of opportune, $\frac{1}{n} - \frac{1}{m} > \frac{1}{n+1}$. By 2) of the same definition, $\mu(Y_i) > \frac{1}{n+1}$. Also the Y_i's are clearly pairwise disjoint. We now apply Lemma 2.7. This gives distinct integers $i_1, \ldots, i_r < N$, such that: 1) $Y_j \cap C_{i_j}(s) \neq \emptyset$; 2) if we replace each $C_{i_j}(s)$ by Y_j and leave the other C_i's unchanged, the resulting configuration, say $\langle C'_0, \ldots, C'_{N-1} \rangle$, has property M.

Let $B_i = \{q \in \omega\colon h_i(q) = 1\}$. From 1) and the fact that $\sup A_i(s) < \ell$, we get $A_{i_j}(s) \subseteq B_i$. We now set $A_{i_j}(s+1) = B_i$. For integers $i < N$ not of the form i_j, we let $A_i(s+1) = A_i(s)$. Note that we have $C'_i \subseteq C_i(s+1)$. So property M is preserved. Note also that our proof of Lemma 2.7 gives an effective procedure for determining $i_1, \ldots, i_r$ and we should follow this procedure so that our construction will be effective. We have now completed step s; the construction now proceeds to step $s+1$.

The description of our construction is now complete. Our next task is to show that if k is correctly chosen (cf. Section 1.10), the construction succeeds.

4. THE CONSTRUCTION WORKS.

4.1. We let e, n, N be as in the preceding paragraph, but we now assume that, with notations as in Section 1.10, $\lambda \in I_k$. Our construction, for this k, yields a sequence of sets $A_0, \ldots, A_{N-1}$. Let $g \in {}^{\omega}2$ such that $\rho(g) \geq 1/n$. Let $B = \{n: g(n) = 1\}$. We must prove that for some i, $B = A_i$.

Our proof will proceed as follows: (a) We show that infinitely many stages are opportune. This will follow from the continuity properties of Section 1.5 and the fact that $\lambda \in I_k$. (b) The next step is to analyze what happens at a large opportune stage. Our method for doing this is to take an increasing sequence of opportune stages and continually simplify the situation by passing to a subsequence. The final upshot is that for a certain sequence $g_1, \ldots, g_r$ of elements of ${}^{\omega}2$, we have:

1) $\rho(g_i) \geq \frac{1}{n}$.

2) $\sum_{i=1}^{r} \rho(g_i) \geq a_k$.

3) For any integer m, there are infinitely many stages s at which $g_i | m$ is the characteristic function of some $A_j(s) \cap m$.

It is crucial, of course, that g be among the g_i's. But, if not, $\lambda \geq \rho(g) + \sum_{i=1}^{r} \rho(g_i) \geq a_k + \frac{1}{n} > b_k$. This contradicts $\lambda \in I_k$. (c) An easy argument now shows that for some $i < N$, $A_i = B$.

4.1. We now commence part (a) of the proof. Towards a contradiction, assume that there is a largest opportune stage, say t. (Recall that there is at least one opportune stage, at least formally, namely $t = -1$.) Our argument will construct a series of very large integers, $n_0, \ldots, n_4$. By the time we are done, it will be clear that some integer $\leq 2^{n_0} 3^{n_3} 5^{n_4}$ is opportune. This will complete part (a) of the proof.

4.3. To start things off, let $g_1, \ldots, g_r$ be those elements at ${}^{\omega}2$ such that $\rho(g_i) \geq \frac{1}{n}$. Since at most $n+1$ functions g have $\rho(g) \geq \frac{1}{n+1}$, we can find an $\varepsilon > 0$ so that if $\rho(g) < \frac{1}{n}$, $\rho(g) < \frac{1}{n} - \varepsilon$. We pick n_0 so large that 1) $n_0 \geq t+2$; 2) $n_0 > n^2 + n$; 3) $\frac{1}{n_0} < \varepsilon$; 4) the functions $g_1|n_0, \ldots, g_r|n_0$ are distinct; 5) $\rho(g_i|n_0) > \frac{1}{n} - \varepsilon$.

LEMMA. *There is an* $n_1 \geq n_0$ *so that whenever* $h \in {}^m2$, $m \geq n_1$, $h|n_1 = g_i|n_1$, *and* $h \neq g_i|m$, *then* $\rho(h) < \frac{1}{n} - \varepsilon$.

PROOF: By Section 1.5, $\rho(g_i) = \lim_{n\to\infty} \rho(g_i|m)$. An inspection of the definition of ρ (cf. Section 1.5) shows that $\rho(g_i|m)$ is monotone nonincreasing in m. Pick $n_1 \geq n_0$ so large that $\rho(g_i|n_1) - \rho(g_i) < \frac{1}{n} - \varepsilon$. Then if h is as in the statement of the lemma, $\rho(h)$ is the measure of a subset of

$$\{f\colon W(f)|n_1 = g_i|n_1 \text{ and } W(f) \neq g_i\}.$$

By our choice of n_1 this latter set has measure $< \frac{1}{n} - \varepsilon$.

4.4. We shall need to cite the next step of the argument again, so we isolate it out as a lemma.

LEMMA. *Let* $\langle h_i : i \in \omega \rangle$ *be a sequence of finite sequences whose length go to infinity with* i. *Suppose*

$$\lim \rho(h_i) \geq \gamma.$$

Then, by passing to a subsequence, we can arrange that for some $g \in {}^\omega 2$,

(1) $\lim_{i\to\infty} h_i(m) = g(m)$, *all* $m \in \omega$.

If g *satisfies* (1), *then* $\rho(g) \geq \gamma$.

PROOF: By the usual Cantor diagonal argument, we may arrange for (1) to hold. It remains to show $\rho(g) \geq \gamma$. For that it suffices, by 1) of Lemma 1.5, to show $\rho(g|m) \geq \gamma$, for any $m \in \omega$.

Suppose, towards a contradiction, that $\rho(g|m) < \gamma$. Then for i sufficiently large, h_i has length $\geq m$, $\rho(h_i) > \rho(g|m)$, and $h_i|m = g|m$. But then,

$$\rho(g|m) < \rho(h_i) \leq \rho(h_i|m) = \rho(g|m).$$

Contradiction!

4.5. LEMMA. *There is an* $n_2 \geq n_1$ *such that if* $m \geq n_2$, *and* $\rho(h|m) \geq \frac{1}{n} - \varepsilon$, *then* $h = g_i|m$ *for some* $i \leq r$.

PROOF: Deny the lemma. Let $\langle h_i : i \in \omega \rangle$ be a sequence of finite binary sequences whose length tend to ∞ such that if m_i = length (h_i), $h_i|m_i \neq g_j|m_i$, any $j \leq r$, and $\rho(h_j) \geq \frac{1}{n} - \varepsilon$. Apply Lemma 4.4. For some g, after passing to a subsequence, we have $\rho(g) \geq \frac{1}{n} - \varepsilon$, and

(2) $\lim_{i\to\infty} h_i(m) = g(m)$.

Thus $g = g_i$, some $i \leq r$ (since $\rho(g) \geq \frac{1}{n} - \varepsilon$). (Cf. Section 4.3.) If j is large, $h_j|n_1 = g_i|n_1$, by (2). But by assumption $h_j \neq g_i|m_j$. This contradicts Lemma 4.3, since $\rho(h_j) \geq \frac{1}{n} - \varepsilon$

4.6. By our choice of k,

$$a_k < \sum_{i=1}^{r} \rho(g_i) < b_k.$$

Let $n_3 > n_2$ be chosen so large that $a_k < \sum_{i=1}^{r} \rho(g_i|n_3) < b_k$. (Lemma 1.5 1).)

Next select $n_4 \geq n_3$ large enough so that the following are true:

(3) $a_k < \sum_{i=1}^{r} \rho(g_i|n_3, m) < b_k$, if $m \geq n_4$.

(Use Lemma 1.5 2) to prove n_4 exists.). Next, we want for $h \in {}^{n_3}2$ and $\rho(h) < \frac{1}{n} - \varepsilon$, that $\rho(h,m) < \frac{1}{n} - \varepsilon$, if $m \geq m_4$. Since there are only finitely many h's of length n_3 we may arrange this by 2) of Lemma 1.5. Finally, we require

$$\rho(g_i|n_3, m) > \frac{1}{n} - \frac{1}{n_0}, \text{ if } m \geq n_4.$$

Since $\frac{1}{n} \leq \rho(g_i) \leq \rho(g_i|n_3)$, we may clearly arrange this as well.

Let now $s = 2^{n_0} 3^{n_3} 5^{n_4}$. Then $s \geq n_4$. If we put $m = (s)_0$ $(= n_0)$, and $\ell = (s)_1$, then since $s \geq n_4$, we have for $h \in {}^{\ell}2$, $\rho(h,s) > \frac{1}{n} - \frac{1}{m}$ iff

$h = g_i | \ell$. Equation (3) now shows that if no s' with $t < s' < s$ is opportune, then s is opportune. This completes our proof that infinitely many stages are opportune.

4.7. We now enter the second phase of our proof. We have an increasing sequence $< s_i : i \in \omega >$ of opportune stages. By passing to a subsequence, repeatedly, we shall be able to arrive at a fairly cogent picture of "what happens at stage s_i". So as not to drown in a sea of subscripts, we refer, by a relabeling, to each new subsequence as s_i.

To start matters off, let $m_i = (s_i)_0$, $\ell_i = (s_i)_1$, and let t_i be the largest opportune stage $< s_i$. Let r_i be the number of distinct binary sequences h of length ℓ_i with $\rho(h|s_i) \geq \frac{1}{n} - \frac{1}{m_i}$. Then $r_i \leq n+1$. Thus by passing to a subsequence and relabeling, we may assume r_i has some constant value r.

Let $h_{i,1}, \ldots, h_{i,r}$ be the binary sequences of length ℓ_i with $\rho(h_{i,j}) \geq \frac{1}{n} - \frac{1}{m_i}$. By passing to a subsequence, r times, and relabeling, we may assume, by Lemma 4.4 that for certain functions $g_1, \ldots, g_r$ in ${}^{\omega}2$, we have

$$\lim_{i \to \infty} h_{i,j}(m) = g_i(m), \text{ all } m \in \omega .$$

Notice that we do not claim the g_i's are distinct.

Since $m_{i+1} > t_{i+1} \geq s_i > m_i$, m_i goes to infinity with i. Thus

$$\lim \rho(h_{i,j}) \geq \lim [\frac{1}{n} - \frac{1}{m_i}] = \frac{1}{n} .$$

By Lemma 4.4, $\rho(g_j) \geq \frac{1}{n}$, for $1 \leq j \leq r$.

4.8. By a suitable relabeling of the $h_{i,j}$'s, we may assume that $g_1, \ldots, g_s$ are distinct and each g_j is $=$ to some g_i with $i \leq s$. We wish to prove next that

$$(4) \quad a_k \leq \sum_{i=1}^{s} \rho(g_i).$$

If not, by Lemma 1.5 1), we can find ℓ such that

$$\sum_{i=1}^{s} \rho(g_i | \ell) < a_k ,$$

and the functions $g_1|\ell,\ldots,g_s|\ell$ are distinct. Now by Lemma 1.5 2), we can find t_0 so that

$$(5)\quad \sum_{i=1}^{s} \rho(g_i|\ell,t) < a_k\ , \text{ if } \quad t \geq t_0\ .$$

Now select j so large that 1) $\ell_j \geq \ell$, 2) for $i \leq r$, $h_j|\ell = g_j|\ell$, and 3) $s_j \geq t_0$. The sum in (5) is, if t is replaced by s_j, the measure of

$$(6)\quad \{f\colon W(f,s_j)|\ell = g_i|\ell, \text{ for some } i \leq s\}.$$

The set (6) is a superset of

$$(7)\quad \{f\colon W(f,s_j)|\ell_j = h_i, \text{ some } i \leq r\}\ .$$

But this set (7) has measure $> a_k$ since s_j is opportune. This contradicts (5), and proves (4) as desired.

4.9. Let now $g \in {}^{\omega}2$ with $\rho(g) \geq \frac{1}{n}$. I say that g is one of the g_i with $i \leq s$. Suppose not. Then since $\rho(g_i) \geq \frac{1}{n}$ for $1 \leq i \leq s$, we have:

$$\lambda \geq \rho(g) + \sum_{i=1}^{s} \rho(g_i) \geq \frac{1}{n} + a_k > b_k\ .$$

(Cf. Section 1.10.) But by our choice of k, $\lambda < b_k$. This contradiction shows that g is one of the g_i's, say g_j.

4.10. Let $B = \{n\colon g(n) = 1\}$. We now show that $B = A_p$ for some $p < N$.

At each opportune stage of our sequence, s_i, there is an integer p_i so that $A_{p_i}(s_i + 1) = \{m < \ell_i\colon h_{i,j}(m) = 1\}$. By passing to a subsequence, we may assume, since $p_i < N$, that p_i has some constant value p. We shall show that $A_p = B$.

Suppose not. Pick n_0 so large that $A_p \cap n_0 \neq B \cap n_0$. Pick $n_1 \geq n_0$ so that $A_p(m) \cap n_0 = A_p \cap n_0$, for $m \geq n_1$. Let s_i be an opportune stage such that $s_{i-1} \geq n_1$, and such that $h_{i,j}|n_0 = g_j|n_0$. Then

$$B \cap n_0 = \{m < n_0 : h_{i,j}(m) = 1\} = A_p(s_i + 1) \cap n_0 = A_p \cap n_0$$

(since $\ell_i \geq s_{i-1} \geq n_1$).

This contradicts our choice of n_0. The upshot is that $B = A_p$, and our theorem is proved.

5. H vs. I.

5.1. The purpose of this section is largely expository. We review enough of the work of Chaitin 1976 and 1975 to motivate the notions $H(A)$, $I(A)$ and make the derivation of the inequality $I(A) \leq 3H(A) + O(\log H(A))$ from our main theorem comprehensible.

In this section, we do *not* identify the integer n with $\{m \in \omega: m < n\}$. Rather we identify each integer n with a binary string. The sequence of binary strings that corresponds to the integers 0, 1, 2, 3,...,is then 0, 1, 10, 11,... (I. e. if $n > 1$, n is identified with the dyadic expansion of n.) Occasionally, we will have to encode all the integers $< 2^m$ into strings of the fixed length m. When we have to do this we fill out the binary string with leading zeros. (E. g., 3 will be encoded by '0011' if $m = 4$.)

5.2. Our first goal is to define the '*information content*' of a finite binary string, s. We shall give a provisional definition (of the number $K(s)$) and then present the refinement (of Chaitin 1975) which we shall, by analogy with Chaitin 1976, refer to as $I(s)$.

The intuition behind the definition is as follows. We have a Turing machine M which acts as a decoder. We give the Turing machine a description of the string s. M decodes the description and outputs s. M will be chosen universal (or 'almost optimal') in a certain precise sense. The number of bits in the shortest description of s is the "*information content*" of s.

Note that a long string may have few bits of information. For example the string of length 10^{100} consisting of zeros will certainly have less than 10^3 bits on any reasonable encoding of Turing machines.

5.3. For our provisional definition we use the following variant of Turing machines. A machine will have three tapes, an input tape, an output tape, and a scratch tape. M determines a partial function, u_M from the set of binaty strings, Σ^*, into itself. $u_M(s) = t$ if M started in its initial state, with scratch and output tapes blank, and s on its input tape, writes t on its output tape and then halts.

Let $|s|$ be the length of the binary string s. (Our convention identifying integers with binary strings makes $|n| = \log_2 n + O(1)$ for $n \geq 1$.) We put

$$K_M(s) = \min\{|t| : u_M(t) = s\}.$$

A machine, M_0, is *universal* if for each machine, M, there is a prefix, π_M, so that for each binary string x,

$$(1) \quad u_{M_0}(\pi_M\hat{\ }x) \simeq u_M(x).$$

(Here $s\hat{\ }t$ is the concatenation of s and t. As usual, the symbol $\simeq$ for partially defined expressions indicates that the left side is defined iff the right side is, and if defined, both are equal.)

It follows from (1) that

$$(2) \quad K_{M_0}(s) \leq K_M(s) + O(1).$$

In particular, if M_0 and M_1 are universal machines,

$$(3) \quad K_{M_0}(s) = K_{M_1}(s) + O(1).$$

It is easy to modify the usual construction of universal Turing machines so as to construct a machine M_0 universal in the precise sense just described. For example, suppose we have Gödel numbered all Turing machines in some standard way. Let M_g be the machine with Gödel number g. (This is a very temporary piece of notation.) Then there is a Turing machine M such that

$$u_M(0^g 1\hat{\ }x) \simeq u_{M_g}(x).$$

(Here $0^g 1$ is g zeros followed by a 1.) Clearly this M is universal.

We fix a universal machine, M_0, and put $K(s) = K_{M_0}(s)$. (This definition is due to various authors including Chaitin and Kolmogoroff. Cf. the paper Chaitin 1975, and its bibliography.)

5.4. The K notion has the unfortunate property that various intuitively true formulae are true only up to a "log" error term. We give one example. Let

$$h\colon \omega \times \omega \simeq \omega$$

be a recursive isomorphism. (For example, $h(<a,b>) = 2^a(2b+1)-1$.) Put $K(<a,b>) = K(h(<a,b>))$. Intuitively, one would expect $K(<a,b>) \leq K(a) + K(b) + O(1)$. But this can be shown to be false. One can prove the weaker result:

$$K(<a,b>) \leq K(a) + K(b) + \min\left[\log_2 K(a),\ \log_2 K(b)\right] + O(1).$$

(The difficulty is this. Let s, t be the minimal length programs for a, b. Then from the string $s^\frown t$, we can not recover s and t unless we know the length of one of the strings s and t.

Thus knowing not only the string s but when it ends means that s has in addition to $|s|$, up to $\log_2 |s|$ bits conveyed in knowing when it ends.)

Chaitin (and independently the Russian mathematician Levin) have figured out how to take this extra information into account. Roughly speaking they require that as we are reading the code word s, we are able to tell when we have read the entire word. The precise concept is captured in the notion of an end-detecting Turing machine, which we now describe.

5.5. We now present revised definitions of the notion of Turing machine and of the function U_M. We require that the input tape reading head can not move to the left. At the start of the calculation, the input tape is positioned at the leftmost binary digit of s. At the end of the computation, for $U_M(s)$ to be defined, we now require that the input head be positioned on the last digit of s. Thus, while reading s, M was able to detect at which point the last digit of s occurred.

The notion of a universal machine is essentially that of Section 5.3. (modulo the new definition of U_M). Once again it is easy to see that universal machines exist; we fix a universal machine M_0, and put

$$I(s) = \min\{|t| : U_{M_0}(t) = s\}.$$

Again, to within $O(1)$, this does not depend on the choice of universal machine .

Following Chaitin we can give the following probabilistic interpretation to $I(s)$. (The proof is non-trivial. Cf. Chaitin 1975.) Suppose we start our machine M_0 with $f \in {}^{\omega}2$ written on the input tape (and the input reading head of M_0 positioned on the leftmost digit of f). Say that $U^{M_0}(f) = t$, if M_0 halts after writing the string t (and nothing else) on its output tape.

Put $P(t) = \mu(\{f: U_{M_0}(f) = t\})$. Put $H(t) = -\log_2 P(t)$. Then

$$I(t) = H(t) + O(1).$$

(The corresponding result when t is replaced by an r.e. set A is not true, though our results will provide the weaker version: $I(A) = O(H(A))$, $H(A) = O(I(A))$.)

5.6. We mention briefly what is known about the relationship between $I(s)$ and $K(s)$. It is trivial to show $K(s) \leq I(s) + O(1)$. The following unpublished result of the author measures the cost required to make the code for s "*self-delimiting*".

(4) $$I(s) = K(s) + K[K(s)] + O(\log_2 K[K(s)]).$$

(This formula allows $I(s)$ to be "computed" from $K(s)$. One can show there is no way of "computing" $I(s)$ from $K(s)$ with an error term which is $O(\log_2 K[K(s)])$.) (We put "computing" in quotes since the function K is not recursive.))

(4) has the following intuitive content. In order to make the minimal K-style program for s self-delimiting, we must prefix it with an encoding of its length. This can be done in $K(K(s)) + O(\log_2 K[K(s)])$ bits in a self-delimiting fashion. The content of (4) is that, to within the error term cited, this simple procedure is optimal. We remark that unlike K , I does satisfy

$$I(<n,m>) \leq I(n) + I(m) + O(1).$$

This follows readily from the arguments used to prove 3) of Lemma 5.7.

5.7. LEMMA. (Chaitin 1975)

1) $I(n) \leq \log_2 n + O(\log_2 \log_2 n)$;

2) $I(n) \leq 2\log_2 n + O(1)$;

3) $I(n \pm m) = I(n) + O(\log_2 m)$;

4) $I(s) \leq |s| + I(|s|) + O(1)$.

REMARKS: 1) is the principal result for us. 2) through 4) are way-stations on the way to the proof.

In order to make 1) through 4) literally true it is useful to interpret $\log_2 0$ and $\log_2 1$ as 1. (Alternatively, we could just introduce $\log^+ n = \log_2(\max(n,2))$ and state the lemma in terms of $\log^+$.)

PROOF: First note that 1) follows easily from 2) to 4). Let s be the binary string identified with n. Then $|s| = \log_2 n + O(1)$. Thus $I(n) = I(s) \leq |s| + I(|s|) + O(1) = \log_2 n + O(\log_2 \log_2 n)$.

The general approach to proving upper bounds on I is to construct special purpose machines M and using $I(s) \leq \min\{|t|: U_M(t) = s\} + O(1)$, which follows from the universality of M_0.

For example, to prove 2), use a machine M that works on the following plan. It reads the digits of t two at a time. It then interprets them as follows: '0x' means "print x on output tape and shift one square to the right" ($x = 0,1$); '11' means "halt". Then if s is the binary string corresponding to n, $U_M(t) = s$ for some t of length $2|s| + 2 = 2\log_2 n + O(1)$.

To prove 3) consider a machine which proceeds as follows. It reads the first digit of the number t to find out whether to add or subtract. It then simulates the universal machine M_0 reading a segment t_1 of t that encodes a number m. It then simulates M_0 again, reading a segment t_2 of t. If $t_0 = 0$, it outputs $n + m$. If $t_0 = 1$ it outputs $n - m$. Thus $I(n \pm m) \leq I(n) + I(m) + O(1) \leq I(n) + O(\log_2 m)$.

The proof of 4) is similar. The auxilliary machine M simulates M_0 and reads an initial segment of t that encodes a number m. It then reads the next m digits of t and writes them on the output tape. If s is our given string and s' is a string of length $I(|s|)$ such that $U_{M_0}(s') = |s|$, then $U_M(s'^\frown s) = s$. This proves 4).

We remark that Chaitin shows (in Chaitin 1975) that

$$\sum_{i=0}^{\infty} 2^{-I(i)} < 1.$$

Whence it follows easily that

$$I(n) \geq \log_2 n + \log_2 \log_2 n \quad ,$$

for infinitely many n.

5.8. We now recall the definitions of $I(A)$ and $H(A)$, for A an r.e. set, given in Chaitin 1976.

The definition of H is similar to that of $H(n)$ given above. Let us envision the Monte-Carlo machines of the introduction as being provided their random string f of 0's and 1's on a read-only input tape. We define a universal machine as before: M_0 is *universal* iff for every Monte-Carlo machine M, there is a finite string π_M so that M_0 on input $\pi_M{}^\frown f$ simulates M's action on input f:

1) $M_0[\pi_M{}^\frown f]$ will enumerate the same set A as $M[f]$.

2) If $M[f]$ only reads the initial segment s of f, then $M_0[\pi_M{}^\frown f]$ reads the initial segment $\pi_M{}^\frown s$.

We put $P(A) = \mu(\{f\colon M_0\,[f]$ enumerates $A\})$. We put $H(A) = -\log_2 P(A)$.

Next, we define $I(A)$. The definition presented here is that of Chaitin 1976. It is *not* the same (to within $O(1)$) as the definition presented in the introduction. We say that a finite binary string s is an M_0-program for A, if M_0 with the string $s^\frown f$ on its input tape will enumerate A on its output tape and *not read beyond* s on its input tape. (Of course it follows that $M_0\,[s^\frown g]$ will also enumerate A for any $g \in {}^{\omega}2$, since the contents of f can not affect the history of $M_0\,[s^\frown f]$.)

Put $I(A) = \min\{|s|\colon s$ is an M_0-program for $A\}$.

We can formulate the concept referred to as $I(A)$ in the introduction, call it now $I^*(A)$ by

$$I^*(A) = \min\{I(j): W_j = A\}.$$

We remark that all the results referred to in this paper are equally valid for $I^*(A)$ in place of $I(A)$.

LEMMA. $I(A) \leq I^*(A) + O(1)$.

PROOF: Let us refer to the universal machine employed in the definition of $I(s)$ as M_1.

We construct an auxilliary machine M as follows. M simulates the action of M_1. M_1 will read an initial segment s of its input and output a number j. M then proceeds to enumerate W_j on its output tape.

It is clear that if s is a minimal M_1-style program for j, then $\pi_M{}^\frown s$ is an M_0-program for enumerating W_j. This proves the lemma.

5.9. LEMMA. $I(A) \leq 3H(A) + I(H(A)) + O(1)$.

Of course it follows from this lemma and Lemma 5.7 that $I(A) \leq 3H(A) + O(\log H(A))$. Also, by Lemma 5.8, it suffices to prove

$$I^*(A) \leq 3H(A) + I(H(A)) + O(1).$$

Suppose that $H(A) \leq n$. Then $P(A) \geq 2^{-n}$. We note that if $N = 2^n$,

$$\frac{1}{2}(N^2(N + 1) < 2^{3n+1} .$$

Thus our main result implies the following. There is a recursive function $f(n,m)$ so that whenever A is an r.e. set with $H(A) \leq n$, then

$$(\exists j)(\exists m)\ f(n,m) = j, \quad m < 2^{3n+1}, \text{ and } A = W_j.$$

To complete the proof we now describe a machine M so that if $H(A) \leq n$, then $U_M(s) = j$, and $A = W_j$ for some s with $|s| = I(n) + 3n + 1$.

M proceeds as follows, on input s: It first simulates M_1 and reads an initial segment s_0 of s. It computes $n = U_{M_1}(s_0)$. It next reads the next $3n + 1$ digits of s, interpreting it as a number $m < 2^{3n+1}$. Finally, M computes $f(n,m)$ and writes it on the output tape. In view of our main result, as recalled two paragraphs ago, M is easily seen to have the desired properties: If $H(A) \leq n$, let s_0 be a string of length $I(n)$ such that $U_{M_1}(s_0) = n$. Let m, j be such that $f(n,m) = j$, $A = W_j$, and $m < 2^{3n+1}$. Let s_1 be a binary string of length $3n + 1$ that gives the binary expansion of m, preceded if necessary by zeros. Then $U_M(s_0 {}^\frown s_1) = j$ and $|s_0 {}^\frown s_1| = I(n) + 3n + 1$.

References.

Bishop, E.

1967, Foundations of Constructive Analysis, McGraw-Hill.

Chaitin, G. J.

1975, *A theory of program size formally identical to information theory*, J. ACM, vol. 22, 329-340

1976, Algorithmic entropy of sets, IBM Research Report RC5799, 36 pages, IBM Watson Lab., Yorktown Heights, N. Y.

de Leeuw, K., E. F. Moore, C. E. Shannon, and N. Shapiro

1956, *Computability by probabilistic machines*, in Automata Studies, C. E. Shannon and J. McCarthey, (Eds.), Princeton University Press, Princeton, New Jersey, 183-212.

Halmos, P. R.

1964, Measure Theory, van Nostrand, Princeton, New Jersey.

Department of Mathematics
University of California
Berkeley, California, U.S.A.